Lecture Notes in Mathematics

Edited by A. Dold and B. Eckmann

606

Mathematical Aspects of Finite Element Methods

Proceedings of the Conference
Held in Rome, December 10–12, 1975

Edited by
I. Galligani and E. Magenes

Springer-Verlag
Berlin Heidelberg New York 1977

Editors
Ilio Galligani
Istituto per le Applicazioni
del Calcolo „Mauro Picone", C.N.R.
Viale del Policlinico 137
00161 Roma/Italia

Enrico Magenes
Laboratorio di Analisi Numerica, C.N.R.
Via Carlo Alberto, 5
Pavia/Italia

Library of Congress Cataloging in Publication Data

Meeting on Mathematical Aspects of Finite Element
Methods, Rome, 1975.
Mathematical aspects of finite element methods.

(Lecture notes in mathematics ; 606)
"Organized by the Istituto per le applicazioni
del calcolo "Mauro Picone" and Laboratorio di
analisi numerica."
Bibliography: p.
Includes index.
1. Numerical analysis--Congresses. 2. Finite
element method--Congresses. I. Galligani, Ilio.
II. Magenes, Enrico. III. Istituto per le
applicazioni del calcolo. IV. Laboratorio di
analisi numerica. V. Title. VI. Series:
Lecture notes in mathematics (Berlin) ; 606.
QA3.L28 no. 606 [QA297] 510'.8s [519.4]
77-21425

AMS Subject Classifications (1970): 35F25, 35F30, 65N30, 65N36, 49A20

ISBN 3-540-08432-0 Springer-Verlag Berlin Heidelberg New York
ISBN 0-387-08432-0 Springer-Verlag New York Heidelberg Berlin

Printing and binding: Beltz Offsetdruck, Hemsbach/Bergstr.
2141/3140-543210

FOREWORD

The contents of this book are based on lectures given at the Meeting on Mathematical Aspects of Finite Element Methods, held in Rome, December 10-12 1975, at the "Consiglio Nazionale delle Ricerche" (C.N.R.) organized by the *Istituto per le Applicazioni del Calcolo "Mauro Picone"* and *Laboratorio di Analisi Numerica*.

The subject of this meeting is of particular interest owing to the importance that the Finite Element Method has in many fields of engineering, not only from the point of view of research but also in the industrial routine. It is well known that this method has been developed by engineers as a concept of structural analysis.

When there was discovered the connection between the Finite Element Method and the Ritz-Galerkin-Faedo procedure applied to the spaces of piecewise polynomial functions, the interest of mathematicians to this method increased enormously: now, in this field, the interests and ideas of engineers and mathematicians converge and overlap and the cooperation between them has become more and more essential.

To this aim, the Istituto per le Applicazioni del Calcolo "Mauro Picone" (IAC) in Roma and Laboratorio di Analisi Numerica (LAN) in Pavia have considered the opportunity of organizing this meeting, by emphasizing the mathematical aspects of the Finite Element Method.

Twenty-five papers were presented and discussed at the Meeting; but only twenty-two lectures have been made available for publication on time

We should like to conclude by thanking the members of IAC for their help in the management of the meeting.

I. Galligani - E. Magenes

Roma, July 1976

CONTENTS

Mathematical Problems of Computational Decisions in the Finite Element Method

I. Babuška
Department of Mathematics and
Institute for Physical Science and Technology
University of Maryland, College Park

and

W. C. Rheinboldt
Computer Science Center
and Department of Mathematics
University of Maryland, College Park

Abstract. Present programs for finite element analysis require the user to make numerous, critical, a-priori decisions. They often represent difficult mathematical problems and may influence strongly the accuracy and reliability of the results, the cost of the computation, and other related factors. This paper discusses some of these decisions and their mathematical aspects in the case of several typical examples. More specifically, the questions addressed here concern the effect of different mathematical formulations of the basic problem upon the results, the influence of the desired accuracy on the efficiency of the process, the selection and comparison of different types of elements, and, for nonlinear problems, the choice of efficient methods for solving the resulting finite dimensional equations. In all cases a consistent use of self-adaptive techniques is strongly indicated.

Acknowledgment. This work was supported in part under Grant AT(40-1)-3443 from the U.S. Energy Research and Development Administration and Grant GJ-35568X from the National Science Foundation.

* * * * *

1. Introduction

The finite element method has advanced rapidly in the past two decades. The most far-reaching progress probably occurred in the practical application of the method in various fields and especially in continuum mechanics. Numerous, often large, general and special-purpose programs for finite element analysis have been built and are widely applied to increasingly complex problems (see, e.g., [1]).

The mathematical analysis of the method began somewhat later but is also progressing at a quick pace. Without question, the method has now been placed on a firm mathematical foundation.

However, in looking over these advances, it is surprising to notice the relatively weak interaction between the mathematical progress and the practical application of the finite element method. Often, in practice, the method is not interpreted as an approximate solution process of a differential equation of, say, continuum mechanics. On the other hand, the theoretical analysis has principally addressed the mathematical basis of the method and of the related approximation problems. There appears to be an urgent need to extend now this theoretical analysis to all phases of the solution process and their interactions. This involves the selection of the mathematical formulation of the original problem and the characterization of the desired type of solution. It also includes the variety of questions about the numerical procedures and last, but not least, the many computer science problems arising in the overall implementation.

Today's finite element programs require the user to make numerous, very critical, a-priori decisions which, in fact, often represent difficult mathematical questions. This includes decisions about the mathematical model, as, for example, whether a plate or shell may be considered thin, or whether nonlinear behavior may be disregarded. It also includes the questions of the selection of the elements and the meshes, the specification of the time steps and of various other process parameters, as well as the decisions when updates or refinements are to be used, etc. The architecture of all present--and probably many of the future--programs incorporates the need for all these options. Any of the decisions required from the user may influence strongly the accuracy of his results, the cost of the computation, and so on. It appears that only a consistent use of self-adaptive techniques can significantly alter this situation.

The mathematical problems involved in all this are wide-ranging, and in part, novel in nature, especially when it comes to the computer science questions. Some starting points for such studies may well be the many questions raised by the often-startling results reported by experienced practitioners of the finite element method. Our aim here is to delineate some such questions for several typical examples. More specifically, in Section 2 we show the effect of different formulations of the basic mathematical problem upon the results. Then Section 3 addresses the influence of the desired accuracy of the solution upon the efficiency of the solution process and the need for further types of asymptotic analyses. Section 4 considers some aspects related to the theoretical comparison of different types of elements, and finally Section 5 shows that, especially in the nonlinear case, the methods for solving the resulting finite dimensional equations depend once again strongly on the selection of the mathematical formulation of the problem.

2. Formulation of the Mathematical Model

Most physical problems may be formulated mathematically in a variety of more or less simplified forms, and a numerical method applied to any one such mathematical model introduces a further transformation. Clearly, a principal mathematical question must be the analysis and estimation of the errors resulting from the various simplifications and transformations. This, however, requires a decision as to which formulation is to be considered as the reference model. Here, usually, attention is only focused on the approximation errors introduced by the numerical method, although sometimes these errors are much smaller than those caused by earlier simplifications of an original mathematical model.

This situation arises, in particular, in continuum mechanics where theoretical advances now allow for the formulation of very general mathematical models. Because of their complexity, numerical procedures are in most cases only applied to considerably simplified formulations which then are also used as the reference models in the error analysis. In this section we illustrate how much we may have to adjust our assessment of the numerical results, if some of the earlier simplifications are taken into account.

As an example, we consider a bending analysis of a simply supported plate for which either a two- or three-dimensional formulation may be used. Let $\bar{\Omega} \subset R^2$ denote the (compact and Lipschitzian) domain of the plate and d its thickness. The material is assumed to be homogeneous and isotrop with Young's modulus E and, for simplicity, Poisson's ratio $\sigma = 0$.

The three dimensional formulation involves the solution of a system of strongly elliptic equations for the unknown vector $\underline{u} = (u_1,u_2,u_3)$ on

(2.1) $$\Omega_d = \{(x_1,x_2,x_3) \in R^3;\ (x_1,x_2) \in \Omega,\ |x_3| < \tfrac{d}{2}\},$$

subject to certain boundary conditions on $\partial\Omega_d$. In its weak form this boundary value problem requires the determination of

$$u_1 \in H^1(\Omega_d),\quad u_2 \in H^1(\Omega_d),\quad u_3 \in H^1_{[0]}(\Omega_d)$$

such that

(2.2) $$E\int_{\Omega_d} \sum_{i,k=1}^{3} \frac{1}{4}\left(\frac{\partial u_i}{\partial x_k} + \frac{\partial u_k}{\partial x_i}\right)\left(\frac{\partial v_i}{\partial x_k} + \frac{\partial v_k}{\partial x_i}\right) dx_1dx_2dx_3 = \int_\Omega v_3(x_1,x_2,\tfrac{d}{2}) f(x_1,x_2)dx_1dx_2$$

holds for any

$$v_1 \in H^1(\Omega_d),\quad v_2 \in H^1(\Omega_d),\quad v_3 \in H^1_{[0]}(\Omega_d).$$

Here $f \in L_2(\Omega)$ is given, $H^1(\Omega_d)$ denotes the usual Sobolev space and

$$H^1_{[0]}(\Omega_d) = \{u \in H^1(\Omega_d) \mid u(x) = 0,\ \ (x_1,x_2) \in \partial\Omega,\ |x_3| < \tfrac{d}{2}\}.$$

The two-dimensional formulation leads to the well-known (see, e.g., [2],[3]) biharmonic problem of finding $w \in H^2(\Omega) \cap H^1_0(\Omega)$ such that

(2.3) $$\frac{Ed^3}{12}\int_\Omega \left[\frac{\partial^2 w}{\partial x_1^2}\frac{\partial^2 v}{\partial x_1^2} + 2\frac{\partial^2 w}{\partial x_1\partial x_2}\frac{\partial^2 v}{\partial x_1\partial x_2} + \frac{\partial^2 w}{\partial x_2^2}\frac{\partial^2 v}{\partial x_2^2}\right] dx_1dx_2$$

$$= \int_\Omega v\, f\, dx_1dx_2,\ \forall\ \ v \in H^2(\Omega) \cap H^1_0(\Omega)\ .$$

Physically, we expect that approximately $w(x_1,x_2) = u_3(x_1,x_2,0)$ for $(x_1,x_2) \in \Omega$. The formulation (2.3) may be derived from (2.2) by requiring that $\underline{u}$ (and correspondingly $\underline{v}$) satisfies

(2.4) $$u_1 = -x_3\frac{\partial w}{\partial x_1},\ u_2 = -x_3\frac{\partial w}{\partial x_2},\ u_3 = w,$$

that is, by restricting the space $H^1(\Omega_d)$. Accordingly, the finite element method for (2.3) may be interpreted as a method for (2.2) with special elements which incorporate two "small" parameters, namely, the size h of the elements in the x_1,x_2-directions, and the thickness d with $d \ll h$, in most cases.

A direct three-dimensional finite element solution of (2.2) is most likely inefficient for small d. At the same time, (2.4) is certainly not the only possible restriction. For example, we may use

$$(2.5)\qquad \begin{aligned} u_1(x_1,x_2,x_3) &= -x_3\varphi_1(x_1,x_2) \\ u_2(x_1,x_2,x_3) &= -x_3\varphi_2(x_1,x_2) \\ u_3(x_1,x_2,x_3) &= \varphi_3(x_1,x_2) , \end{aligned}$$

where now $\varphi_1,\varphi_2 \in H^1(\Omega)$, $\varphi_3 \in H_0^1(\Omega)$. This leads to a system of three equations of second order involving a small parameter, in contrast to the one equation of fourth order in the case of (2.4) without such a small parameter. Hence, the finite element discretization may now involve C^0-elements, instead of the C^1-elements needed before. Of course, the design of the elements has to account for the small parameter (d). A direct use of, say, piecewise linear elements would lead to very inaccurate results when $d \ll h$ is small.

However, there is a significant difference between the restrictions (2.4) and (2.5), namely, their dependence upon the domain. Let $\mathcal{B} \subset R^2$ be the open unit ball in R^2 and $\mathcal{B}^n \subset \mathcal{B}$, $n \geq 3$, the regular n-sided (open) polygon inscribed in $\mathcal{B}$. The corresponding three-dimensional domain $\mathcal{B}_d$ is defined as in (2.1). Then we have the following result for (2.2).

Theorem 2.1: Let $\underline{u}$ and $\underline{u}^{(n)}$ denote the solutions of (2.2) (with $f \in L_2(\Omega)$) for $\Omega = \mathcal{B}$ and $\Omega = \mathcal{B}^n$, respectively. Then $\lim_{n\to\infty} \underline{u}^{(n)} = \underline{u}$ in $H^1(\mathcal{D}) \times H^1(\mathcal{D}) \times H^1(\mathcal{D})$ for any compact $\bar{\mathcal{D}} \subset \mathcal{B}_d$.

On the other hand, for (2.3), that is, the restriction (2.4) of (2.2), the following theorem holds:

Theorem 2.2: Let w and $w^{(n)}$ be the solutions of (2.3) (with $f \equiv 1$) for $\Omega = \mathcal{B}$ and $\Omega = \mathcal{B}^n$, respectively. Then

$$\lim_{n\to\infty} w^{(n)} = W = \frac{3}{16d^3E}(3-4r^2+r^4), \quad (r^2=x_1^2+x_2^2) ,$$

in $H^2(\mathcal{D})$ for any compact $\bar{\mathcal{D}} \subset \mathcal{B}$, and

$$w = \frac{3}{16d^3E}(5-6r^2+r^4) \neq W .$$

This result was proved in [4] (see also [5], [6]) and its meaning discussed in various papers (see, e.g., [7]) under the name "Babuška's paradox."

In the case of the two-dimensional problem (2.2)/(2.5), obtained from (2.2) by applying the restriction (2.5), we have once again a result of the type of Theorem 2.1 (see [8]).

Theorem 2.3: Let $\underline{u}$ and $\underline{u}^{(n)}$ be the solutions of the two-dimensional problem (2.2)/(2.5) (with $f \in L_2(\Omega)$) for $\Omega = B$ and $\Omega = B^n$, respectively. Then $\lim_{n\to\infty} u^{(n)} = u$ on $H^1(D) \times H^1(D) \times H^1(D)$ for any compact $\bar{D} \subset B_d$.

The questions discussed here belong to the general range of problems of dimension reduction. There exists a large literature in this area especially for problems related to plates and shells. We mention here, for instance, [9], [10], [11], [12], [33], [34], [35], [36], [37], where additional references may be found. It should be noted, however, that all these presentations assume a smooth solution and hence cannot distinguish between the reductions (2.4) and (2.5) that led to the Theorems 2.1 and 2.2.

Some of the results of these three theorems may be summarized as follows:

(1) For small d and $\Omega = B$ the formulations (2.2),(2.3), and (2.2)/(2.5) essentially give the same results. However, if Ω has corners, as does B^n, then, even for large n, the results for (2.3) may be very different from those of (2.2) and (2.2)/(2.5) if h is small.

(2) In order to overcome the effect of Theorem 2.2 we may combine (2.4) and (2.5). More specifically, (2.5) is used in a neighborhood of the boundary and (2.4) elsewhere in the domain. Of course, in doing so we need to take account of the fact that $d \ll h$.

(3) The restriction (2.5) may be generalized to include higher order polynomials in x_3. This may be desirable when d is not sufficiently small and the resulting two-dimensional formulations are still less costly to solve than (2.2).

(4) Dimensional reduction may be treated as a special selection principle for the elements. Asymptotic analysis alone is insufficient to determine the influence of this type of reduction upon the solution. It is an open question how these problems may be approached theoretically. From a computational viewpoint, the only realistic way may be the use of self-adaptive techniques.

Although we discussed here only one particular example, it should be evident that similar situations may arise in connection with various other problems. We certainly encounter them in shell theory, but analogous questions also occur, for instance, when linearizations are introduced. In practice, the decision about the choice of the specific mathematical model is almost entirely left up to "experience". There appears to be a need for new theoretical and computational approaches.

3. Accuracy and Asymptotic Behavior

In practical applications of the finite element method, the accuracy required of the solution is rarely very high; in fact, an error of 10-20% is often fully acceptable. This means that the standard asymptotic error analysis may not provide us with sufficient insight. Since asymptotic approaches are hardly to be avoided, this suggests that we should consider a variety of different asymptotic analyses for characterizing more completely the computational process.

As an example, we consider a simple version of a type of problem occurring in reactor computations in the presence of interfaces. More specifically, let the following boundary value problem be given:

(3.1a)
$$\frac{\partial}{\partial x_1}\left(a \frac{\partial u}{\partial x_1}\right) + \frac{\partial}{\partial x_2}\left(a \frac{\partial u}{\partial x_2}\right) = 1, \ \forall\ (x_1,x_2) \in \Omega$$

$$u = 0 \text{ on } \partial\Omega$$

where

(3.1b)
$$\Omega = \{(x_1,x_2) \in R^2;\ |x_1| < \tfrac{1}{2},\ |x_2| < \tfrac{1}{2}\}$$

and

(3.1c)
$$a = \begin{cases} \Phi \text{ on } \Omega_1 = \{(x_1,x_2);\ 0 < x_1 < \tfrac{1}{2},\ 0 < x_2 < \tfrac{1}{2}\} \\ 1 \text{ on } \Omega - \Omega_1 \end{cases}$$

The solution has a singularity of the type $r^\beta \varphi(\theta)$ at the origin[1)] with β and φ depending on Φ. For an analysis of the solution of such interface problems, see [13] and [14].

Singularities of this type influence considerably the convergence of the finite element or finite difference methods. We refer to [15], [16], [17], and [18] for thorough studies of these problems. Suppose that a regular triangular finite element mesh with grid size h is used, as shown in Figure 3.1. Theoretically, the error then satisfies

$$\|e\|_{L_\infty} \geq ch^{\beta+\varepsilon},$$

with arbitrary $\varepsilon > 0$, and this result is independent of the order of the elements.

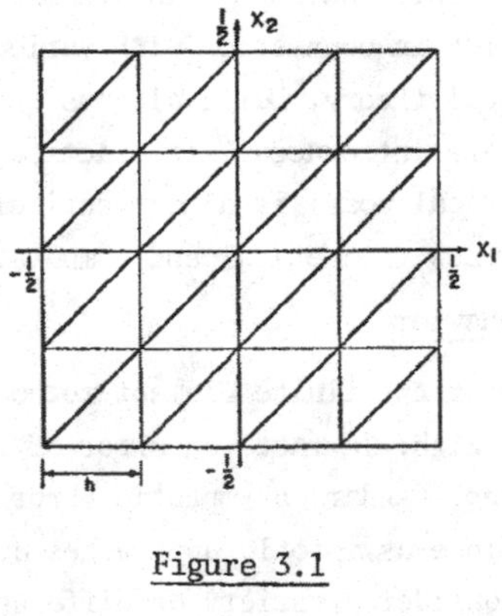

Figure 3.1

1) Here, and further below, r and θ represent polar coordinates.

However, this error behavior has not been observed in reactor computations. In the case of problem (3.1), the experimentally-observed rate of convergence for 1% accuracy as a function of Φ is indicated in Figure 3.2. For small $\Phi > 1$ the rate of convergence is better than the expected asymptotic value $\alpha(\Phi) < 1$; while for large Φ it essentially equals the predicted rate. The reason is that for small Φ the effect of the singularity is negligible in comparison to the desired 1% accuracy. The higher the required accuracy, the more the effect of the singularity becomes visible.

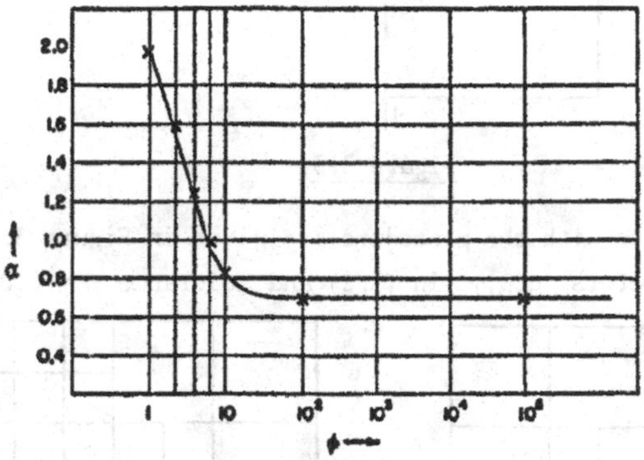

Figure 3.2

Generally, when a singularity of this type is present, it is known that for a regular mesh its effect on the accuracy will be felt throughout the region, and not just in some neighborhood. On the other hand, it has been shown (see, e.g., [19]) that there exist refinements of the mesh such that the resulting rate of convergence is the same as if no singularity were present.

These observations together indicate that for efficient computation a mesh should be constructed which incorporates a proper degree of refinement commensurate with the effect of the singularity at the desired accuracy. Such a mesh can hardly be designed a priori; instead, it must be evolved adaptively during the course of the computation.

In [20] (see also [21]) a procedure has been described for such a self-adaptive mesh refinement. More specifically we considered the numerical solution of the Dirichlet problem for Laplace's equation on an L-shaped domain (see Figure 3.3). The Dirichlet boundary conditions were chosen such that the exact solution has the form

(3.2) $$u = \alpha r^{2/3} \sin 2/3\, \theta + \gamma e^{\beta x_1} \cos \beta x_2 \,.$$

A piecewise regular triangular mesh was used analogous to that shown in Figure 3.1.

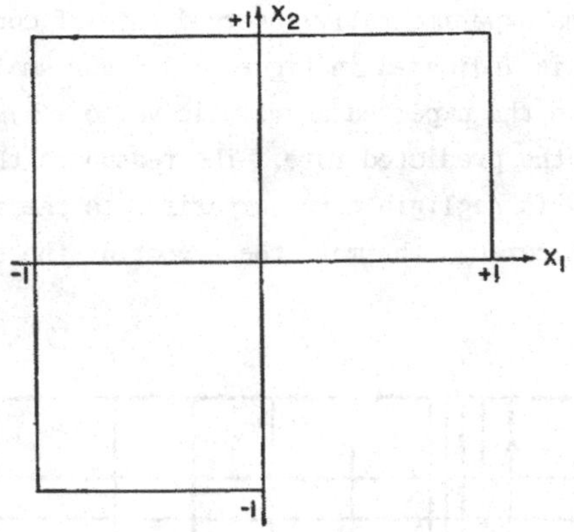

Figure 3.3

Some computational results with the procedure are given in Figure 3.4 for three different sets of the parameters α,β,γ in (3.2) and tolerance $\tau = 0.050$. Every

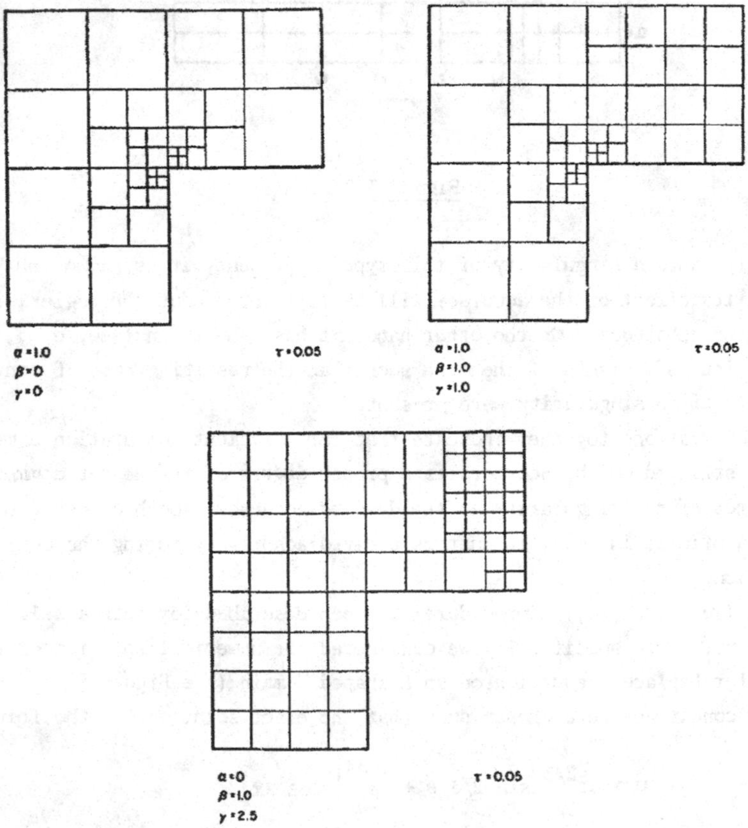

Figure 3.4

square represents here a block of 32 equally sized triangles. Figure 3.5 shows the dependence of the error in the C-norm on the number of unknowns N (that is, the number of nodal points) obtained for different tolerances τ. The dashed lines correspond to the use of a regular mesh while the solid lines give the results obtained with the adaptive mesh generator. The behavior is analogous when the L_2-norm or the H^1-norm is used. It is interesting that before the onset of the asymptotic behavior the rate of convergence is actually better than its theoretical, asymptotic bound. The explanation is that far from the singularity the mesh is already much too fine for the desired accuracy. Therefore local refinements around the corner, which increase only moderately the number of unknowns, provide for a large increase in accuracy.

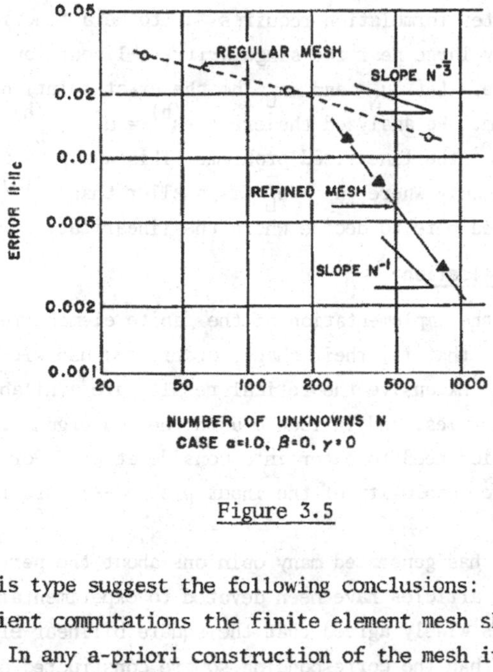

Figure 3.5

Studies of this type suggest the following conclusions:

(1) For efficient computations the finite element mesh should correspond to the desired accuracy. In any a-priori construction of the mesh it is difficult, if not impossible, to avoid over- or under-refinements in some parts of the domain resulting in decreased efficiency or accuracy or both.

(2) A self-adaptive procedure should be based on some asymptotic analysis. In our case, the behavior for $\tau \to 0$ was used, and the numerical experiments indicate that the results of this type of analysis may have a wider range of applicability than those of the standard asymptotic analysis in terms of element size.

(3) The accuracies used in the computations were relatively high and are probably not achieved in practical problems. This indicates that the discrepancy between the predictions of today's asymptotic theories and the results of practical computations should be larger than those shown here.

(4) Any self-adaptive mesh-refinement procedure depends critically on the complexity of the data structures it entails. There appears to be considerable need for studies of the data management problems of such mesh-refinements.

(5) For large problems in which the size and structure of the elements is significantly influenced by the geometry, the question remains how to obtain reasonable estimates of the reliability of computational results. Once again asymptotic results are needed here, since rigorous and realistic a-posteriori estimates are not likely to be obtainable. Some mathematical problems related to this question are addressed in the next section.

(6) In problem (3.1) the coefficient function was assumed to be constant inside the two subdomains Ω_1 and $\Omega - \Omega_1$. If, say, (3.1) represents a torsion problem, then u is the stress function and a describes the constitutive law. In this case, a theoretically better formulation requires a to be a function of grad u, and since grad u is very large near the singularity, a linear constitutive law is clearly unacceptable there. Let u_N and u_L be the exact solutions of the nonlinear and the linearized problem. We analyzed the error $e^{(h)} = u_L - u_L^{(h)}$ between u_L and the approximate solution of the linearized problem. This analysis remains relevant for the nonlinear problem only where $u_N - u_L$ is smaller than $e^{(h)}$. Once again an adaptive approach is needed here to decide where the linearization may be used.

4. Problems of Element Selection

A major question in the implementation of the finite element method is the best selection of the elements, that is, their shape, order, and use within the subdivision of the given domain. Extensive theoretical results are available about various element types, their properties, and influence upon the convergence, etc. But there are many other factors which need to enter into consideration. For instance, we should take account of the complexity of the input problem--a most laborious part of the method.

Practical experience has generated many opinions about the performance of different elements, and many articles have been devoted to experimental results on this topic. For example, it is widely agreed that the square bilinear element in R^2 performs slightly better than the corresponding square constructed of two triangular linear elements.

The performance of an element differs with the context and we should distinguish whether the element is used (i) in the interior of the domain, (ii) at its boundary, or (iii) in the presence of irregularities such as singularities, etc. We shall restrict ourselves here to some results about case (i), although some of the ideas are easily generalized to (ii). The third category (iii) requires special approaches (see, e.g., [18]).

Recent results [22], [23], and [24] about interior estimates for elliptic equations show that the error has two essential parts, a global and a local one. The global error is well understood and is generally of higher order than the local one.

Hence different elements of the same order have to be compared in terms of their local performance.

For computational ease there is good reason for the elements to be distributed locally regularly. Under this assumption, we may concentrate on the performance of meshes in R^n with translation properties. As an example, we present an analysis of a simple case which lends itself easily to considerable generalizations.

We denote by $H^k \equiv H^k(R^2)$ the standard Sobolev space over R^2, and assume that its norm is written in the form

$$(4.1) \qquad \|u\|^2_{H^k} = \int_{R^2} |Fu|^2 (1+(x_1^2+x_2^2))^k \, dx$$

where F is the Fourier transform of u. Let $S_h \subset H^t$, $t \geq 0$, be a family of functions depending on the real parameter $h \in [h_0,h_1]$. Then, for any two spaces H^{k_1},H^{k_2} with $k_2 \geq k_1$ and $t \geq k_1$, we define the approximation bound

$$(4.2) \qquad \Phi(H^{k_1},H^{k_2},S_h) = \sup_{\|u\|_{H^{k_2}} \leq 1} \|u-P_h u\|_{H^{k_1}}$$

where P_h is the orthogonal projection (in the sense of H^{-k_1}) of H^{k_1} onto S_h. It allows for a comparison of different sets $S_h^{[1]},S_h^{[2]}$ over the same parameter interval $h_0 \leq h \leq h_1$. More specifically, we call $S_h^{[1]}$ superior to $S_h^{[2]}$ with respect to H^{k_1},H^{k_2} if

$$(4.3) \qquad \Phi(H^{k_1},H^{k_2},S_h^{[1]}) < \Phi(H^{k_1},H^{k_2},S_h^{[2]}) .$$

In the case of equality, the sets are said to be equivalent with respect to the two spaces.

As examples, we consider the sets:

(i) $S_h^{[1]}$, the space of continuous, piecewise bilinear functions on squares of size h as shown in Figure 4.1;

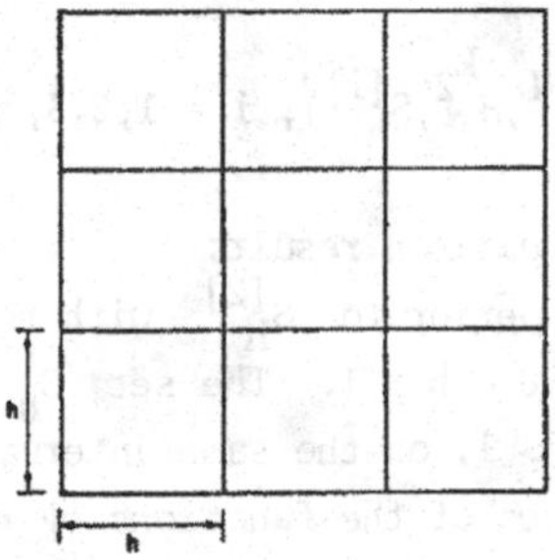

Figure 4.1

(ii) $S_h^{[2]}$, the space of continuous, piecewise linear functions on right triangles as shown in Figure 4.2;

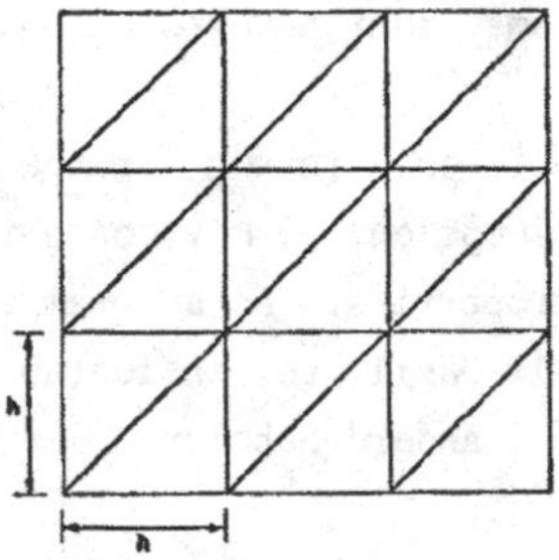

Figure 4.2

(iii) $S_h^{[3]}$, the space of continuous, piecewise linear functions on triangles as shown in Figure 4.3.

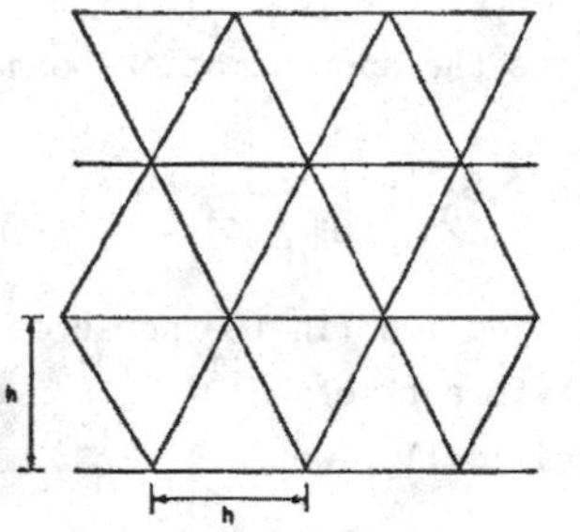

Figure 4.3

All three spaces have the same density of nodal points. They represent so-called $S_h^{t,k}$-spaces (see, e.g., [25]) with $t = 2$, $k = 3/2 - \varepsilon$, and any $\varepsilon > 0$. Then for any pair $k_1 \le k_2$, $k_1 \le k$ we have

$$\Phi(H^{k_1}, H^{k_2}, S_h^{t,k}) \le C(k_1, k_2, S_h^{t,k}) h^{\mu}, \quad \mu = \min(k_2 - k_1, t - k_1) \ . \tag{4.4}$$

This suggests in our case the definition

$$\psi(H^{k_1}, H^{k_2}, S_h^{[i]}) = h^{-\mu}\Phi(H^{k_1}, H^{k_2}, S_h^{[i]}), \ i = 1,2,3, \ \mu = \min(k_2 - k_1, 2 - k_1). \tag{4.5}$$

Now we have the following comparison result:

Theorem 4.1: The set $S_h^{[1]}$ is superior to $S_h^{[2]}$ with respect to H^1 and H^{k_2}, $k_2 > 1$, on the parameter interval $0 < h \le 1$. The set $S_h^{[1]}$ is equivalent to $S_h^{[3]}$ with respect to H^1 and H^{k_2}, $k_2 > 1$, on the same interval $0 < h \le 1$.

Table 1 below gives some values of the functions ψ of (4.5) for $i = 1,2$. All numbers are rounded to two digits.

Table 1

Values of the Function ψ of (4.5)

i = 1, bilinear elements

k_2	h = 1.0	h = 0.2	h = 0.02
1.5	.56	.57	.57
2	.31	.32	.32
3	.12	.11	.11

i = 2, triangular elements of Figure 4.2

k_2	h = 1.0	h = 0.2	h = 0.02
1.5	.58	.59	.59
2	.35	.44	.46
3	.18	.17	.17

The function Φ is closely related to the spectrum of certain operators which in turn were analyzed by means of the Fourier transform.

Table 1 is in complete agreement with the experience mentioned at the outset of this section. At the same time, Theorem 4.1 shows that, in general, we may not claim the superiority of the bilinear elements, since the spaces $S_h^{[1]}$ and $S_h^{[3]}$ are equivalent. It is interesting that ψ depends only weakly on h, which offers promise for asymptotic considerations.

The same analysis can be carried through for other classes of elements.

We turn now to some other aspects of comparing elements. In many problems the required output includes the values of certain derivatives at a number of points. This occurs, for instance, when stress data are desired in structural analysis. Often only the density of these stress-evaluation points is of importance and not their exact location. The problem then arises how to make best use of the earlier obtained data in locating these points and in calculating the desired quantities, such as the derivatives. In this connection, it is interesting to note some recent results [24] where for certain regular meshes superconvergence was attained by the use of some averaging process with coefficients that are independent of the given differential equation.

In order to bring the question into the framework of the discussion of the earlier part of this section, let $S_h \subset H^t$ be a one-parameter space of functions, and H^{k_1}, H^{k_2}, $k_2 \geq k_1$, $t \geq k_1$, given Sobolev spaces. In addition, suppose that we have some sets Z and T of linear functionals over H^{k_2} and S_h, respectively. The elements of Z are denoted by ζ and those of T by τ. In extension of (4.2) we may then define the quantities

(4.6) $$\gamma(H^{k_1},H^{k_2},S_h,\zeta,\tau) = \sup_{\|u\|_{H^{k_2}}\le 1} |\zeta u-\tau P_h u|$$

(4.7) $$\Gamma(H^{k_1},H^{k_2},S_h,\zeta) = \inf_{\tau\in T} \gamma(H^{k_1},H^{k_2},S_h,\zeta,\tau)$$

(4.8) $$N(H^{k_1},H^{k_2},S_h,Z) = \inf_{\zeta\in Z} \Gamma(H^{k_1},H^{k_2},S_h,\zeta)$$

The connection with the stated problem is evident. The quantity Γ indicates how best to recover the value ζu if only $P_h u$ is known, and N suggests which functional ζ is most advantageous, that is, where we might locate the stress points.

From a practical viewpoint, it may be difficult to evaluate any one of the three quantities. Nevertheless, they provide very useful theoretical information for comparing different elements.

As an example, we consider a simple one-dimensional problem. For this, let H^k now denote the Sobolev spaces $H^k(R_1)$ and S_h the space of continuous, piecewise linear functions on R^1 with nodal points at $x_j = jh$, $j = 0,\pm1,\pm2,\ldots$. We choose the spaces H^1 and H^{k_2} with $k_2 > 3/2$, and define T as the space of all linear functionals over S_h, and Z as

$$Z = \{\zeta_{x_0};\ \zeta_{x_0}(u) = \left.\frac{\partial u}{\partial x}\right|_{x=x_0},\ 0 \le x_0 \le h\}\ .$$

Then it follows that

(4.9) $$\Gamma(H^1,H^{k_2},S_h,\zeta_{x_0}) \le c(k_2,x_0)h^{k_2-3/2}$$

(4.10) $$N(H^1,H^{k_2},S_h,Z) \le \hat{c}(k_2)h^{k_2-3/2}\ .$$

The exponent on the right side is independent of the order of S_h. This is not surprising since this part of the estimate does not reflect the global part of the error which restricts the useful range of k_2.

Table 2 below gives some data for (4.10) rounded to two digits.

Table 2

Values of $h^{-(k_2-3/2)}N(H^1,H^{k_2},S_h,Z)$

k_2	h=1	h=0.2
2	.71	.67
3	.087	.088
4	.016	.016

Once again we note the very weak dependence on h.

Table 3 gives some data for (4.9) rounded to two digits.

Table 3

Values of $h^{-(k_2-3/2)}\Gamma(H^1,H^{k_2},S_h,\zeta_{x_0})$

	h=1.0			h=0.2		
k_2	$x_0 = 0$	$x_0=.25h$	$x_0=.5h$	$x_0=0$	$x_0=.25h$	$x_0=.5h$
2	.88	.75	.71	.94	.80	.67
3	.18	.14	.087	.21	.16	.088
4	.046	.034	.016	.055	.040	.016

Clearly, for any k_2 the center point $x_0 = .5h$ is best for the computation of the derivative. This choice brings a nonnegligible gain in accuracy.

Table 4 below concerns the optimal functional τ for approximating ζ_0 and $\zeta_{1/2h}$. We have

$$\tau(P_h u) = \frac{1}{2h} \sum_{j=-\infty}^{\infty} \alpha_j (P_h u)(jh)$$

where

$$\alpha_j = -\alpha_{-j} \text{ for } x_0 = 0, \quad \alpha_j = -\alpha_{-j+1} \text{ for } x_0 = \frac{1}{2} h$$

Table 4

Coefficients α_j, j = 0,...,5, for the Optimal Functional

<u>Case A</u>. $x_0 = 0$, $\alpha_j = -\alpha_{-j}$

	$k_1=2$		$k_1=3$		$k_1=4$	
j	h=1.0	h=.2	h=1.0	h=.2	h=1.0	h=.2
0	0	0	0	0	0	0
1	1.51	1.60	1.69	1.79	1.79	1.87
2	-.38	-.43	-.60	-.66	-.72	-.78
3	.11	.11	.26	.28	.37	.40
4	-.027	-.030	-.11	-.12	-.19	-.21
5	.0082	.0082	.046	.052	.099	.11

Case B. $x_0 = \frac{1}{2}h,\ \alpha_j = -\alpha_{-j+1}$

	$k_1=2$		$k_1=3$		$k_1=4$	
j	h=1.0	h=.2	h=1.0	h=.2	h=1.0	h=.2
1	2.50	2.50	2.51	2.57	2.48	2.57
2	-.30	-.29	-.32	-.32	-.31	-.31
3	.092	.070	.14	.11	.15	.11
4	-.020	-.021	-.050	-.049	-.062	-.059
5	.0070	.0057	.023	.022	.034	.031

The coefficients α_j show considerable variations in size and are sensitive to k_2. Therefore, it is worthwhile not to look for the best functional but to choose one which is independent of k_2. Evidently, different choices may be compared on the basis of the quotient γ/Γ, that is, the function

(4.11) $$\omega(H^{k_1},H^{k_2},S_h,\zeta,\tau) = \frac{\gamma(H^{k_1},H^{k_2},S_h,\zeta,\tau)}{\Gamma(H^{k_1},H^{k_2},S_h,\zeta)}$$

As an example, consider the functionals τ_{x_0} with the following coefficients:

(4.12)
$$x_0 = 0,\ \alpha_0 = 0;\ \ \alpha_1 = 1;\ \ \alpha_j = 0,\ \ j \geq 2;\ \ \alpha_j = -\alpha_{-j}$$
$$x_0 = \frac{1}{2}h,\ \alpha_1 = 2;\ \ \alpha_j = 0,\ \ j \geq 2;\ \ \alpha_j = -\alpha_{-j+1}$$

Table 5 gives some values of ω for this case.

Table 5

Values of ω for the Functionals (4.12)

	$x_0=0$		$x_0=\frac{1}{2}h$	
k_2	h=1	h=.2	h=1	h=.2
2	1.04	1.08	1.04	1.06
3	1.27	1.70	1.18	1.46
4	1.89	2.50	1.60	4.31

Another plausible functional at the mid-point might be given by

(4.13) $$x_0 = \frac{1}{2}h,\ \ \alpha_1 = \frac{27}{12},\ \ \alpha_2 = -\frac{1}{12},\ \ \alpha_j = 0,\ j \geq 3,\ \ \alpha_j = -\alpha_{-j+1}\ ,$$

for which some values of ω are given in Table 6.

Table 6

Values of ω for the Functionals (4.13)

k_2	h=1	h=.2
2	1.01	1.03
3	1.07	1.13
4	1.39	3.05

A comparison of Tables 5 and 6 shows that (4.13) is more advantageous. On the other hand, the spread of values is now larger (namely 3h instead of h). This could become critical when only locally uniform meshes are used.

It should be evident how this type of analysis may be carried out in two or three dimensions and for different types of elements. Another extension concerns the condition that the mesh is regular with the translation property. This assumption may be relaxed by requiring only the existence of a smooth mapping of the elements onto a regular mesh, as, for instance, in the case of isoparametric elements.

We end the section again with some observations and conclusions.

(1) On regular, or at least locally sufficiently regular meshes it is possible to develop theoretically sound comparisons of different elements with respect to various properties.

(2) Along the same line it is possible to analyze the best choice of certain functionals for the approximation of specific quantities computed from element data. In particular, it turns out that there are special points inside the elements where the computation of derivatives is more advantageous than at others.

(3) The approach presented here may well provide a tool for adaptive error controls. For example, in the evaluation of a derived quantity, say, the stress, several formulae of different asymptotic order may be compared to obtain some error estimates. The outcome of the comparison might then trigger appropriate modifications in the course of the computation. This type of approach has become standard in the numerical solution of initial value problems for ordinary differential equations.

(4) In general, programming considerations suggest a preference for meshes with regularity properties of the type needed here. However, in practice it is often necessary to introduce abrupt changes into a mesh. Such a change creates a boundary layer disturbance in the error near the "artificial interface". Experience has shown that disturbances of this kind tend to dampen out quickly, often within a distance of about 2h if the accuracy requirement is not too severe. There is a need for further theoretical studies of this aspect.

(5) Ideas of the type discussed in this section may also be useful for the comparison of the behavior of different elements at the boundary and possibly even near singularities.

5. The Solution of the Nonlinear Problem in R^n

The practical application of the finite element method leads to linear or nonlinear equations in R^n with large dimension n, and these equations are, in general, sparse, that is, each component equation depends only on a few of the n independent variables. Methods for solving large, sparse linear equations in R^n are beginning to be reasonably well understood, although there is certainly still room for further improvements. On the other hand, the computational solution of large, sparse nonlinear equations is as yet a rather poorly developed topic.

For any such nonlinear equations, the choice of an efficient solution process depends critically on our knowledge of the properties of the equation and their desired solution. Here, once again, it is important to select carefully the most advantageous formulation for the infinite dimensional model that is to be discretized. In addition, it appears that adaptive strategies are among the most promising tools for the design of effective procedures for solving large, sparse nonlinear equations in R^n. In this section we discuss a typical problem which illustrates these points.

Among the most widely used processes for solving equations arising in finite element applications are the so-called continuation methods. They are based on the fact that frequently the given equation depends on one or several real parameters and hence has the form

$$H(x,t) = 0, \quad t \in [0,t_{max}], \quad x \in R^n. \tag{5.1}$$

If, for each t, a solution $x(t)$ of (5.1) exists that varies continuously with t, then the function

$$x : [0,t_{max}] \to R^n \tag{5.2}$$

constitutes a curve in R^n between the--assumed to be given--point $x^0 = x(0)$ and an unknown solution $\hat{x} = x(t_{max})$. The continuation processes use the curve as a guide and channel their iterates in its proximity from x^0 to $\hat{x}$.

This approach often has a clear physical meaning as, for instance, in the case of quasi-static behavior as a function of time. In this connection, it is also important to note that the curve (5.2) may not be defined by an equation of the form (5.1) but instead by some initial value problem. The so-called Davidenko processes create such a differential equation artifically by differentiating (5.1). In practice, however, such an equation often arises in a more natural way.

In order to focus the discussion, we consider now a quasi-static torsion problem for an infinite rod with cross-section $\bar{\Omega} \subset R^2$. In its formulation, we use the so-called single-curve hypothesis in the theory of plasticity (see, e.g., [26]). The basic relations between the stress components τ_1,τ_2 and strain-components γ_1,γ_2 as functions of x_1, x_2 and t, are as follows:

$$\text{(5.3)}\quad \begin{aligned}
&\text{(a)}\quad \frac{\partial}{\partial x_1}\tau_1 + \frac{\partial}{\partial x_2}\tau_2 = 0 \\
&\qquad\qquad\qquad\qquad\qquad \text{on } \Omega,\ t \in R^1 \\
&\text{(b)}\quad -\frac{\partial}{\partial x_2}\gamma_1 + \frac{\partial}{\partial x_1}\gamma_2 = 2\omega(t) \\
&\text{(c)}\quad \tau_1 \cos(nx_1) + \tau_2 \cos(nx_2) = 0 \text{ on } \partial\Omega \ (n \text{ the outward normal}) \\
&\text{(d)}\quad \gamma_i = g(\Gamma)\tau_i,\ i = 1,2,\ \Gamma = \gamma_1^2 + \gamma_2^2
\end{aligned}$$

Alternately, the constitutive laws (5.3d) may be re-written in the form

$$\text{(5.4)}\qquad \gamma_i = \bar{g}(T)\tau_i,\ i = 1,2,\ T = \tau_1^2 + \tau_2^2 .$$

With (5.4) and the stress-function u, that is,

$$\text{(5.5)}\qquad \tau_1 = \frac{\partial u}{\partial x_2},\quad \tau_2 = -\frac{\partial u}{\partial x_1},$$

we obtain the standard boundary value problem

$$\text{(5.6)}\qquad \begin{aligned}
&-\sum_{i=1}^{2} \frac{\partial}{\partial x_i}\, \bar{g}(u_{x_1}^2 + u_{x_2}^2)\, \frac{\partial u}{\partial x_i} = 2\omega \text{ on } \Omega \\
&u = 0 \text{ on } \partial\Omega
\end{aligned}$$

The load function ω of t usually satisfies $\omega(0) = 0$, and then we also have the initial condition

$$\text{(5.7)}\qquad u(x,0) = 0,\ x \in \Omega.$$

A different formulation is obtained if we do not eliminate the strain components γ_1, γ_2. For this, suppose that--as for hypoelastic materials, (see, e.g., [27])--the constitutive laws (5.3d) are only available in "incremental" form involving the derivatives $\dot{\tau}_1, \dot{\tau}_2$ and $\dot{\gamma}_1, \dot{\gamma}_2$ of the stresses and strains with respect to t, respectively. In the simplest case--using already (5.5)--we have then

$$\text{(5.8)}\qquad \dot{\gamma}_i = (-1)^i \sum_{j=1}^{2} h_{ij}(\gamma_1, \gamma_2)\, \frac{\partial \dot{u}}{\partial x_j},\ i = 1,2$$

and, more generally, the h_{ij} may also depend on the stresses and the sign of the t-derivatives of the so-called yield functions. From the derivative form of (5.3b) and (5.8) we obtain now the equation

$$\text{(5.9)}\qquad \sum_{i,j=1}^{2} \frac{\partial}{\partial x_i}\, \hat{h}_{ij}(\gamma_1, \gamma_2)\, \frac{\partial \dot{u}}{\partial x_j} = 2\dot{\omega} \text{ on } \Omega$$

where $\hat{h}_{11} = h_{21}$, $\hat{h}_{12} = h_{22}$, $\hat{h}_{21} = h_{11}$, $\hat{h}_{22} = h_{12}$. The system of differential equations (5.8)/(5.9) has to be taken together with the boundary and initial conditions

$$\text{(5.10)} \qquad \begin{aligned} &u = 0 \text{ on } \partial\Omega \\ &u(x,0) = 0,\ \gamma_1(x,0) = \gamma_2(x,0), \text{ on } \Omega. \end{aligned}$$

In the case of the first formulation (5.6), a standard finite element approximation leads to a system of equations on R^n,

$$\text{(5.11)} \qquad A(x)x = c(t),\quad x \in R^n,\quad t \in [0,t_{max}],\quad c(0) = 0\ ,$$

which is of the form (5.1). On the other hand, for the system (5.8)/(5.9) we obtain the approximating system of differential equations

$$\text{(5.12)} \qquad \begin{aligned} &\text{(a)}\quad B(y)\dot{x} = \dot{c} \\ &\text{(b)}\quad \dot{y} = K(y)\dot{x} \end{aligned} \Bigg\} \quad x \in R^n,\quad y \in R^m,\quad t \in [0,t_{max}],\ c(0) = 0\ .$$
$$\text{(c)}\quad x(0) = 0,\ y(0) = 0$$

The components of x represent nodal values while those of y are associated with Gaussian-quadrature points in the interior of the elements. Hence m is considerably larger than n. For the computation, it is useful to discretize also the (redundant) equation (5.3b). This gives the additional equation

$$\text{(5.12)} \qquad \text{(d)}\quad My = c.$$

The equations (5.12) are essentially those used in [28].

The evaluation of the matrix functions A and B involves a full assembly of the elemental stiffness matrices and hence is costly. On the other hand, (5.12b) is a direct discretization of (5.8) and thus K is very sparse and simpler to evaluate. Similarly, M is sparse and relatively cheap to compute.

When the formulation (5.6) and hence the equation (5.11) is used, then the derivative of the operator $x \to A(x)x$ on R^n is practically inaccessible. In fact, even a discrete approximation of this derivative, based on several values of A, would be extremely costly to compute. Therefore, it is desirable to consider only processes that use evaluations of A itself. But this is a very restrictive condition. In fact, most such processes show only a slow rate of convergence which may even deteriorate with increasing n (see, e.g., [29]), and there may also be limitations to the possible size of $c(t)$. As an example, consider the simple algorithm

$$\text{(5.13)} \qquad A(x^j)x^{j+1} = c(t_k),\quad j = 0,1,\ldots$$

for solving (5.11) with fixed $t = t_k$. The following convergence result was proved in [30] (see also [31]).

Theorem 5.1: Suppose that $A : D \subset R^n \to L(R^n)$ satisfies

$$\|A(x)-A(y)\| \leq \beta\|x-y\|, \ \forall \ x,y \in D$$

and that $x^* \in \text{int}(D)$ is a solution of (5.11) for $t = t_k$ such that $A(x^*)$ is non-singular and

$$\beta\eta^2\|c(t_k)\| < 1, \ \eta = \|A(x^*)^{-1}\| . \tag{5.14}$$

Then for any starting point x^0 in the ball

$$\bar{B}(x^*,r) \subset D, \quad r < \frac{1}{\beta\eta} \ (1-\beta\eta^2\|c(t_k)\|)$$

the process (5.13) converges to x^*.

The result indicates that the attraction radius r decreases with increasing $\|c(t_k)\|$. The proof also shows that the convergence factor deteriorates accordingly. This has been widely observed in connection with structural problems. Numerical experiments show that already for relatively small values of $\|c(t_k)\|$ the process (5.13) fails to converge. Analogous results hold for several variations of this method (see, e.g., [31]).

In cases where the derivative is available, it is readily seen that Newton's method and its modified forms do not share this behavior. This represents a strong reason against the formulation (5.6) and its finite analog (5.11), whenever the necessary derivative is too costly to evaluate.

But we need not remain with the standard formulation (5.6) and, in fact, the equations (5.12) provide sufficient information for Newton-type processes without excessive computational costs.

In order to see the relation between (5.11) and (5.12), suppose that (5.12b) has a total differential and hence that

$$y = G(x). \tag{5.15}$$

This corresponds to the constitutive law in the form (5.4). Then (5.12a) and (5.12b) reduce to

$$\begin{aligned} &\text{(a)} \quad B(G(x))\dot{x} = \dot{c}, \quad x(0) = 0 \\ &\text{(b)} \quad MG(x) = c(t). \end{aligned} \tag{5.16}$$

Now note that (5.16b) represents the discretization of (5.3b)--after substitution of (5.4) and (5.5)--and hence that the equations (5.16b) and (5.11) are identical. On the other hand, (5.16a) is the discretization of the derivative of (5.3b)--after substitution of the derivatives of (5.4) and (5.5). Under certain conditions, the operations of differentiation and discretization commute (see [32]), and then it follows that BG must be the derivative of the operator $x \to A(x)x$ on R^n. In this case, Newton's method for the solution of (5.16b) (with fixed t) is given by

$$B(G(x^j))(x^{j+1}-x^j) = c(t) - MG(x^j), \; j = 0,1,\ldots . \tag{5.17}$$

In general, BG need not be the exact derivative of MG. But then MG and BG are still approximations of an infinite dimensional operator and of its derivative, respectively, and hence it may be expected that the method (5.17) continues to converge well. This and related mathematical problems will be discussed elsewhere.

The method (5.17) solves, of course, only the equation (5.16b) for one fixed t. Moreover, we need an appropriate starting point for the iteration. This is a critical problem, especially since the convergence domains of Newton-type methods often shrink drastically with growing n (see, e.g., [29]). Here we have to depend on the continuation approach to progress along the curve (5.2).

However, in considering this approach it is necessary to decide first whether the entire curve is of interest and should be approximated closely, or whether the curve itself is of lesser interest and only a limited number of target points on it should be computed. In general, the literature on continuation methods addresses the first objective; but in high-dimensional problems only the second objective is a reasonable one. Accordingly, it appears that here the so-called numerical continuation methods are the suitable choice. They use a locally convergent iterative method--such as (5.17)--for solving (5.16b) at an increasing sequence of discrete parameter values $0 = t_0 < t_1 < \ldots < t_m < t_{max}$ corresponding to the target points. The locations of these points are rarely given; instead, only a reasonable density is specified. Hence, some effective step-algorithm for the t_i-values is needed which meets our objective. An adaptive method of this type was described in [30], [31]. It allows the process to move more rapidly along the curve whenever the local iterations are less sensitive to the choice of their initial values.

Briefly, suppose that we have progressed through the parameter values $0 = t_0 < t_1 < \ldots < t_k < t_{max}$ and that $\bar{x}^j$, $j = 1,\ldots,k$, $(\bar{x}^0 = x^0)$, are the terminal iterates of the local method at these t-values. From these data we obtain by suitable extrapolation an approximation $x = \hat{x}(t)$ of the desired curve (5.2) as well as an upper bound of the error $\varepsilon(t) = \|x(t)-\hat{x}(t)\|$ for all t in some interval $[t_k, t_k+\delta_k]$. In addition, during the local iterations estimates of the convergence domains are established which then allow--again by extrapolation--for a prediction of the radius $r(t)$ of a possible convergence ball $\bar{B}(x(t),r(t))$ at any $t \in [t_k, t_k+\delta_k]$. The next step $h = t_{k+1} - t_k$ may then be chosen such that

$$\varepsilon(t_k+h) \leq \theta r(t_k+h)$$

where $\theta \to 0$ is a suitable factor to adjust for the uncertainty of the various estimates.

As in the case of modern methods for initial value problems for ordinary differential equations, the type and degree of the extrapolation may be adaptively controlled. The critical part of the process is, of course, the estimation of the local convergence radii. For this computable bounds are hardly available. But exact bounds are not needed; in fact, we require only estimates which are neither too pessimistic nor too expensive and which usually lead to convergence. If convergence is not obtained, the prediction can be adaptively corrected. In other words, we need estimates which are heuristic in nature but are based on a sound mathematical analysis, for instance, of an asymptotic type.

The adaptive continuation process has been applied to a variety of problems with excellent success. Some computational results are given in the cited articles [30], [31].

In the case of (5.16), we noted already that B is fairly costly to evaluate, while M and G are relatively simple to compute. This suggests the use of a modified Newton method in place of (5.17), where BG is not re-evaluated at each step.

Evidently, it is also possible to solve (5.16a) directly as an ordinary differential equation using, of course, a method applicable to such stiff equations. Once again, because BG is costly to evaluate and also because it changes much less than x, we are led to use different steps for computing x and BG. Here, as in the case of the modified Newton method above, an adaptive technique is needed to select the different time steps.

Clearly, whenever (5.15) is available, a direct solution of the differential equation (5.16a) is likely to be less effective and not in line with our continuation objective. But when (5.12b) does not have a total differential, then we cannot avoid the solution of the initial value problem (5.12a-c). As before, because of the differences between x,y and B(y) in physical context and evaluation costs, different time steps for computing these quantities are highly advisable. In addition, we have the auxiliary equation (5.12d), which can be used advantageously to correct the process, and this equilibration iteration may be scheduled more or less frequently. Hence, we are faced with the simultaneous use of four different time steps, namely (i) for x, (ii) for y, (iii) for updating B, and (iv) for the iteration involving (5.12d). In practice, the steps (i) and (ii) are often the same. Experience has shown that the accuracy of the solution, cost of computation and other related factors are often rather sensitive to the selection of the various t-steps. Clearly then, an a-priori selection cannot be made by the user and again adaptive procedures are the only reasonable answer for this problem.

As before, we end this section with several concluding remarks.

(1) The selection of the appropriate formulation of a problem about to be discretized is important for the design of efficient methods for solving the resulting finite dimensional equations. Although we discussed here only one particular example, it should be evident that similar situations arise in connection with many other problems, as, for example, in the case of geometrical nonlinearities.

(2) Continuation approaches for solving large nonlinear equations in R^n possess very attractive properties. However, care must be taken in selecting the defining relations for the continuation curve, in order not to introduce undesirable computa-

tional restrictions. Moreover, the objective of the continuation approach must be clearly delineated. For large equations, the numerical continuation technique appears to offer advantages over the frequently-discussed Davidenko methods.

(3) Adaptive steplength algorithms for numerical continuation are now available; the open problem is their extension to the more general problems of the type of (5.12). Here the interesting question arises under what condition the correction of the process by means of the Newton-type method for solving (5.12d) does indeed lead the iterates back onto the curve. When (5.15) holds, this is clearly the case, but this is certainly not a necessary condition. These problems will be addressed elsewhere.

References

[1] W. Pilkey, K. Saczalski, H. Schaeffer (editors), Structural Mechanics Computer Programs, Surveys, Assessments, and Availability, University of Virginia Press, Charlottesville, Va., 1974.

[2] L. D. Landau, M. E. Lifshitz, Theory of Elasticity, Pergamon Press, 1970, pp. 44-53.

[3] M. Filonenko-Borodich, Theory of Elasticity, P. Noordhoff, Groningen, The Netherlands, 1964.

[4] I. Babuška, The stability of the domain of definition with respect to basic problems of the theory of partial differential equations especially with respect to the theory of elasticity, I, II (in Russian), Czechosl. Mat. J., 1961, pp. 76-105, 165-203.

[5] I. Babuška, The theory of small change in the domain of definition in the theory of partial differential equations and its applications, in Proceedings of the EQUADIFF Conference , Prague, 1962, pp. 13-26.

[6] I. Babuška, The continuity of the solutions of elasticity problems on small deformation of the region (in German), ZAMM, 1959, pp. 411-412.

[7] G. Birkhoff, The numerical solution of elliptic equations, Reg. Conf. Series in Appl. Math., Vol. 1, SIAM Publications, Philadelphia, Pa., 1971.

[8] I. Babuška, The stability of domains and the question of the formulation of the plate problems (in German), Apl. Mat., 1962, pp. 463-467.

[9] D. Morgenstern, Bernoulli hypotheses for beam and plate theories (in German), ZAMM, 1963, pp. 420-422.

[10] D. Morgenstern, Mathematical foundations of shell theory (in German), Archive for Analysis 3, 1959, pp. 91-96.

[11] I. Babuška, M. Prager, Reissnerian algorithms in the theory of elasticity, Bul. de l'Academia Pol., Ser. Sci. Techn., 1960, pp. 411-417.

[12] I. Babuška, R. Babušková, I. Hlaváček, B. Kepr, L. Pachta, J. Švejdová, Algorithms of Reissnerian type in mathematical theory of elasticity (in Czech), Sbornik ČVUT, 1962, pp. 15-47.

[13] R. B. Kellogg, Higher order singularities for interface problems, in Mathematical Foundations of the Finite Element Method with Applications to Partial Differential Equations , A. K. Aziz (editor), Academic Press, New York, 1972, pp. 589-602.

[14] R. B. Kellogg, Singularities in interface problems, in Numerical Solution of Partial Differential Equations II, B. Hubbard (editor), Academic Press, New York, 1971, pp. 351-400.

[15] A. H. Schatz, Private communication, to appear.

[16] E. A. Volkov, Method of composite meshes for bounded and unbounded domains with piecewise smooth boundary, Proc. Steklov Inst. Math. 96, 1968, pp. 117-148.

[17] P. Laasonen, On the degree of convergence of discrete approximations for the solution of the Dirichlet problem, Ann. Acad. Sci. Finn. Ser. AI, 1957, p. 246.

[18] I. Babuška, R. B. Kellogg, Mathematical and computational problems in reactor calculations, in Proceedings of the 1973 Conference on Mathematical Models and Computing Techniques for Analysis of Nuclear Systems University of Michigan, Ann Arbor, Mich., 1973, pp. VII-67-VII-94.

[19] I. Babuška, Finite element method for domains with corners, Computing 6, 1970, pp. 264-273.

[20] I. Babuška, W. Rheinboldt, C. Mesztenyi, Self-adaptive refinements in the finite element method, University of Maryland, Computer Science Technical Report TR-375, May 1975.

[21] I. Babuška, Self-adaptiveness and the finite element method, in The Mathematics of Finite Elements and Applications II, J. Whiteman (editor), Academic Press, London, to appear.

[22] J. A. Nitsche, A. H. Schatz, Interior estimates for Ritz-Galerkin methods, Math. Comp. 28, 1974, pp. 937-958.

[23] A. H. Schatz, L. B. Wahlbin, Interior maximum norm estimates for finite element methods, Private preprint, 1975.

[24] J. H. Bramble, A. H. Schatz, Higher order local accuracy by averaging in the finite element method, Private preprint, 1975.

[25] I. Babuška, A. K. Aziz, Survey lectures on the mathematical foundations of the finite element method, in The Mathematical Foundations of the Finite Element Method with Applications to Partial Differential Equations, A. K. Aziz (editor), Academic Press, New York, 1972, pp. 3-359.

[26] L. M. Kachanov, Foundations of the Theory of Plasticity, North Holland Publ. Co., Amsterdam, The Netherlands, 1971.

[27] Y. C. Fung, Foundations of Solid Mechanics, Prentice-Hall, Englewood Cliffs, N.J., 1965.

[28] K. J. Bathe, An assessment of current finite element analysis of nonlinear problems in solid mechanics, in Proceedings of the Symposium on the Numerical Solution of Partial Differential Equations 1975, B. Hubbard (editor), Academic Press, New York, 1976, pp. 117-164.

[29] W. Rheinboldt, On the solution of large, sparse sets of nonlinear equations, in Computational Mechanics, J. T. Oden (editor), Lecture Notes in Mathematics, Vol. 461, Springer Verlag, Heidelberg, Germany, 1975, pp. 169-194.

[30] W. Rheinboldt, An adaptive continuation process for solving systems of nonlinear equations, University of Maryland, Computer Science Technical Report TR-393, July 1975.

[31] W. Rheinboldt, On the solution of some nonlinear equations arising in the applications of the finite element method, in The Mathematics of Finite Elements and Applications II , J. Whiteman (editor), Academic Press, London, to appear.

[32] J. Ortega, W. Rheinboldt, On discretization and differentiation of operators with application to Newton's method, SIAM J. Num. Anal., 1966, pp. 143-156.

[33] D. Morgenstern, I. Szabo, Vorlesungen über theoretische Mechanik, Springer-Verlag, 1961, pp. 120-128.

[34] W. T. Koiter, On the foundation of the linear theory of thin shells, I, II, Proc. of Koninkl Nederl. Akad. Wetenschapen, Ser. B, 73, 1970, pp. 169-182, 183-195.

[35] I. N. Vekua, On two ways of constructing the theory of elastic shells, Proc. of International Congress of Theor. and Appl. Mech. 1972, Springer-Verlag, 1973, pp. 322-339.

[36] I. N. Vekua, On one version of the consistent theory of elastic shells, Proc. of International Congress of Theor. and Appl. Mech. 1967, Springer-Verlag, 1969, pp. 59-84.

ESTIMATIONS D'ERREUR DANS L^∞ POUR LES INEQUATIONS A OBSTACLE

C. BAIOCCHI
Istituto di Matematica dell'Università
et L.A.N. del C.N.R.
Pavia (Italie)

RESUME: Soit w_h (resp. u_h) la solution approchée obtenue en discrétisant par éléments finis du premier ordre une équation (resp. une inéquation) variationnelle dont la solution est u. On compare les quantités $\|u-u_h\|_{L^\infty}$ et $\|u-w_h\|_{L^\infty}$ (cf. (4.2) suivante); on en déduit une estimation "presque optimale" pour $\|u-u_h\|_{L^\infty}$ (cf. (4.3) suivante).

NOTATIONS: $W^{k,p}(\Omega)$ est l'espace des fonctions dont les dérivées jusqu'à l'ordre k sont dans $L^p(\Omega)$; $H^k(\Omega)=W^{k,2}(\Omega)$; $H^1_o(\Omega)$ est l'espace des fonctions de $H^1(\Omega)$ nulles sur $\partial\Omega$; $H^{-1}(\Omega)$ est le dual de $H^1_o(\Omega)$; $C^{k,\alpha}(\bar{\Omega})$ est l'espace des fonctions dont les dérivées jusqu'à l'ordre k sont höldériennes de exposant α sur $\bar{\Omega}$. Toutes les fonctions considérées sont à valeurs réelles.

1. DESCRIPTION DU PROBLEME.

Il est bien connu (cf. p.ex. [5],[18]) que pour les problèmes aux limites du type:

(1.1) $\quad Au = f$ dans Ω ; $u = 0$ sur $\partial\Omega$

(Ω ouvert borné de $\mathbb{R}^m$, $m \geq 1$; A opérateur différentiel linéaire du deuxième ordre, elliptique, à coefficients réguliers (1)),pour tout $k \in \mathbb{N}$, quitte à discrétiser en éléments finis d'ordre suffisamment élevé, on a (2):

(1.2) $\quad \|u-u_h\|_{L^\infty(\Omega)} \leq C_k \|u\|_{H^{k+1}(\Omega)} h^k;$

et il s'agit d'estimations "bien adaptées" car pour le problème (1.1) on a le résultat de régularité: $f \in H^{k-1}(\Omega) \rightarrow u \in H^{k+1}(\Omega)$.

---

(1) Pour préciser le signe (essentiel dans (1.3) suivante) on peut p. ex. choisir $A = -\sum_{i=1}^{m} \partial^2/\partial x_i^2$.

(2) h étant le pas de discrétisation, u_h la correspondente solution discrète; cf. toujours [5],[18], pour les hypothèses précises.

Au contraire, pour la discrétisation d'inéquations, du type:

$$(1.3)\quad \begin{cases} u\geq\psi \text{ et } Au\geq f \text{ dans } \Omega \text{ ; } u = 0 \text{ sur } \partial\Omega \\ Au = f \text{ où } u>\psi \end{cases}$$

une estimation du type (1.2) (si elle était valable!)ne serait utilisable que pour k=1 ([3]) car en général, même pour ψ,f très régulières, on n'a pas $u\in H^3(\Omega)$.

Pour le problème (1.3) on a toutefois des résultats de régularité, p.ex. (cf. [8],[3],[13]):

$$(1.4)\quad f\in L^p(\Omega), \psi\in W^{2,p}(\Omega) \longrightarrow u\in W^{2,p}(\Omega)$$

et donc il serait intéressant d'étendre à la approximation de (1.3) la validité d'estimations du type ([4]):

$$(1.5)\quad \text{si}\quad u\in W^{2,p}(\Omega)\ \forall p<+\infty, \text{ on a } \|u-u_h\|_{L^\infty(\Omega)} = O(h^{2-\varepsilon})\ \forall \varepsilon>0.$$

On va montrer qu'en effet, discrétisant (1.3) en éléments finis linéaires, et imposant la contrainte $u\geq\psi$ uniquement aux noeuds de la triangulation ([5]), sous des conditions géométriques simples sur la triangulation, (1.5) est valable.

Avant de passer aux détails on veut souligner l'intérêt d'estimations L^∞ pour les problèmes à obstacle. Posons, u étant la solution de (1.3):

$$\Omega^+ = \{x\in\Omega \mid u(x)>\psi(x)\};$$

il est évident ([6]) qu'aucune convergence de u_h à u ne peut assurer la convergence à Ω^+ des $\Omega_h^+ = \{x\in\Omega \mid u_h(x)>\psi(x)\}$ ([7]); toutefois on peut montrer ([8]) que si ε_h satisfait:

$$\varepsilon_h>0;\ \lim_{h\to 0^+} \varepsilon_h = \lim_{h\to 0^+} \| u-u_h \|_{L^\infty(\Omega)}/\varepsilon_h = 0,$$

posant $\tilde{\Omega}_h^+ = \{x\in\Omega \mid u(x)>\psi(x)+\varepsilon_h\}$ on a que $\tilde{\Omega}_h^+$ est une approximation de Ω

([3]) Pour k=1 (1.3) est valable (et utilisable!);cf. [7],[12]. En employant les espaces de Sobolev fractionnaires on peut arriver jusqu'à k<3/2; cf. [4]. Pour des généralités sur ce sujet on renvoye à la conférence de Mosco [10] de ce Symposium.

([4]) Pour la validité de (1.5) dans le cas des équations cf.[14],[15] pour le problème de Dirichlet et [16] pour le problème de Neumann.

([5]) Ce qui fournit un problème discret "plus simple" à résoudre (cf.le n. 3 suivant). Un résultat analogue, dans le quel on impose la contrainte $u_h\geq\psi$ partout, a été présenté à ce Symposium par Nitsche [15].

([6]) Il suffit de choisir $u\equiv\psi\equiv 0$; $u_h\equiv h$ (donc $u_h\to u$ "au mieux") pour avoir $\Omega_h^+\equiv\Omega$, $\Omega^+ = \emptyset$.

"convergeante " (i.e. $\lim_{h\to 0^+} \tilde{\Omega}_h^+ = \Omega^+$, limite au sens des ensembles) et "par défaut" (i.e. $\tilde{\Omega}_h^+ \subset \Omega^+$ pour h petit). En particulier, si (1.5) est remplie, une telle approximation de Ω^+ est donnée, pour tout $\varepsilon\in]0,2[$, par:

$$\Omega^+_{h,\varepsilon} = \{x\in\Omega \mid u_h(x) > \psi(x) + h^{2-\varepsilon}\}.$$

2. HYPOTHESES SUR LA TRIANGULATION.

Soit $\mathcal{T}_h$ une triangulation ([9]) de Ω ; pour fixer les idées (et les notations) on va supposer que $\forall T\in\mathcal{T}_h$ on a $T\subset\bar{\Omega}$; $\{P_i\}_{i=1,\dots,N}$ désigneront les sommets intérieurs ([10]); et on supposera que $\forall i$, $\forall T\in\mathcal{T}_h$, si $P_i\in\partial T$ on a $T\subset\mathcal{B}(P_i,h/2)$ ([11]).

Pour tout i on notera $\phi_i(x)$ la fonction continue sur $\bar{\Omega}$, nulle hors de $\bigcup_{T\in\mathcal{T}_h} \bar{T}$, affine sur chaque $T\in\mathcal{T}_h$, qui vaut 1 sur P_i et 0 sur les autres sommets de $\mathcal{T}_h$; on remarquera que:

(2.1) posant S_i = intérieur du support de ϕ_i, on a $S_i\subset\mathcal{B}(P_i,h/2)$.

On notera $\vec{u}\equiv(u_i)$, $\vec{v}\equiv(v_i)$ etc. les vecteurs de $\mathbb{R}^N$; on posera $(\vec{u}|\vec{v}) = \sum_{i=1}^N u_i v_i$.

On définit p_h, r_h par:

$$(2.2)\quad \begin{cases} (p_h\vec{v})(x) = \sum_{i=1}^N v_i\phi_i(x); \quad \overrightarrow{r_h(v)} \equiv (r_{h,i}(v)) \text{ où} \\ r_{h,i}(v) = \dfrac{1}{\text{mes } S_i}\displaystyle\int_{S_i} v(x)dx \text{ ([12])}; \end{cases}$$

([7]) Et toutefois, dans des nombreux problèmes concrets, la "vraie" inconnue du problème (1.3) est $\partial\Omega^+$; cf. p.ex. [1] où l'on a donné un premier résultat de convergence de Ω_h^+ à Ω^+ rélatif à une discrétisation de (1.3) en différences finies.

([8]) Cf. [2] où l'on a entroduit cet artifice pour l'approximation d'un problème de frontière libre parabolique. Quelques compléments à ce résultat, ainsi qu'à la validité de (1.5), seront donnés dans un travail en préparation.

([9]) Au sens usuel; cf. p.ex. [5],[18]. On remarquera que seulement la terminologie est bidimensionnelle, les résultats restant valables dans $\mathbb{R}^m$, $m\geq 1$ quelconque.

([10]) A savoir P_i est un sommet d'un triangle de $\mathcal{T}_h$, et P_i est intérieur à $\bigcup_{T\in\mathcal{T}_h}\bar{T}$. On note N au lieu de N_h par brévité.

([11]) $\mathcal{B}(P,r)$ désignant la boule ouverte de rayon r centrée en P; en particulier on a $\max_{T\in\mathcal{T}_h} \text{diam}(T) \leq h$.

et on va expliciter quelques propriétés (immédiates!) de p_h, r_h dont on aura bésoin [13]:

(2.3) $r_h : L^2(\Omega) \to \mathbb{R}^N$ est linéaire continue non décroissante; $\overline{r_h(1)} = \vec{1}$ [14];

(2.4) $p_h : \mathbb{R}^N \to H_o^1(\Omega)$ est linéaire continue injective; $\|p_h \vec{v}\|_{L^\infty(\Omega)} = |\vec{v}|_\infty$ [15];

(2.5) p_h est non décroissante, donc $p_h^* : H^{-1}(\Omega) \to \mathbb{R}^N$ l'est aussi [16];

finalement on rappelle que (cf. p.ex. [5],[18]) sous des consitions géométriques simples sur $\mathcal{T}_h$ on a:

$$(2.6)\quad \begin{cases} \text{si, } p<+\infty,\ v \in H_o^1(\Omega) \cap W^{2,p}(\Omega), \text{ on a:} \\ \forall \varepsilon>0,\ \|u - p_h r_h u\|_{L^\infty(\Omega)} = O(h^{2-\varepsilon}). \end{cases}$$

Une dernière propriété, reliant p_h à l'opérateur différentiel A(cf. (3.19) (3.10) suivantes) sera imposée plus loin. Dégageons maintenant une conséquence de (2.1). Posons, pour $\alpha \in]0,1]$ (ce sont les sémi normes dans $C^{0,\alpha}(\bar{\Omega})$ et $C^{1,\alpha}(\bar{\Omega})$ resp.):

$$(2.7)\quad [v]_{0,\alpha} = \sup_{x,y \in \Omega} \frac{|v(x)-v(y)|}{|x-y|^\alpha};\ [v]_{1,\alpha} = [|\text{grad} v|]_{0,\alpha}.$$

On a évidemment:

(2.8) s'il existe $x \in S_i$ avec $v(x)=0$, on a $r_{h,i}(v) \leq [v]_{0,\alpha} h^\alpha$ [17].

<u>LEMME</u>: Soit $v \geq 0$. On a:

(2.9) s'il existe $x \in S_i$ avec $v(x)=0$, on a $r_{h,i}(v) \leq [v]_{1,\alpha} h^{1+\alpha}$.

Dém.: On remarque d'abord que $|(\text{grad} v)(x)| = 0$ (si non v déviendrait négative sur un segment sortant de x) donc $|(\text{grad} v)(y)| \leq [v]_{1,\alpha} h^\alpha$ pour tout

---

[12] On peut bien entendu remplacer cette formule par $r_{h,i}(v) = = (\int_\Omega \phi_i(x)dx)^{-1} \int_\Omega \phi_i(x) v(x) dx$; si v est "regulier", en changeant quelques constantes on peut aussi choisir $r_{h,i}(v) = v(P_i)$.

[13] Les monotonies sont par rapport aux ordres "naturels": $\vec{u} \leq \vec{v}$ signifie $u_i \leq v_i\ \forall_i$; $u \leq v$ dans $L^2(\Omega)$ (ou $H_o^1(\Omega)$) signifie $u(x) \leq v(x)$ p.p.; dans $H^{-1}(\Omega)$ on a l'ordre dual de $H_o^1(\Omega)$ (qui est "compatible": pour $u,v \in L^2(\Omega)$ on a $u \leq v$ au sens de $H^{-1}(\Omega) \leftrightarrow u(x) \leq v(x)$ p.p.).

[14] Pour $\lambda \in \mathbb{R}$ on note $\vec{\lambda}$ l'élément de $\mathbb{R}^N$ dont toutes les composantes sont λ .

[15] $|v|_\infty = \max_i |v_i|$.

[16] $p_h^* \equiv p_{h,i}^*$ étant la transposée de p_h. On remarquera que $f_{|S_i} = g_{|S_i}$ entraine $p_{h,i}^*(f) = p_{h,i}^*(g)$.

$y \in \mathcal{B}(x,h)$, en particulier $\forall y \in S_i$ (cf. [17]).

On en tire, $\forall y \in S_i$:

$$v(y) = v(y)-v(x) =^{(18)} \int_0^1 \frac{d}{dt} v(x+t(y-x))\, dt =$$

$$= \int_0^1 ((\text{grad} v)(x+t(y-x))|y-x)dt \leqslant [v]_{1,\alpha} h^\alpha |y-x| \leqslant [v]_{1,\alpha} h^{1+\alpha};$$

d'où la thèse, grâce à (2.3).

Par un simple procédé par l'absurde, de (2.8), (2.9), on déduit:

$$r_{h,i}(v) > [v]_{\ell,\alpha} h^{\ell+\alpha},\ v \geqslant 0 \rightarrow v_{|S_i} > 0\ (\ell=0,1);$$

en particulier, si l'on cherche $\sigma_h(v)$ tel que:

$$(2.10) \qquad v \geqslant 0,\ r_{h,i}(v) > \sigma_h(v) \rightarrow v_{|S_i} > 0$$

on pourra prendre $\sigma_h(v) = h^\alpha . \min(h[v]_{1,\alpha}, [v]_{0,\alpha})$; et, d'après les inclusions de Sobolev [19]:

$$(2.11) \quad \begin{cases} \text{soit } v \in W^{2,p}(\Omega)\ \forall p<+\infty.\ \forall \varepsilon>0\ \exists C_{\varepsilon,v} \\ \text{tel que, posant } \sigma_h(v) = C_{\varepsilon,v} h^{2-\varepsilon}, \text{ on a (2.10).} \end{cases}$$

3. DISCRETISATION.

Suivant [17], pour donner une formulation de (1.3), on introduit la forme a par:

$$a(u,v) = \langle Au,v \rangle \quad \forall u,v \in H_0^1(\Omega) \quad [20];$$

on supposera (cf. toujours [17]):

$$(3.1) \qquad a:\{u,v\} \longmapsto a(u,v) \text{ est bilinéaire, continue et coercitive sur } H_0^1(\Omega).$$

On se donne aussi f, ψ avec:

$$(3.2) \qquad f \in H^{-1}(\Omega);\ \psi \in H^1(\Omega),\ \psi \leqslant 0 \text{ sur } \partial\Omega$$

de façon que, posant:

$$(3.3) \qquad K = \{v \in H_0^1(\Omega) \mid v \geqslant \psi \text{ dans } \Omega\}$$

---

[17] (2.1) entraine $S_i \subset \mathcal{B}(z,h)\ \forall z \in S_i$; donc $v(y) = v(y)-v(x) \leqslant [v]_{0,\alpha} h^\alpha\ \forall y \in S_i$; ensuite on fait usage de (2.3).

[18] On suppose ici Ω convexe. Si Ω n'était pas convexe il suffirait de avoir un opérateur π de prolongement de $C^{1,\alpha}(\bar\Omega)$ dans $C^{1,\alpha}(\bar{\mathcal{B}})$, $\mathcal{B}$ boule conténant Ω; avec notations évidentes il suffirait alors de remplacer dans (2.9) $[v]_{1,\alpha,\Omega}$ par $[\pi v]_{1,\alpha,\mathcal{B}}$.

[19] Qui, toujours pour Ω régulier, assurent que, pour $p>m$, $[v]_{1,1-m/p} \leqslant C_p \|v\|_{W^{2,p}(\Omega)}$. Naturellement dans la constante $C_{\varepsilon,v}$ de (2.11) on peut englober la constante liée à l'opetateur π introduit dans [18].

[20] On note $\langle , \rangle$ la dualité entre $H^{-1}(\Omega)$ et $H_0^1(\Omega)$.

K est un sousensemble de $H^1_o(\Omega)$ fermé, convexe, non vide; d'après [17] le problème:

(3.4) trouver $u\in K$ tel que $a(u,u-v)\leq\langle f,u-v\rangle\ \forall v\in K$

admet donc une solution unique; et (3.4) est la formulation précise de (1.3).

A côté du problème (3.4) on considère le problème discret:

$$(3.5)\quad \begin{cases} \text{soit } K_h = \{\vec{v}\in\mathbb{R}^N \mid \vec{v}\geq\overrightarrow{r_h\psi}\};\ \text{trouver } \vec{u}_h \in K_h \text{ tel que} \\ a(p_h\vec{u}_h,\ p_h\vec{u}_h-p_h\vec{v})\leq \langle f,p_h\vec{u}_h-p_h\vec{v}\rangle\ \forall\vec{v}\in K_h \end{cases}$$

(existence et unicité dans (3.5) suivent toujours de [17] grâce à l'injectivité de p_h); on remarquera que, posant:

(3.6) $A_h\equiv(a_{ij})$ où $a_{ij} = a(\phi_j,\phi_i)$ ([21])

le problème (3.5) équivant au "système de complémentarité" ([22]):

$$(3.7)\quad \begin{cases} \vec{u}_h\geq\overrightarrow{r_h(\psi)};\ A_h\vec{u}_h\geq p_h^*(f);\ (A_h\vec{u}_h)_i=p_{h,i}^*(f) \text{ pour les i tels que} \\ u_{h,i}>r_{h,i}(\psi); \end{cases}$$

on remarquera que, grâce à (1.3),(2.5),(2.10) (cf. aussi ([16])) on a:

(3.8) $p_h^*(f)\leq p_h^*(Au);\ p_{h,i}^*(f)=p_{h,i}^*(Au)$ si $r_{h,i}(u-\psi)>\sigma_h(u-\psi)$.

On aura bésoin que A_h satisfasse un "principe du maximum", à savoir (cf. ([14])):

(3.9) $A_h\ \vec{1}\geq\vec{0}$;

(3.10) si $\vec{\delta}\in\mathbb{R}^N$ vérifie $\vec{\delta}\nleq\vec{0}$, il existe $\bar{i}$ avec $\delta_{\bar{i}}>0$, $(A_h\vec{\delta})_{\bar{i}}\geq 0$;

on remarquera que (3.9), (3.10) sont remplies par example si A_h est une M-matrice ([23]); et on connait (cf. [6]) des conditions géométriques simples sur $\mathcal{T}_h$ suffisantes à impliquer cette proprieté de A_h.

<u>REMARQUE</u>: La fonction u_h du Résumé correspond évidemment à $p_h\vec{u}_h$; analoguement la fonction w_h du résumé est donnée par $p_h\vec{w}_h$ où $\vec{w}_h$ résoud:

(3.11) $\vec{w}_h\in\mathbb{R}^N;\ \forall\vec{v}\in\mathbb{R}^N\ a(p_h\vec{w}_h,p_h\vec{v}) = \langle Au,p_h\vec{v}\rangle$

ou, ce qui revient au même ([24]):

(3.12) $\vec{w}_h = A_h^{-1}p_h^*Au$.

([21]) Equivalemment $A_h=p_h^*Ap_h$; on remarquera que (cf. (3.1),(2.4)) A_h est définie positive.

([22]) Pour les systemes de complémentarité et leur utilisation dans la résolution de problèmes du type (1.3) on renvoye à [11].

4. ESTIMATIONS.

Soit u la solution du problème (3.4); $\vec{u}_h$ et $\vec{w}_h$ désignant les solutions de (3.5),(3.11) resp. on posera ([25]):

(4.1) $\varepsilon_{h,Eq}(u) = u - p_h\vec{w}_h; \quad \varepsilon_{h,Inéq}(u) = u - p_h\vec{u}_h.$

On va démontrer que, sous les hypothèses faites, on a l'estimation:

$$(4.2)\quad \begin{cases} \|\varepsilon_{h,Inéq}(u)\|_{L^\infty(\Omega)} \leq 2\,\|\varepsilon_{h,Eq}(u)\|_{L^\infty(\Omega)} + \\ +3\,\|u - p_h r_h u\|_{L^\infty(\Omega)} + \sigma_h(u-\psi). \end{cases}$$

En particulier, lorsqu'on a (1.4) pour la résolution des équations, grâce à (2.6),(2.11) on aura (1.4) aussi pour la résolution des inéquations, i.e.:

$$(4.3)\quad \begin{cases} \text{si dans (3.2) on a } f\in L^p(\Omega) \text{ et } \psi\in W^{2,p}(\Omega)\ \forall p<+\infty, \text{ pour les solu} \\ \text{tions } u,\vec{u}_h \text{ de (3.4),(3.5) on a } \|u-p_h\vec{u}_h\|_{L^\infty(\Omega)} = O(h^{2-\varepsilon})\ \forall\varepsilon>0. \end{cases}$$

Pour demontrer (4.2) on pose (cf. ([25])):

(4.4) $\vec{v}_h = \overrightarrow{r_h u} - A_h^{-1} p_h^*(Au)$

et on montrera que (cf. ([14]), ([15])):

(4.5) $-\overrightarrow{|\vec{v}_h|_\infty} - \sigma_h(u-\psi) \leq \vec{u}_h - \overrightarrow{r_h u} + \vec{v}_h \leq \overrightarrow{|\vec{v}_h|_\infty}$;

de (4.5) on tire évidemment:

$$|\vec{u}_h - \overrightarrow{r_h u}|_\infty \leq 2|\vec{v}_h|_\infty + \sigma_h(u-\psi)$$

d'où, passant aux p_h (cf. (2.4)):

$$\|p_h\vec{u}_h - p_h\overrightarrow{r_h u}\|_{L^\infty(\Omega)} \leq 2\,\|p_h\vec{v}_h\|_{L^\infty(\Omega)} + \sigma_h(u-\psi)$$

et donc, étant $\vec{v}_h = \overrightarrow{r_h u} - \vec{w}_h$ (cf. (4.4), (3.12)), (4.2) en suivra.

Il nous reste seulement à montrer (4.5). Soit:

$$\vec{\delta} = \vec{u}_h - \overrightarrow{r_h(u)} + \vec{v}_h - \overrightarrow{|\vec{v}_h|_\infty} ;$$

pour tout i tel que $\delta_i > 0$ on a $u_{h,i} > r_{h,i}(u) \geq r_{h,i}(\psi)$ (monotonie de r_h, et $u\geq\psi$), donc, d'après (3.7), $(A_h\vec{u}_h)_i = p^*_{h,i}(f)$; (3.9) entraine alors $(A_h\vec{\delta})_i \leq p^*_{h,i}(f) -$

([23]) Ce qui entraine aussi la convergeance de méthodes itératives pour la résolution effective du système (3.7); cf. [9].

([24]) L'existence de A_h^{-1} suit de (3.9), (3.10); cf. d'ailleurs ([21]).

([25]) Il s'agit évidemment des erreurs de discrétisation que l'on fait en resolvant l'équation (resp. l'inéquation) dont la solution est u.

$p^*_{h,i}(Au) \leqslant 0$ (grâce à (3.8)). On a donc $(A_h \vec{\delta})_i \leqslant 0$ $\forall_i$ tel que $\delta_i > 0$, ce qui (cf. (3.10)) implique $\vec{\delta} \leqslant \vec{0}$.

Analoguement, posant

$$\vec{\delta} = \overrightarrow{r_h u} - \vec{u}_h - \vec{v}_h - |\vec{v}_h|_\infty - \overrightarrow{\sigma_h (u-\psi)}$$

pour les i tels que $\delta_i > 0$ on a $r_{h,i}(u) > u_{h,i} + \sigma_h(u-\psi)$, à savoir $r_{h,i}(u-\psi) > \sigma_h(u-\psi)$; d'où (cf. (3.8) ,(3.9),(3.7)):

$$(A_h \vec{\delta})_i \leqslant p^*_{h,i}(Au) - (A_h \vec{u}_h)_i = p^*_{h,i}(f) - (A_h \vec{u}_h)_i \leqslant 0$$

ce qui, toujours d'après (3.10), implique $\vec{\delta} \leqslant \vec{0}$; (4.5) est donc démontrée.

BIBLIOGRAPHIE

[1] C. BAIOCCHI, E. MAGENES: "Problemi di frontiera libera in idraulica". Accad. Naz. Lincei, Quad. 217 , Roma (1975) 394-421.

[2] C. BAIOCCHI, G.A. POZZI: Travail en préparation .

[3] H. BREZIS, G. STAMPACCHIA: "Sur la régilarité de la solution d'inéquations elliptiques". Bull. Soc. Math. France 96(1968) 153-180.

[4] F. BREZZI, G. SACCHI: "A Finite Element Approximation of a Variational Inequality Related to Hydraulics". A paraître sur Calcolo.

[5] P.G. CIARLET, P.A. RAVIART: "General Lagrange and Hermite interpolation in $\mathbb{R}^n$, with applications to finite element methods". Arch. Rat. Mech. Anal. 46 (1972) 177-199.

[6] P.G. CIARLET, P.A. RAVIART: "Maximum principle and uniform convergence for the finite element method". Comp. Mat. Appl. Mech. Eng. 2 (1971) 17-31 .

[7] R.S. FALK: "Error estimates for the approximation of a class of variational inequalities". Math. of Comp. 28 (1974) 963-971 .

[8] H. LEWY, G. STAMPACCHIA: "On the regularity of the solution of a variational inequality". Comm. P.A.M. 22 (1969) 153-188.

[9] J.C. MIELLOUX: "Méthodes de Jacobi, Gauss-Seidel, sur (sous)-relaxation par blocs, appliquées à une classe de problèmes non linéaires". C.R.A.S. Paris, 273 (1971) 1257-1260.

[10] U. MOSCO: Conférence de ce Symposium .

[11] U. MOSCO, F. SCARPINI: "Complementarity systems and approximation of variational inequalities" R.A.I.R.O., 9 (1975) 83-104.

[12] U. MOSCO, G. STRANG: "One side approximation and variational inequalities". Bull. Amer. Math. Soc. 80 (1974) 308-312.

[13] U. MOSCO, G.M. TROIANIELLO: "On the smoothness of solution of the unilateral Dirichlet problem". Boll. U.M.I. 8(1973) 56-67.

[14] J. NITSCHE: "L^∞ convergence of finite element approximation". 2ème conférence sur les éléments finis, Rennes (France) (1975).

[15] J. NITSCHE: Conférence de ce Symposium.

[16] R. SCOTT: "Optimal L^∞ estimates for the finite element method on irregular meshes". A paraître.

[17] G. STAMPACCHIA: "Formes bilinéaires coercitives sur les ensembles convexes". C.R.A.S. Paris, 258 (1964) 4413-4416.

[18] G. STRANG, G. FIX: "An analysis of the finite element method". Prentice-Hall, Englewood Cliffs, N.J. (1973).

HYBRID METHODS FOR FOURTH ORDER ELLIPTIC EQUATIONS

Franco BREZZI
Università di Pavia e Laboratorio
di Analisi Numerica del C.N.R.

ABSTRACT.- A "displacement version" of the assumed stresses hybrid method is proposed for solving fourth order elliptic equations with Dirichlet boundary conditions.

0.- THE STARTING PROBLEM.

Let A be a fourth order elliptic operator; for sake of simplicity we shall suppose that A is reduced to its principal part (i.e. A contains only derivatives of order four), self adjoint and with constant coefficients. Let Ω be a convex polygon in $\mathbb{R}^2$ and consider the problem:

$$(0.0)\qquad \begin{cases} Aw = p \text{ in } \Omega \\ w = \dfrac{\partial w}{\partial n} = 0 \text{ on } \partial\Omega \end{cases}$$

where n denotes the outward normal derivative and p is an element of $L^2(\Omega)$.

Suppose that there exists a linear continuous operator T from $W_2 = H^2_0(\Omega)$ into some functional Hilbert space H (identified with its dual space H') such that

$$(0.1)\qquad A = T^*T,$$

with T^*= formal adjoint operator of T, i.e. $T^*: H\equiv H' \to W'$ and

$$\langle T^*v,\varphi\rangle = [v,T\varphi] \quad \forall v\in H \quad \forall \varphi\in W_2,$$

where [,] denotes the scalar product in H and < , > denotes the duality between W_2' and W_2.

We suppose moreover that there exists a positive constant α such that

$$(0.3)\qquad ||T\varphi||_H \geq \alpha||\varphi||_{W_2} \quad \forall \varphi \in W_2$$

It is well known that, under all these hypotheses, problem (0.0) can be written in the "variational form":

$$(0.4)\qquad \begin{cases} \underline{\text{find}}\ w \in W_2\ \underline{\text{such that}}: \\ [Tw,\ T\varphi] = \int_\Omega p\varphi dx \qquad \forall\ \varphi \in W_2 . \end{cases}$$

We shall deal, in the following, with the approximation of problem (0.1) in the form (0.4). The approach here proposed will coincide, in the case of plate bending problems, with the "assumed stresses hybrid method", and therefore we can think at it as an hybrid method itself. We end section 0 with some examples of choices for A,T,H.

Example 1.- $A=\Delta^2$; many possible choices for T and H are admissible. One of them is $T=T^*= -\Delta$ and $H=L^2(\Omega)$; another one is

$$(0.5)\qquad T:\varphi \to \left(\frac{\partial^2\varphi}{\partial x_1^2},\ \frac{\partial^2\varphi}{\partial x_1\partial x_2},\ \frac{\partial^2\varphi}{\partial x_2\partial x_1},\ \frac{\partial^2\varphi}{\partial x_2^2}\right) = \varphi_{/ij}$$

with:

$$H=\left\{\underline{v} \mid v_{ij} \in L^2(\Omega)\ (i,j=1,2),\ v_{12}=v_{21}\right\},$$

$$[\underline{u},\underline{v}] = \int_\Omega (u_{11}v_{11}+u_{12}v_{12}+u_{21}v_{21}+u_{22}v_{22})dx = \int_\Omega u_{ij}v_{ij}dx,$$

and

$$T^*:\underline{v} \to \frac{\partial^2 v_{11}}{\partial x_1^2} + 2\frac{\partial^2 v_{12}}{\partial x_1\partial x_2} + \frac{\partial^2 v_{22}}{\partial x_2^2} = v_{ij/ij}$$

Example 2.- $A= \frac{\partial^4}{\partial x_1^4} + \frac{\partial^4}{\partial x_2^4}$; many possible choices are also admissible: for instance

$$T:\varphi \to \left(\frac{\partial^2\varphi}{\partial x_1^2},\ \frac{\partial^2\varphi}{\partial x_2^2}\right)$$

with $H=(L^2(\Omega))^2$ and

$$T^*:\ \underline{v} \to \frac{\partial^2 v_1}{\partial x_1^2} + \frac{\partial^2 v_2}{\partial x_2^2}.$$

1.- THE ASSUMED STRESSES HYBRID APPROACH.

The assumed stresses hybrid approach has been introduced first by T.H.H. PIAN and P. TONG (cfr. for instance [1]), for elasticity and plate bending problems, starting from the so called "modified minimum complementary energy principle". Although the final linear system can be treated in such a way that the only remaining unknowns have in fact the physical meaning of "displacements" (cfr. always PIAN [1] and also the more recent [2] for more details) the method has always been considered as a "dual method" (or "stress method", cfr. for instance [3]). From the mathematical point of view the first proof of the convergence for the plate bending case has been given in [4] and, after that, a more general study has been done in [5] for second order problems and in [2] for the plate bending case. Similar results have been obtained in [6] for the "assumed displacement hybrid methods" for the second order case and, more recently, also in [7] but in a more general context including also mixed methods, always for second order problems.

In the present paper I will introduce the assumed stresses hybrid method directly as a "displacement method", that is starting from the variational formulation (0.4) which translates, in some sense, the "minimum potential energy principle". While the resulting mathematical approach seems quite different with respect to the previous one [2], the final "convergence condition" (obtained here by means of two very simple lemmas about the projection operators in finite dimensional spaces) is exactly the same, but in a more general context. Since the true difficulties consist in verifying such condition in the real examples (which has been done in details in [2] for the plate bending problems), the results here reported are essentially contained also in [2]. On the other hand, in the author's opinion, the following "displacement approach" is of simpler presentation and also suggestful for new developements, and therefore of some interest itself.

Let us now consider problem (0.4) and define approximate problems in the following way. Let r,s be integer numbers with $r\geq 3$ and $s\geq 1$. For each decomposition $\mathcal{T}_h$ of Ω into convex subdomains K (for simplicity let us suppose that each K is a triangle), consider the following finite dimensional subspace of $W_2=H^2_0(\Omega)$.

$$W_h=W(r,s,\mathcal{T}_h)=\left\{\varphi \mid \varphi\in W_2,\ \varphi_{|\partial K}\in P_r(\partial K),\ \frac{\partial}{\partial n}_{|\partial K}\ P_s(\partial K) \text{ and } A\varphi=0 \text{ in } K, \text{ for each } K \text{ in } \mathcal{T}_h\right.$$

where $P_m(\partial K)$ ($m\in\mathbb{N}$) denotes the set of functions defined on ∂K which are polynomials of degree m on each edge of ∂K.

<u>Remark</u> 1.- Note that the condition "$\varphi\in W_2$", in the definition of W_h, implies, in some sense, that φ and $\varphi_{/i}$ (i=1,2) must be continuous. Therefore, in the choice of the degrees of freedom in W_h, the degrees of freedom "values of φ, $\frac{\partial\varphi}{\partial x_1}$, $\frac{\partial\varphi}{\partial x_2}$ at the corners" should always be used. This justifies the requirement $r\geq 3$, $s\geq 1$. After that the choice of the d.o.f. for each given values of r and s is quite easy. For instance for r=4 and s=2 we can choose as d.o.f. the values of φ and $\varphi_{/i}$ (i=1,2) at the corners and the values of φ and $\varphi_{/n}$ at the midpoint of each edge.

<u>Remark</u> 2.- The condition "$A\varphi=0$ in each K" in the definition of W_h, is, in some sense, unessential. In facts we shall try to compute the functions φ of W_h only at the interelement boundaries (and the solution w itself will be approximated at the interelement boundaries). Therefore the condition "$A\varphi=0$" has the only goal of reducing W_h to a finite dimensional space still preserving, as an essential data, the values of φ and $\varphi_{/n}$ at

the interelement boundaries.

We consider now the following discrete problem:

$$(1.1)\quad \begin{cases} \underline{\text{Find}}\ \psi \in W_h\ \underline{\text{such that}}: \\ [T\psi,\ T\varphi] = \int_\Omega p\varphi dx \qquad \forall\, \varphi \in W_h. \end{cases}$$

It is immediate to verify that (1.1) has a unique solution. In order to compute "the error" between ψ and w, we shall not compare ψ and w directly, but, since we are interested essentially in the values they assume on the interelement boundaries, we introduce first an auxiliary function $\mathring{w}$ defined as

$$(1.2)\quad \left.\begin{array}{l} \mathring{w} = w \\ \mathring{w}_{/i} = w_{/i} \quad (i=1,2) \end{array}\right\} \text{ on } \partial K,\ A\mathring{w}=0 \text{ in } K, \quad \forall\, K \in \mathcal{T}_h$$

and finally compare the distance between ψ and $\mathring{w}$.

For this we remark first that, for any given A and for any given choi ce of T and H, a "Green formula" can be found of the type:

$$(1.3)\ [T\varphi, v] - \sum_{K\in\mathcal{T}_h} \int_K T^* v\varphi\, dx = \sum_{K\in\mathcal{T}_h} <\varphi_{/n}, G_o v>_{\partial K} - <\varphi, G_1 v>_{\partial K}$$

where $<,>_{\partial k}$ denotes the duality pairing between suitable distributional spaces defined on ∂K, and G_o, G_1 are "trace operators" on ∂K of order 0 and 1 respectively. It is possible to give explicit representation of such G_o, G_1 in order to precise (1.3) mathematically. For simplicity we shall show only a few examples.

Ex. 1.- $A=\Delta^2$, $T=-\Delta$, $H=L^2(\Omega)$

$$(1.4)\quad \int_\Omega -\Delta\varphi v dx - \sum_{K\in\mathcal{T}_h} \int_K \varphi(-\Delta v)\, dx = \sum_{K\in\mathcal{T}_h} \int_{\partial K} (\varphi v_{/n} - \varphi_{/n} v)\, d\partial\sigma$$

(for $\varphi \in H^2(\Omega)$, $v \in L^2(\Omega)$ $v \in H^2(K)\ \forall K$)

Ex.2.- $A=\Delta^2$, $T: \varphi \to (\varphi_{/ij})$

$$(1.5)\quad \int_\Omega \varphi_{/ij} v_{ij} dx - \sum_{K\in\mathcal{T}_h} \int_K \varphi v_{ij_{/ij}} dx = \sum_{K\in\mathcal{T}_h} \int_{\partial K} (\varphi_{/i} v_{ij} n_j - \varphi v_{ij_{/i}} n_j)\, d\partial\sigma$$

(for $\varphi \in H^2(\Omega)$, $v_{ij} \in H^2(K)\ \forall K$);

setting:

$$(1.6)\quad M_n(\underline{v}) = v_{ij} n_j n_j,\ M_{nt}(\underline{v}) = v_{ij} n_i t_j,\ Q_n(\underline{v}) = v_{ij_{/i}} n_j$$

the right hand term of (1.5) can be written:

$$(1.7)\quad \sum_{K\in\mathcal{T}_h} \int_{\partial K} (M_n(\underline{v})\, \varphi_{/n} + M_{nt}(\underline{v})\, \varphi_{/t} - Q_n(\underline{v})\, \varphi)\, d\sigma =$$

$$= \sum_{K \in \mathcal{T}_h} \int_{\partial K} \left(\varphi_{/n} M_n(\underline{v}) - \varphi\left(\frac{\partial M_{nt}(\underline{v})}{\partial t} + Q_n(\underline{v}) \right) \right) d\sigma.$$

and has, in that way, the "right form" used in (1.3).

By means of a Green formula of the type (1.3) it is possible to see that, for each $\varphi \in W_h$, we have

$$(1.8) \qquad [T\mathring{w}, T\varphi] = [Tw, T\varphi] = \int_\Omega p\varphi dx.$$

Therefore the solution ψ of (1.1) verifies

$$(1.9) \qquad [T\psi, T\varphi] = [T\mathring{w}, T\varphi] \quad \forall \varphi \in W_h$$

and we immediately get the following result.

<u>PROPOSITION 1.- The distance between</u> ψ , <u>solution of</u> (1.1) <u>and</u> $\mathring{w}$, <u>defined in</u> (1.2) <u>is given by</u>:

$$(1.10) \qquad ||\psi - \mathring{w}||_{W_2} \leq \frac{1}{\alpha} \operatorname*{Inf}_{\varphi \in W_h} ||T\mathring{w} - T\varphi||_H .$$

It is also quite straight-forward to prove (cfr.e.g. [2]) the following result.

<u>PROPOSITION 2.- If</u> $\mathring{w}$ <u>is sufficiently regular in each</u> K (<u>that is if</u> w <u>is sufficiently regular</u>) <u>then</u>

$$(1.11) \qquad \operatorname*{Inf}_{\varphi \in W_h} ||T\mathring{w} - T\varphi||_H < c|h|^\mu$$

<u>with</u> $\mu = \min(r-1, s)$, $|h|$ <u>=mesh size</u>, c <u>independent of</u> h.

Proposition 1 and 2 seem to solve any problem, but, on the other hand, it is easy to see that the practical computation of terms like

$$(1.12) \qquad [T\varphi^{(r)}, T\varphi^{(s)}]$$

or even like

$$(1.13) \qquad \int_\Omega p\varphi^{(s)} dx,$$

when $\varphi^{(r)}$ and $\varphi^{(s)}$ vary over a basis for W_h, is, if not impossible, at least very difficult.

To overcome this difficulty, consider the following procedure: choose a finite dimensional subspace V_h of H, project the elements $T\varphi$ of $T(W_h)$ on V_h and compute, instead of (1.12),

$$(1.4) \qquad [\pi_h T\varphi^{(r)}, \pi_h T\varphi^{(s)}]$$

where π_h is the projection operator, in H, on V_h.

Terms of the type (1.13) can be treated in a similar manner; for this, suppose for the moment that a particular solution f of the equation

(1.15) $T^{*}f = p$ in Ω, $f \in H$.

is explicitely known. We have

(1.16) $\int_{\Omega} p \; \varphi \, dx = \langle T^{*}f, \varphi \rangle = [f, T\varphi] \quad \forall \varphi \in W_h$.

In the numerical computation, then, (1.13) will be substituted by

(1.17) $[f, r_h \, T\varphi^{(s)}]$

and finally problem (1.1) can be substituted by the following:

(1.18) $\begin{cases} \underline{\text{Find}} \; \bar{\psi} \; \underline{\text{in}} \; W_h \; \underline{\text{such that}}: \\ [\pi_h T \, \bar{\psi}, \pi_h T\varphi] = [f, \pi_h T\varphi] \quad \forall \varphi \in W_h. \end{cases}$

Ii is clear that the choice of V_h cannot be made arbitrarly, but it must satisfy the following requirements:

1) Functions $\pi_h T\varphi$ must be easily computable.
2) Problem (1.18) must have a unique solution
3) The solution $\bar{\psi}$ of (1.18) must converge to $\overset{\circ}{w}$ with the same rate of ψ.

In the pratical cases, V_h will be chosen as a subspace of H whose elements are "polinominals" (with a certain degree) inside each element K; usually no continuity requirements are given on the elements of V_h (that is they are independently assumed in each K). The operator r_h will be computable if any term of the type

(1.19) $[T\varphi, v] \quad \varphi \in W_h, \; v \in V_h$

is computable. By means of a Green formula like (1.3), we can write (1.19) as follows

(1.20) $[T\varphi, v] = \sum_{K \in \mathcal{C}_h} \{ \int_K (T^{*}v)\varphi dx + \langle \varphi_{/n}; G_0 v \rangle_{\partial K} - \langle \varphi, G_1 v \rangle_{\partial K} \}$

Since φ and $\varphi_{/n}$ are computable on each ∂K we immediately get the following result.

<u>PROPOSITION 3</u>.- <u>If the elements</u> v <u>of</u> V_h <u>verify</u>

(1.21) $T^{*}v = 0$ <u>inside each K</u>

<u>then the projection operator</u> π_h <u>is computable</u>.

We look now for sufficient conditions in order that (1.18) has a unique solution. It is very easy to see that (1.18) has always a solution; uniqueness will hold if the following hypothesis is added.

<u>PROJECTION HYPOTHESIS</u>. <u>There exists a constant</u> $\gamma(h) > 0$ <u>such that</u>:

(1.22) $||\pi_h T\varphi|| \geq \gamma(h) ||T\varphi|| \quad \forall \varphi \in W_h$.

It is immediate to prove the following proposition.

PROPOSITION 4.- If the projection hypothesis is satisfied then (1.18) has a unique solution.

In order to study the convergence of the solution $\bar{\psi}$ of (1.18) we shall need the following lemmas

Lemma 1.- Let $\mathcal{H}$ be an Hilbert space and let $\mathcal{V}$, $\mathcal{W}$ be finite dimensional subspaces of $\mathcal{H}$. We denote by $P_{\mathcal{V}}$ and $P_{\mathcal{W}}$ the projection operators (in $\mathcal{H}$) on $\mathcal{V}$ and $\mathcal{W}$ respectively. We define

(1.23) $$\mathcal{Q}=P_{\mathcal{V}}(\mathcal{W})$$

and we denote by P_Q the projection operator on $\mathcal{Q}$ and by π the restriction of $P_{\mathcal{V}}$ to $\mathcal{W}$.

We suppose that π is such that:

(1.24) $$\exists k>0;\quad k||\pi\varphi||\geq||\varphi||\quad \forall\varphi\in W.$$

Under these hypotheses we have that for any $g\in\mathcal{H}$:

(1.25) $$||P_{\mathcal{W}}g-\pi^{-1}P_Q g||\leq k(2||(I-P_{\mathcal{W}})g||+||(I-P_{\mathcal{V}})g||).$$

Proof.- Let us set

(1.26) $$\lambda=P_{\mathcal{W}}g;\qquad \chi=\pi^{-1}P_Q g.$$

We remark that the vector $\pi\chi-r\lambda$ belongs to $\mathcal{Q}$ and therefore it is orthogonal to $r\chi-g$. Thus

$$\begin{aligned}
&||\pi\chi-\pi\lambda||\leq||\pi\lambda-g||\leq||\pi\lambda-\lambda||+||\lambda-g||\leq\\
&\leq||(P_{\mathcal{V}}-I)\lambda||+||(P_{\mathcal{W}}-I)g||\leq\\
(1.27)\quad&\leq||(P_{\mathcal{V}}-I)(\lambda-g)||+||(P_{\mathcal{V}}-I)g||+||(P_{\mathcal{W}}-I)g||\leq\\
&\leq||\lambda-g||+||(P_{\mathcal{V}}-I)g||+||(P_{\mathcal{W}}-I)g||\leq\\
&\leq2||(P_{\mathcal{W}}-I)g||+||(P_{\mathcal{V}}-I)g||
\end{aligned}$$

and condition (1.24) gives us the final result.

Lemma 2.- Under the same hypotheses of Lemma 1, we have that if f_1 and f_2 are two elements in $\mathcal{H}$ such that

(1.28) $$P_{\mathcal{W}}(f_1-f_2)=0,$$

then:

(1.29) $$||\pi^{-1}P_Q(f_1-f_2)||\leq k^2||(I-P_{\mathcal{V}})(f_1-f_2)||.$$

Proof.- We set

(1.30) $$\delta=f_1-f_2$$

and we have, from (1.24)

(1.31) $$||\pi^{-1}P_Q\delta||\leq k||P_Q\delta||.$$

on the other hand

$$||P_Q\delta||^2=[P_Q\delta,\delta]=[P_Q\delta-\pi^{-1}P_Q\delta,\delta]=$$

(1.32) $$=[P_{\mathcal{V}}\pi^{-1}P_Q\delta-\pi^{-1}P_Q\delta,\delta]=[\pi^{-1}P_Q\delta,(P_{\mathcal{V}}-I)\delta]\leq$$

$$\leq k||P_Q\delta||\ ||(P_{\mathcal{V}}-I)\delta||$$

and therefore

(1.33) $$||P_Q\delta||\leq k||(I-P_{\mathcal{V}})\delta||$$

and from (1.31),(1.33) we get (1.29).

In order to apply the projection lemmas to our situation we choose

(1.34) $$\mathcal{H}=H \qquad \mathcal{W}=T(W_h) \qquad \mathcal{V}=V_h.$$

The projection hypothesis (1.22) guarantees that (1.24) is satisfied with $k=\gamma(h)$. By means of formulas (1.8) and (1.16) we have that $T\psi$ (where ψ is the solution of (1.1)) can be regarded as

(1.35) $$T\psi=P_{\mathcal{W}}(T\mathring{w})=P_{\mathcal{W}}(f)$$

On the other hand, if $\bar{\psi}$ is the solution of (1.18) we have

(1.36) $$T\bar{\psi}=\pi^{-1}P_Q f.$$

Setting now

(1.37) $$\tilde{\psi}=\pi^{-1}P_Q T\mathring{w}$$

we have by lemma 1

(1.38) $$||T(\psi-\tilde{\psi})||\leq\gamma(h)(2||(I-P_{\mathcal{W}})T\mathring{w}||+||(I-P_{\mathcal{V}})T\mathring{w}||).$$

On the other hand it is easy to see that, if (1.21) is satisfied, by Green formula (1.3) we get

(1.39) $$[T\mathring{w},v]=[Tw,v] \quad \forall v\in V_h$$

and therefore

(1.40) $$\pi^{-1}P_Q T\mathring{w}=\pi^{-1}P_Q Tw$$

By means of lemma 2 we get thus

(1.41) $$||T\tilde{\psi}-T\bar{\psi}||=||\pi^{-1}P_Q(Tw-f)||\leq$$
$$\leq\gamma(h)^2||(I-P_{\mathcal{V}})(Tw-f)||.$$

Therefore the following theorem:

THEOREM 1.- _If the projection hypothesis_ (1.22) _and the condition_ (1.21) _are satisfied, then the distance between the solution_ $\bar{\psi}$ _of_ (1.18) _and_ $\mathring{w}$ _can be bounded as follows._

(1.42) $$||T\bar{\psi}-T\mathring{w}||\leq||T\bar{\psi}-T\tilde{\psi}||+||T\tilde{\psi}-T\psi||+||T\psi-T\mathring{w}||\leq$$

$$\leq \gamma(h)^2 \operatorname*{Inf}_{v \in V_h} ||(Tw-f)-v||+$$

$$+\gamma(h)(2\operatorname*{Inf}_{v \in V_h} ||T\mathring{w}-v||+\operatorname*{Inf}_{\varphi \in W_h} ||T\mathring{w}-T\varphi||)+$$

$$+\operatorname*{Inf}_{\varphi \in W_h} ||T\mathring{w}-T\varphi||.$$

It can be proved, by means of an abstract generalisation of the Bramble--Hilbert lemma, (cfr.[2] lemma 4.1) that if V_h contains all the "polinomials" of degree$\leq$m which verify (1.21), then for each element z in H which satisfies also (1.21) and which is sufficiently smooth,

(1.43) $$\operatorname*{Inf}_{v \in V_h} ||z-v|| \leq c|h|^{m+1}$$

From theorem 1, and formulae (1.11),(1.43) we have finally the following result.

<u>THEOREM 2.-</u> <u>If</u> w <u>and</u> f <u>are sufficiently regulars, and if the following condition is satisfied:</u>

(1.44) $$\gamma(h) \geq \gamma \geq 0 \qquad \forall h$$

<u>then the distance between</u> $\bar{\psi}$ <u>and</u> $\mathring{w}$ <u>is bounded as follows:</u>

(1.45) $$||\bar{\psi}-\mathring{w}||_{W_2} \leq \alpha^{-1} ||T\bar{\psi}-T\mathring{w}|| \leq c|h|^q$$

<u>with</u> q=min(r-1,s,m+1) <u>and</u> c <u>constant independent of h.</u>

<u>Remark</u> 3.- The projection condition (1.44) is essentially equivalent to the "compatibility condition" of [4] and [2] which can be regarded, itself, as a special case of the abstract conditions that can be deduced from the results reported in [8] or, in a formulation equivalent but more explicitely referred to these kind of problems, reported in [9]. The importance of a condition of this kind has been pointed out very clearly by Strang [10] who interpreted it as a "stability condition". In many particular cases, like the present one, the condition (1.44), which is essentially a global condition, can be "localised" and studied on a fixed element (reference element) (cfr. e.g. [4]). In its "localised form" the condition has been called also "rank condition" (see e.g. [7]).

<u>Remark</u> 4.- Although an abstract result like the one of theorem 2 could seem satisfactory, in the concrete cases the true difficulties arise often from the research of sufficient conditions on m,r,s in order that (1.44) be satisfied. For the present case the only concrete example that can be mentioned has been obtained in [2] for the particular choice: $A=\Delta^2$, $T:\varphi \to (\varphi_{/ij})$. In this case it has been proved that sufficient con

ditions in order that (1.44) is satisfied, are:

$$(1.46)\qquad m \geq \begin{cases} r-2 & \underline{\text{if}}\ r>s+1 \\ r-2 & \underline{\text{if}}\ r=s+1\ \underline{\text{and}}\ r\ \underline{\text{is even}} \\ r-1 & \underline{\text{if}}\ r=s+1\ \underline{\text{and}}\ r\ \underline{\text{is odd}} \\ s-1 & \underline{\text{if}}\ r<s+1 \end{cases}$$

In view of (1.45),(1.46) the optimal case seems to be: $r=4$, $s=3$, $m=2$, which has given also very good results in the numerical experiments. We point out that in this case the d.o.f. in W_h can be chosen as follows: values of $\varphi, \varphi_{/i}$ $(i=1,2)$ at the corners and values of $\varphi, \varphi_{/n}, \varphi_{/nt}$ at the midpoint of each edge.

Until now we assumed the knowledge of a "particular solution" f of the equation

$$(1.47)\qquad T^{*}f = p \text{ in } \Omega$$

In many concrete cases this can be a real difficulty. We shall see that, in fact, we could suppose only the knowledge of a solution $\tilde{f}$ of the equation

$$(1.48)\qquad T^{*}\tilde{f} = p \text{ in each } K$$

In this case we will have, by (1.3)

$$(1.49)\qquad [\tilde{f},T\varphi]=\int_{\Omega} p\varphi\,dx+\sum_{K\in\mathcal{T}_h} (\langle\varphi_{/n},G_0\tilde{f}\rangle_{\partial K} -\langle\varphi,G_1\tilde{f}\rangle_{\partial K}).$$

Let us define, for simplicity

$$(1.50)\qquad b(\tilde{f},\varphi)=\sum_{K\in\mathcal{T}_h} (\langle\varphi_{/n},G_0\tilde{f}\rangle_{\partial K} -\langle\varphi,G_1\tilde{f}\rangle_{\partial K})$$

and consider, as discrete problem, the following one.

$$(1.51)\qquad \begin{cases} \underline{\text{Find}}\ \chi\in W_h\ \underline{\text{such that:}} \\ [\pi_h T\chi,\pi_h T\varphi]= [\tilde{f},\pi_h T\varphi]-b(\tilde{f},\varphi). \end{cases}$$

We want to find a bound for the distance between $\bar{\psi}$, solution of (1.18) and χ.

For this we consider

$$||\pi_h T(\bar{\psi}-\chi)||^2= [\pi_h T(\bar{\psi}-\chi),\pi_h T(\bar{\psi}-\chi)] =$$

$$(1.52)\qquad = [f,\pi_h T(\bar{\psi}-\chi)]-[\tilde{f},\pi_h T(\bar{\psi}-\chi)] +b(\tilde{f},\bar{\psi}-\chi)$$

Setting $\delta=\bar{\psi}-\chi$ we have:

$$||\pi_h T\delta||^2= [f-\tilde{f},\pi_h T\delta-T\delta]+ [f-\tilde{f},T\delta]+b(\tilde{f},\delta) =$$

$$(1.53)\qquad = [f-\tilde{f},(\pi_h-I)T\delta]\leq||(I-P_{\nu})(f-\tilde{f})||\ .\ ||T\delta||\leq$$

$$\leq\gamma(h)\,||(I-P_{\nu})(f-\tilde{f})||\ ||\pi_h T\delta||$$

and therefore

$$(1.54)\quad ||T\bar{\psi}-T\chi|| \leq \gamma(h)^2 \operatorname*{Inf}_{v\in V_h} ||f-\tilde{f}-v||.$$

We can canclude that, under the same hupotheses, an"error bound" identical to the one of theorem 2 holds even for the distance between χ and $\overset{\circ}{w}$.

We end the paper by showing that a close approximation of Tw can also be deduced by the proposed method; in fact this is very reasonable since the method must "coincide" with a "dual method".

We shall deal for simplicity only with the case of problem (1.12), but a similar result can also be proved for the problem (1.51).

For this let us consider

$$(1.55)\quad u = \pi_h T\bar{\psi}-\pi_h f$$

and note that, for each v in V_h we have

$$(1.56)\quad [u,v]-[Tw-f,v]=[T\bar{\psi},v]-[Tw,v]=[T\bar{\psi}-T\overset{\circ}{w},v].$$

Therefore

$$(1.57)\quad ||u-\pi_h(Tw-f)||=||\pi_h(T\bar{\psi}-T\overset{\circ}{w})||$$

and

$$(1.58)\quad \begin{aligned} ||u+f-Tw|| &\leq ||u-\pi_h(Tw-f)|| + \operatorname*{Inf}_{v\in V_h}||(Tw-f)-v|| \leq \\ &\leq \operatorname*{Inf}_{v\in V_h}||(Tw-f)-v|| + ||T\bar{\psi}-T\overset{\circ}{w}|| \end{aligned}$$

We can conclude with the following theorem

THEOREM 3.- <u>Under the same hypotheses of theorem 2 we have:</u>

$$(1.59)\quad ||(\pi_h T\bar{\psi}-\pi_h f+f)-Tw|| \leq c|h|^q$$

<u>with</u> $q = \min(r-1,s,m+1)$ <u>and</u> c <u>independent of</u> h .

REFERENCES

[1] T.H.H. PIAN, P.TONG: "Basis of finite element methods for solid continua". Internat. Journal Numer. Methods Enginering 1, 3-28, 1969.

[2] F.BREZZI, L.D.MARINI: "On the numerical solution of plate bending problems by hybrid methods". To appear on R.A.I.R.O., 1975.

[3] B.FRAEIJS DE VEUBEKE: "Variational principles and the patch test". International Journal Numerical Methods Engineering, 8, 783-801, 1974.

[4] F.BREZZI: "Sur la méthode des éléments finis hybrides pour le problème biharmonique". Numer. Math. 24, 103-131, 1975.

[5] J.M.THOMAS: "Méthodes des éléments finis hybrides duaux pour les problèmes elliptiques du second ordre". To appear on R.A.I.R.O..

[6] P.A.RAVIART, J.M.THOMAS: "Méthodes des éléments finis hybrides primaux pour les problèmes elliptiques du second ordre". To appear.

[7] I.BABUŠKA, J.T.ODEN, J.K.LEE: "Mixed hybrid finite element approximation in linear elasticity". To appear.

[8] I.BABUŠKA: "Error bound for finite element method". Num. Math., 16, 322-333, 1971.

[9] F.BREZZI: "On the existence, uniqueness and approximation of saddle point problems arising from Lagrangian multipliers". R.A.I.R.O., 8, 129-151, 1974.

[10] G.STRANG: "The finite element methods - Linear and nonlinear applications". Proceeding of the Int. Congress of Math.. Vancouver 1974.

VARIATIONAL TECHNIQUES FOR THE ANALYSIS OF A LUBRICATION PROBLEM

G. Capriz

Istituto di Elaborazione della Informazione, C.N.R.,Pisa.

1. Introduction.

The aim of the paper is an illustration of a mechanical problem, which can be stated in the form of a variational inequality, which can be treated with the attending techniques and whose approximate solution can be obtained through a process of successive approximations, where at each step finite elements procedures can be adopted to advantage.

The problem is relatively simple, but it is exciting to find that analytical devices can provide specific answers to detailed questions from the engineer.

As the problem is relatively little known outside the group of experts in the theory of lubrication, attention is set especially on qualitative aspects, giving, however, specific references to related papers.

2. Reynolds' equation.

The unknown function in the problem is the pressure distribution p in a thin film of lubricant (a viscous fluid of viscosity η) contained in the narrow clearance between two surfaces Σ_a, Σ_b in relative motion. The differential condition satisfied by p is due to O. Reynolds ([1]; see also Sommerfelds contribution [2]); it was obtained through a direct estimation, appropriate to the situation, of the relations between the relevant parameters. Several Authors [3,4,5] have suggested relations between Reynolds' equation and the general Navier-Stokes equations of viscous fluid flow, introducing appropriate power series expansions in a parameter which measures the film thickness. Reynolds' equation is usually quoted for particular films (such as occur, for instance, in cylindrical bearings); here we prefer to sketch rapidly a derivation for a general case.

Inertia effects in the lubricant are disregarded in the direct approach (and they turn out to be irrelevant in the first approximation, when the approach through power series expansions is pursued). Hence any one of the surfaces, say Σ_a can be considered for the present purpose as fixed and the other, Σ_b, as moving. Then one can choose a system of lines of curvature over Σ_a as coordinate lines x_1, x_2 and count a third coordinate x_3, within the film, along the straight lines normal to Σ_a. Here a hypothesis is of the essence: that the clearance $h(x_1, x_2)$ between Σ_a and Σ_b counted along x_3 be much smaller than the radii of curvature of Σ_a, and also small with respect to a characteristic dimension of the portion Γ of Σ_b <u>wetted</u> by the film.

The physical components of the speed of Σ_b are called $V_i(x_1, x_2, t)$ $(i=1,2,3)$ and the flow rate through the film in the direction of $x_1 (x_2)$ per unit width counted along x_2 (x_1) is called $q_1 (q_2)$; the q_i's are assumed to be (or can be shown to be in the first approximation) those which pertain to a Couette flow

$$(2.1) \qquad q_i = \frac{h}{2} V_i - \frac{h^3}{12\eta\sqrt{a_i}} \frac{\partial p}{\partial x_i}, \qquad i = 1,2;$$

here a_1, a_2 are the coefficients of the first fundamental form of Σ_a.

Use is now made of the equation of continuity averaged over the film thickness

$$(2.2) \qquad \frac{\partial}{\partial x_1}(\sqrt{a_2}\, q_1) + \frac{\partial}{\partial x_2}(\sqrt{a_1}\, q_2) = - V_3 \sqrt{a_1 a_2} + \sqrt{a_2}\, V_1 \frac{\partial h}{\partial x_1} + \sqrt{a_1}\, V_2 \frac{\partial h}{\partial x_2}.$$

Introduction of (2.1) in (2.2) leads to Reynolds' equation

$$(2.3)\quad \frac{\partial}{\partial x_1}\left(\frac{h^3}{\eta}\sqrt{\frac{a_2}{a_1}}\,\frac{\partial p}{\partial x_1}\right) + \frac{\partial}{\partial x_2}\left(\frac{h^3}{\eta}\sqrt{\frac{a_1}{a_2}}\,\frac{\partial p}{\partial x_2}\right) =$$

$$= 12\sqrt{a_1 a_2}\; V_3 + 6\,h^2\,\frac{\partial}{\partial x_1}\left(\frac{\sqrt{a_2}\,V_1}{h}\right) + 6\,h^2\,\frac{\partial}{\partial x_2}\left(\frac{\sqrt{a_1}\,V_2}{h}\right).$$

Boundary and side conditions must be added to (2.3). Notice, first of all, that p must be normalized to avoid indetermination; the null value is often assumed to correspond to atmospheric pressure. Because it is found in experiments that, under normal conditions of operation, the lubricating fluid does not support subatmospheric pressures, then (2.3) makes sense only where $p \geq 0$ and everywhere in Γ it must be

$$(2.4)\qquad p \geq 0 .$$

Where the film yields to slight subatmospheric pressure, a bubble full of air at $p \sim 0$ (i.e. <u>a region of cavitation</u>) is formed; thus, we need to state transition conditions at the boundary γ_c of the region of cavitation. There, the flow rate q_n in the direction $\underline{n}$ normal to γ_c equals $\frac{1}{2}h(V_1 n_1 + V_2 n_2)$; beyond the fluid rests on the moving surface leaving on top an empty gap (of initial thickness $\frac{1}{2}$ h). On γ_c the transition conditions are

$$(2.5)\qquad p=0\ ,\quad \frac{\partial p}{\partial n} = 0, \qquad \text{on } \gamma_c .$$

Boundary conditions may involve directly p; for instance, over a portion γ_p of the boundary γ of the lubricated region Γ, p may be supposed to be given

$$(2.6)\qquad p = \hat{p}\ , \qquad \text{on } \gamma_p .$$

This hypothesis is consistent with the earlier remarks only if

$$\hat{p} \geq 0$$

and implies that the lubricant can flow freely in and out of the gap. In particular, no condition is imposed on the input flow rate, although more appropriately one should not exclude conditions of <u>starving feed</u>; i.e. the existence of an upper limit $\bar{q}$ to $-q_n$ (n, outer normal):

$$(2.7)\qquad \frac{h^3}{12}\frac{\partial p}{\partial n} - \frac{h}{2}V_n \leq \bar{q}, \qquad \text{on } \gamma_p;\quad V_n = V_1 n_1 + V_2 n_2 .$$

This question is not discussed here, but attention is directed to it as worthy of further analysis.

As an alternative to (2.6), on a portion $\gamma_q = \gamma - \gamma_p$ of γ, one can attempt to impose the rate $\hat{q}_n$ of lubricant inflow per unit length of γ, so that over γ_q

(2.8) $\frac{\partial p}{\partial n} = \frac{12}{h^3} (\frac{h}{2} v_n + \hat{q}_n)$, on γ_q.

However, this condition might not be always compatible with (2.4).

3. Variational formulation and existence theorem.

Γ is taken here to be an open, bounded, connected set, with a boundary γ which is sufficiently regular to allow everywhere the specification of a normal direction. Our task is that of finding a subset Ω of Γ and a function $p \geq 0$ on Γ, such that: (i) eqn (2.3) is satisfied in Ω; (ii) p and $\frac{\partial p}{\partial n}$ vanish on $\gamma_c = \partial\Omega \cap \Gamma$; (iii) the boundary condition (2.6) is satisfied on γ_p and the boundary condition (2.8) is satisfied on $\gamma_q \cap \partial\Omega$. Here, of course, $\partial\Omega$ is the boundary of Ω; notice the relaxation (iii) of the condition (2.8): through it account is taken of the remark at the end of Sect. 2.

This formulation of our problem implies the acceptance of a condition of regularity of γ_c, so that $\frac{\partial p}{\partial n}$ makes sense there. But such condition is unwarranted a priori ; more appropriate is therefore a weak formulation, which can be expressed in terms of a variational inequality. Thus we can also reach an existence theorem in a way now standard ; for the purpose we need to make a number of points more precise: we list them here.

We assume that γ_p and γ_q are disjoint open subsets of γ, such that $\gamma = \overline{\gamma}_p \cup \overline{\gamma}_q$. We consider the subspace $\mathbb{V}$ of all distributions u of $H^1(\Gamma)$, which vanish on γ_p; thus a Poincaré inequality holds for all $u \in \mathbb{V}$, i.e. a positive constant C exists such that

$$\int_\Gamma u^2 d\Gamma \leq C \int_\Gamma |\text{grad } u|^2 \, d\Gamma, \qquad \forall \; u \in \mathbb{V} .$$

Then, the space $\mathbb{V}$ becomes a Hilbert space if we equip it with the norm

$$\|u\|^2_{\mathbb{V}} = \int_\Gamma |\text{grad } u|^2 \, d\Gamma.$$

Let us call

$$\alpha_1(x_1, x_2) = \frac{h^3}{\eta} \sqrt{\frac{a_2}{a_1}},$$

$$\alpha_2(x_1, x_2) = \frac{h^3}{\eta} \sqrt{\frac{a_1}{a_2}}$$

and assume that α_i are measurable functions such that there exist positive constants m, M for which

$$m|\xi|^2 \leq \alpha_1 \xi_1^2 + \alpha_2 \xi_2^2 \leq M|\xi|^2, \forall \xi \in \mathbb{R}^2,$$

almost everywhere in $\overline{\Gamma}$.

Let us define on $\overline{\Gamma}$ a function $\tilde{p} \in H^1(\Gamma)$, such that its trace on γ_p is $\hat{p}$; let us call

$$f = -\frac{\partial}{\partial x_1}\left(\alpha_1 \frac{\partial \tilde{p}}{\partial x_1}\right) - \frac{\partial}{\partial x_2}\left(\alpha_2 \frac{\partial \tilde{p}}{\partial x_2}\right) + 12\sqrt{a_1 a_2}\, V_3 +$$

$$+ 6h^2 \frac{\partial}{\partial x_1}\left(\frac{\sqrt{a_2}\, V_1}{h}\right) + 6h^2 \frac{\partial}{\partial x_2}\left(\frac{\sqrt{a_1}\, V_2}{h}\right),$$

and let us assume that a choice of $\tilde{p}$ is possible so that $f \in L^r(\Gamma)$, $r \geq 1$. This condition bears on the shape of γ, on the quality of $\hat{p}$, and on the properties of a_1, a_2.

Let us finally assume that

$$g = -\frac{\partial \tilde{p}}{\partial n} + \frac{12\eta}{h^3}\left(\frac{h}{2} V_n + \hat{q}_n\right)$$

belongs to $L^s(\gamma_q)$, $s \geq 1$.

Consider now the bilinear form

$$(3.1) \quad a(u, v) = \int_\Gamma \left(\alpha_1 \frac{\partial u}{\partial x_1} \frac{\partial v}{\partial x_1} + \alpha_2 \frac{\partial u}{\partial x_2} \frac{\partial v}{\partial x_2}\right) d\Gamma, \quad u, v \in \mathbb{V}.$$

and the functional

$$\langle T, v\rangle = \int_\Gamma f\, v\, d\Gamma + \int_{\gamma_q} g\, v\, d\gamma, \qquad v \in \mathbb{V}.$$

In view of the assumptions we have accepted, $\langle T, v\rangle$ belongs to the dual space $\mathbb{V}'$ of $\mathbb{V}$.

Consider now the following closed, connected subset of $\mathbb{V}$

$$\mathbb{K} = \{ v \in \mathbb{V}, \quad v \geq -\tilde{p} \quad \text{a.e. in } \Gamma \};$$

then our problem can be transformed in the problem of finding a function u such that

$$(3.2) \quad \begin{cases} u \in \mathbb{K}, \\ a(u, v-u) \geq \langle T, v-u \rangle, \text{ for all } v \in \mathbb{K}. \end{cases}$$

Since the bilinear form (3.1) is continuous and coercive, it follows from well known-results (see, e.g., [6]) that there exists a unique solution u of (3.2).

Additional requirements (of modest import from a physical point of view) assure ample regularity of the solution of (3.2), so that it can be interpreted also as a solution almost in a classical sense. The proviso refers to the regularity of γ_c, which is a matter of a different order of difficulty, and the behaviour of p in the neighbourhood of $\overline{\gamma}_p \cap \overline{\gamma}_q$, where objective difficulties arise (see, again, [6] for details).

4. A special case : cylindrical bearing.

A special case is now considered of prime interest for engineers: the case of a full cylindrical bearing, where Γ_a and Γ_b are portions of height b of circular cylinders. For simplicity (and also because of prevailing interest), the axes of Γ_a and Γ_b are here taken to be parallel though distinct and it is assumed that the motion of Γ_b is very simple: a rotation around its axis with constant angular speed ω. In such case one can use an obvious system of cylindrical coordinates so that $x_1 = \theta$, $x_2 = z$, $a_1 = R^2$ (R, radius of Σ_a), $a_2 = 1$; $h = c + e\cos\theta$ (c, radial clearance; i.e., difference between the radius of Σ_a and that of Σ_b: $c > 0$. e, eccentricity ; i.e. , distance between the axes of Σ_a and Σ_b: $e < c$);

$$V_1 = \omega R \quad , \qquad V_2 = 0 \quad , \qquad V_3 = -\omega e \sin\theta \, .$$

If we assume also that the viscosity η is constant, then Reynolds' equation becomes

$$(4.1) \qquad \frac{\partial}{R\partial\theta}\left[(c + e\cos\theta)^3 \frac{\partial p}{R\partial\theta}\right] + (c + e\cos\theta)^3 \frac{\partial^2 p}{\partial z^2} = - 6\omega e\eta\sin\theta .$$

The problem is now set in the rectangle Γ : $0 \leq \theta \leq 2\pi$, $-\frac{b}{2} \leq z \leq \frac{b}{2}$; on the boundary $z = \pm\frac{b}{2}$ one can assume $p = 0$ (the lubricant can flow freely at either end of the bearing from conveniently placed reservoirs (flooded bearings) at atmosferic pressure; actually no grave complication would arise in most of the developments below if pressures at either end were assumed to be constant but different from each other and not less than zero).

For $\theta = 0$, $\theta = 2\pi$ one must impose periodicity conditions. These conditions make the problem slightly different from that considered in the previous section. There are various ways of overcoming the difficulty such as the use of the device of changing variables through a conformal mapping so that Γ goes into an annulus. The existence theorem of Sect.2 can be shown to apply because the conditions on γ, on α_1, α_2, and on f and g are now amply satisfied.

The adaptation to the present instance of theorems of regularity is still not easy; but there is one subcase where the analysis becomes less difficult. It is when e/c is small, so that one can presume that p is, in approximation, proportional to e/c

$$p = \frac{e}{c}\pi + o(\frac{e}{c}) \ .$$

Then the differential equation for π is simpler

$$(4.2) \qquad \frac{1}{R^2}\ \frac{\partial^2\pi}{\partial\theta^2} + \frac{\partial^2\pi}{\partial z^2} = - \frac{6\omega\eta}{c^2}\sin\theta \ .$$

In a forthcoming paper [7], Cimatti takes advantage of this simplification to obtain results regarding the regularity of π, the smoothness of γ_c and the extent of Ω.

The results on the last question are far from exhaustive: in particular one would like to know how Ω changes with b/R. Perhaps the problem is better expressed when the quantities in eqn (4.1) are renormalized using a characteristic length L, the parameters

$$a = e/c \ , \quad \alpha = b^2/L^2 \ , \ \beta = R^2/L^2 \ ,$$

the variable $\zeta = z/b$ and the function

$$P = p\, c^3 L^2 / 6\, \omega\, e\, \eta\, R^2 b^2 \ .$$

Then eqn (3.1) becomes

$$\text{(4.3)} \qquad \alpha \frac{\partial}{\partial \theta}\left[(1+ a \cos\theta)^3 \ \frac{\partial P}{\partial \theta}\right] + \beta (1+ a\cos\theta)^3 \frac{\partial^2 P}{\partial \zeta^2} = - \sin\theta$$

and the set Γ is the rectangle $0 \le \theta \le 2\pi$, $-\frac{1}{2} \le \zeta \le \frac{1}{2}$, so that Ω and $P(\theta, \zeta)$ depend only on the choice of the parameters a, α, β.

Numerical results seem to show that:

(i) $\Omega(a, \bar{\alpha}, \bar{\beta}) \supset \Omega(a, \bar{\bar{\alpha}}, \bar{\bar{\beta}})$ if $\bar{\alpha}/\bar{\beta} > \bar{\bar{\alpha}}/\bar{\bar{\beta}}$;

(ii) for $\beta \neq 0$ the <u>region of cavitation</u> $\Gamma - \Omega$ is not empty; for sufficiently small values of α/β, $\Gamma - \Omega$ is connected; whereas for sufficiently large values of α/β, the axis ζ belongs to Ω ; a critical value of α/β exists separating the two possibilities for the shape of Ω ;

(iii) $\Omega(a, \alpha, \beta) \to \Gamma$ if $\beta \to 0$,

$\Omega(a, \alpha, \beta) \to \Gamma'$ if $\alpha \to 0$,

where Γ' is the rectangle $0 < \theta < \pi$, $-\frac{1}{2} < \zeta < \frac{1}{2}$.

The two extreme cases quoted under (iii) are well known to engineers: they are called respectively the asymptotic cases of Sommerfeld and Ocvirk. In engineering analyses it is usually accepted that P is then a solution of the equation obtained from (4.3) inserting respectively $\beta = 0$ and $\alpha = 0$. In the former case the boundary conditions at $\zeta = \pm \frac{1}{2}$ are abandoned and the discrepancy is presumed settled through a <u>partial</u> boundary layer near the boundary. In the latter case the condition of continuity of $\frac{\partial P}{\partial \theta}$ is similarly abandoned, again presuming a boundary layer effect along γ_c .

5. Numerical approximations.

The difficulties met in any approximate numerical analysis of our problem are connected with the non-linear character of the free-boundary condition. The results quoted in Sect. 4 are among those obtained using an appropriate version (see [8]) of a method of approximation proposed by Stampacchia [9] to overcome such difficulties. The results will appear, together with some complementary developments, in a separate paper.

References

1 O. Reynolds, On the theory of lubrication, etc. Phil.Trans. Roy.Soc., A/177/1886, 157-234.

2 A. Sommerfeld, Zur hydrodynamische Theorie der Schmiermittelreibung. Zeitschr. Math. Phys., 50 (1904), 97-155.

3 G.H. Wannier, A contribution to the hydrodynamics of lubrication. Quart. Appl. Math., 8 (1950), 1-32.

4 H.G. Elrod, A derivation of the basic equations for hydrodynamic lubrication with a fluid having constant properties. Quart. Appl. Math., 17 (1960), 349-359.

5 G. Capriz, On some dynamical problems arising in the theory of lubrication. Riv. Mat. Univ. Parma, 1 (1960), 1-20.

6 M.K.V. Murthy and G. Stampacchia, A variational inequality with mixed boundary conditions. Israel J.Math., 13 (1972), 188-224.

7 G. Cimatti, On a problem of the theory of lubrication governed by a variational inequality. To appear in J. Optimization Theory Appl.

8 A. Laratta, O. Menchi, Approssimazione della soluzione di una disequazione variazionale. Applicazione ad un problema di frontiera libera. Calcolo, 11 (1974), 243-267.

9 G. Stampacchia, On a problem of numerical analysis connected with the theory of variational inequalities. Symposia Mathematica, 10 (1972), 281-293.

INTERIOR L^∞ ESTIMATES FOR FINITE ELEMENT APPROXIMATIONS OF SOLUTIONS OF ELLIPTIC EQUATIONS

Jean DESCLOUX
EPFL Dept. Math. of Lausanne

Nabil NASSIF
American University of Beirut

ABSTRACT.- Let $\Lambda \subset\subset \Omega \subset \tilde{\Omega}$ where $\tilde{\Omega}$ is the domain of definition of the solution of an elliptic equation. One assumes certain conditions of regularity on the equation and on the finite elements on Ω. Then one shows that the $L^2(\Omega)$ convergence of the approximate solution towards the exact solution implies the $L^\infty(\Lambda)$ convergence with the same order.

1.- INTRODUCTION

One considers the following situation. $\Omega \subset \tilde{\Omega} \subset R^N$ are open and bounded given sets; V is a subspace of $H^m(\tilde{\Omega})$ (Sobolev space) containing $H_o^m(\Omega)$ (completion in $H^m(\tilde{\Omega})$ of C^∞ functions with support in Ω); b is an elliptic bilinear form on $V \times V$, coercive on $H_o^m(\Omega)$, $b(w,z) = \sum_{0 \leqslant |\alpha|, |\beta| \leqslant m} \int_{\tilde{\Omega}} b_{\alpha\beta} D^\alpha w D^\beta z$ with $b_{\alpha\beta} \in C^\infty(\bar{\Omega})$.

Let us introduce some notations. $\Gamma, \Lambda, A, B, C, \ldots$ will denote open subsets of Ω; $A \subset\subset B$ means that the closure of A is contained in B. For $\Lambda \subset \Omega$, $w \in H^k(\Lambda_i)$ where the Λ_i's are disjoint and open with $\bar{\Lambda} = \bigcup_{i=1}^{n} \Lambda_i$, we set

$$||w||^2_{j,\Lambda} = \sum_{i=1}^{n} \sum_{|s| \leqslant j} \int_{\Lambda_j} (D^s u)^2, \quad 0 \leqslant j \leqslant k; \text{ if } w \in W^{k,\infty}(\Lambda_i) \ i=1,2,\ldots,n, \text{ we write}$$

$$|||w|||_{j,\Lambda} = \max_{|s| \leqslant j} \max_{i=1,\ldots,n} ||D^s w||_{L^\infty(\Lambda_i)} .$$

$\mathcal{S}$ is a family of finite element subspaces $S \subset V$; $h=h(S)$ is the maximum of the diameters of the elements contained in Ω. For $\Lambda \subset \Omega$, $S_o(\Lambda)$ denotes the set of elements of S with support contained in Λ. $\mathcal{S}$ verifies H2, H3, H4 of section 2. In fact these hypotheses are satisfied by all "classical" straight elements with the properties: a) for any $S \in \mathcal{S}$ and any polynomial p of degree r-1 there is $w \in S$ equal to p on Ω; b) there exists a constant $c>0$ independant of $S \in \mathcal{S}$ such that any element contained in Ω contains a sphere a radius ch^N. The best reference for H2a is [2]; H2b is

treated for example in [11]; for H3 and H4, one can consult [5].
Now, let $f\in L^2(\tilde{\Omega})\cap C^\infty(\tilde{\Omega})$, $u\in V$ with $b(u,\phi)=\int_{\tilde{\Omega}} f\phi \ \forall\phi\in V$, $v\in S$ with $b(v,\phi)=\int_{\tilde{\Omega}} f\phi \ \forall\phi\in S$ and $\Lambda\subset\subset\Omega$. Suppose $r\geqslant 2m$. Several interior estimates in quadratic norms have been established in the literature. For example, Nitsche and Schatz [8] have shown under the condition that $\partial\tilde{\Omega}$ is sufficiently regular that $||u-v||_{0,\Lambda}=O(h^r)$ independently of the regularity of f on $\tilde{\Omega}-\Omega$. Descloux [3] has established the pure interior estimate

$$||u-v||_{k,\Lambda} \leqslant c\{||u-v||_{0,\tilde{\Omega}} + h^{r-k}(||u||_{0,\Omega} + ||f||_{r-2m,\Omega})\} \quad 0\leqslant k\leqslant r;$$

this result shows in particular that the $L^2(\Omega)$ convergence of v implies the $L^2(\Lambda)$ convergence of all derivatives of order $<r$.

The purpose of this paper is to use the results contained in [3] for getting L^∞ interior estimates. Several L^∞ estimates are known, for example, [7] and [10]; [4], which concerns L^2 approximations contains likely the oldest result on the subject.

The first result of this paper (theorem 1) states that for the situation described above and $r\geqslant N/2$ one has

$$|||u-v|||_{k,\Lambda} \leqslant c\{||u-v||_{0,\Omega} + h^{r-k}|||u|||_{r,\Omega}\} \quad 0\leqslant k\leqslant r;$$

in some sense this relation justifies the practical use of finite elements However theorem 1 relies on the strong hypothesis H5.

In order to justify H5, consider the particular case of the Dirichlet form $b(w,z)=\int \sum_{|s|=1} D^s w D^s z$ with $V=H_0^1(\tilde{\Omega})$; supposing 1) $\partial\tilde{\Omega}$ regular; 2) the use of convenient curved elements at the boundary, 3) $r\geqslant 3$ (not only $r\geqslant 2$), Nitsche [7] has established for $\tilde{u}\in H_0^1(\tilde{\Omega})$, $\tilde{v}\in S$, $b(\tilde{u}-\tilde{v},\phi)=0 \ \forall\phi\in S$ the inequality $|||\tilde{u}-\tilde{v}|||_{0,\tilde{\Omega}}\leqslant ch^r|||\tilde{u}|||_{1,\tilde{\Omega}}$; now, using the notations of H5, let $\Gamma\subset\subset A\subset\subset\Lambda$ with ∂A regular and $\omega\in C_0^\infty(A)$ with $\omega(x)=1$, $x\in\Gamma$; the finite element space S is modified in a finite element subspace $\tilde{S}\subset H_0^1(A)$ such $\tilde{S}$ satisfies the requirements of Nitsche and $S=\tilde{S}$ on Γ (near ∂A the original elements are replaced by curved elements); then w is defined as the Ritz approximation of ωz. In fact, for the Dirichlet form, using essentially the arguments of [7], the authors have verified H5 under the condition $r\geqslant 3$ without the problematic construction of $\tilde{S}$. Scott's results [10] for $b(w,z)=\int(wz+\sum_{|s|=1} D^s w D^s z)$ can be treated in a similar way. Finally, remark that the restriction $r\geqslant N/2$, due to the use of Sobolev's imbedding theorem has no importance for concrete problems since $r\geqslant 2$ and $N\leqslant 3$.

The second part of this paper concerns "regular" elements as defined by H6 and supposes no hypothesis of type H5. Though different and obtained enterely independently, our results have many connections with those of [1]. Let us first introduce some further notions and notations; a finite difference operator of order M is the product of M operators of the form $v(x)\to h^{-1}(v(x_1,\dots,x_i+\alpha h,\dots,x_n)-v(x_1,\dots,x_i+\beta h,\dots x_n)$ where α and β are integers ; the finite difference operator of order 0 is the identity; a net of step h is a set of the form $R_h=\{a+\gamma h \mid \gamma\in Z^N\}$ where $a\in R^N$; for $\Lambda\subset\Omega$ and w defined on $\Lambda\cap R_h$ we write $||w||^2_{R_h\cap\Lambda}=h^N\sum_{x\in R_h\cap\Lambda} w^2(x)$, $|||w|||^2_{R_h\cap\Lambda}=\max_{x\in R_h\cap\Lambda}|w(x)|$. H6a is simply a definition of regular elements. For the usual elements, w of H6b can be constructed as the interpolant of u; H6b is easily verified when one remarks that the operator of translation by $\gamma h, \gamma\in Z^N$, commutes with the operator of interpolation. For piecewise polynomial elements, H6c is a consequence of the relation $|||v|||_{0,\Pi}\leqslant ch^{-N/2}||v||_{0,\Pi}$ where $v\in S$ and Π is an element (see [5]). The result concerning regular elements is given in theorem 2; it is certainly optimal with respect to orders of convergence; the presence of the term 2M in the norm of f is due to the use of an imbedding theorem. Remark finally that the extension of theorem 2 from "square" elements to "rectangular" elements is possible.

2.- HYPOTHESES

The symbol r denotes a fixed integer which is supposed to be $\geqslant 2m$.

H1. $b_{\alpha\beta}\in C^\infty(\bar\Omega)$, $|\alpha|,|\beta|\leqslant m$; b is coercive on $H_0^m(\Omega)$ i.e. there exists $\gamma>0$ such that $b(v,v)\geqslant\gamma||v||_{m,\Omega}$ for all $v\in H_0^m(\Omega)$; b is elliptic on $\bar\Omega$ i.e. $\sum_{|\alpha|,|\beta|=m} b_{\alpha\beta}(x)\xi^\alpha\xi^\beta\neq 0$ for all $\xi\in R^N, \xi\neq 0$, $x\in\bar\Omega$.

H2. Let $\Gamma\subset\subset\Lambda\subset\Omega$, $u\in C_0^\infty(\Gamma)$; then there exist $v\in S_0(\Omega)$, $w\in S_0(\Omega)$ such that

a) $||u-v||_{k,\Gamma}\leqslant ch^{\ell-k}||u||_{\ell,\Lambda}$, b) $|||u-w|||_{k,\Gamma}\leqslant ch^{\ell-k}|||u|||_{\ell,\Lambda}$

$0\leqslant k\leqslant\ell\leqslant r$,

where c depends on Γ and Λ.

H3. Let $\Lambda\subset\Omega$, $\omega\in C_0^\infty(\Lambda)$, $v\in S$; then there exists $w\in S_0(\Lambda)$ with

$||\omega v-w||_{k,\Lambda}\leqslant ch||v||_{k,\Lambda}$ $0\leqslant k\leqslant r$, where c depends on ω.

H4. Let $\Gamma\subset\subset\Lambda\subset\Omega$, $v\in S$; one has:

a) $||v||_{k+p,\Gamma}\leqslant ch^{-p}||v||_{k,\Lambda}$ $0\leqslant k+p\leqslant r$,

b) $|||v|||_{k+p,\Gamma}\leqslant ch^{-p}|||v|||_{k,\Lambda}$ $0\leqslant k+p\leqslant r$,

c) $|||v|||_{k,\Gamma}\leqslant ch^{-N/2}||v||_{k,\Lambda}$ $0\leqslant k\leqslant r$,

where c depends on Γ and Λ.

H5. Let $\Gamma \subset\subset \Lambda \subset \Omega$, $z \in C^{\infty}(\bar{\Lambda})$; then there exists $w \in S$ such that $b(z-w,\phi)=0$ $\forall \phi \in S_0(\Gamma)$ and $|||z-w|||_{0,\Gamma} \leq ch^r |||z|||_{r,\Lambda}$, where c depends on Γ and Λ.

H6. Let $\Gamma \subset\subset \Lambda \subset \Omega$.

a) There exists $h_0>0$ depending on Γ and Λ such that for any $v \in S_0(\Gamma)$ with $h(S)<h_0$ one has $V_i^{\pm} \in S_0(\Lambda)$ where $V_i^{\pm}(x_1,\ldots,x_N)=v(x_1,\ldots,x_{i-1},x_i \pm h,\ldots,x_N)$, $i=1,2,\ldots,N$.

b) For $u \in C^{\infty}(\bar{\Lambda})$, there exists $w \in S$ such that $|||Tu-Tw|||_{k,\Gamma} \leq ch^{r-k}|||u|||_{r+M}$ $0 \leq k \leq r$, where T is a difference operator of order M, w is independent of T, c depends on Γ, Λ and T.

c) If R_h is a net of step h, then for any $v \in S$ one has $||v||_{R_h \cap \Gamma} \leq c||v||_{0,\Lambda}$ where c depends on Γ and Λ but not on R_h.

3.- RESULTS AND PROOFS

We first recall theorem 2 of [3].

LEMMA 1

One supposes H1,H2,H3,H4. Let $\Gamma \subset\subset \Lambda \subset \Omega$, $f \in C^{\infty}(\bar{\Lambda})$, $w \in S$ with $b(w,\phi) = \int_{\Lambda} f\phi$ $\forall \phi \in S_0(\Lambda)$; then there exists $u \in H_0^m(\Lambda)$ with $b(u,\phi) = \int_{\Lambda} f\phi$ $\forall \phi \in H_0^m(\Gamma)$ and $||u-w||_{k,\Gamma} \leq ch^{r-k}(||w||_{0,\Lambda}+||f||_{r-2m,\Lambda})$, $k = 0,1,\ldots,r$; c depends on Γ and Λ.

Lemma 2 is a consequence of classical interior regularity properties of elliptic equations (see for example [6]) and of imbedding theorems.

LEMMA 2

One supposes H1. Let $\Gamma \subset\subset \Lambda \subset \Omega$, $u \in H^m(\Lambda)$, $f \in C^{\infty}(\bar{\Lambda})$ with $b(u,\phi) = \int_{\Lambda} f\phi$ $\forall \phi \in H_0^m(\Lambda)$; then $|||u|||_{k,\Gamma} \leq c\{||u||_{0,\Lambda}+||f||_{k+M-2m,\Lambda}\}$ $k=0,1,2,\ldots,$ where M is the smallest integer larger than N/2 and where c depends on Γ, Λ and k.

Lemma 3 is a discrete version of lemma 2 for f=0; it should be possible to remove the parasitic term $h^{r-N/2}$.

LEMMA 3

One supposes H1,H2,H3,H4. Let $\Gamma \subset\subset \Lambda \subset \Omega$, $w \in S$ with $b(w,\phi) = 0$ $\forall \phi \in S_0(\Lambda)$; then $|||w|||_{0,\Gamma} \leq c||w||_{0,\Gamma}(1+h^{r-N/2})$ where c depends on Γ and Λ.

PROOF

c will denote a generic constant independent of w and h. Let $\Gamma \subset\subset C \subset\subset B \subset\subset A \subset\subset \Lambda$. By lemma 1, there exists $u \in H^m(\Lambda)$ with $b(u,\phi)= 0$ $\forall \phi \in H_0^m(A)$ and $||u-w||_{0,A} \leq ch^r||w||_{0,\Lambda}$. By lemma 2, $|||u|||_{r,B} \leq c||u||_{0,A} \leq c||w||_{0,\Lambda}$. Let (H2b) $z \in S$ with $|||u-z|||_{0,C} \leq ch^r|||u|||_{r,B} \leq ch^r||w||_{0,\Lambda}$.

By H4c, one gets:

$$|||z-w|||_{0,\Gamma} \leqslant ch^{-N/2}||z-w||_{0,C} \leqslant ch^{-N/2}(||z-u||_{0,C} + ||u-w||_{0,C}) \leqslant ch^{r-N/2}||w||_{0,\Lambda'}$$

$$|||w|||_{0,\Gamma} \leqslant |||w-z|||_{0,\Gamma} + |||z-u|||_{0,\Gamma} + |||u|||_{0,\Gamma} \leqslant c(1+h^{r-N/2})||w||_{0,\Lambda}.$$

Lemma 4 is a classical "trick" based on the "inverse assumption" H4b.

<u>LEMMA 4</u>

One suppose H2 and H4. Let $\Gamma \subset\subset \Lambda \subset \Omega$, $u \in C^{\infty}(\bar{\Lambda})$, $v \in S$; then $|||u-v|||_{K,\Gamma} \leqslant ch^{-k}(|||u-v|||_{0,\Lambda} + h^{r}|||u|||_{r,\Lambda})$ $0 \leqslant k \leqslant r$, where c depends on Γ and Λ.

<u>PROOF</u>

c will denote a generic constant independent of u and v. Let $\Gamma \subset\subset A \subset\subset \Lambda$ and (H2b) $w \in S$ such that $|||u-w|||_{k,A} \leqslant ch^{r-k}|||u|||_{r,\Lambda}$; then H4b implies

$$|||u-v|||_{k,\Gamma} \leqslant |||u-w|||_{k,\Gamma} + |||w-v|||_{k,\Gamma} \leqslant ch^{r-k}|||u|||_{r,\Lambda} + ch^{-k}|||w-v|||_{0,A} \leqslant$$
$$\leqslant ch^{r-k}|||u|||_{r,\Lambda} + ch^{-k}(|||u-w|||_{0,A} + |||u-v|||_{0,A}) \leqslant ch^{r-k}|||u|||_{r,\Lambda} +$$
$$+ ch^{-k}|||u-v|||_{0,\Lambda}.$$

<u>THEOREM 1</u>

One supposes H1,H2,H3,H4,H5 and $r > N/2$. Let $\Lambda \subset\subset \Omega$, $f \in C^{\infty}(\bar{\Omega})$, $u \in C^{\infty}(\bar{\Omega})$ with $b(u,\phi) = \int_{\Omega} f\phi \quad \forall \phi \in H_0^m(\phi)$, $v \in S$ with $b(v,\phi) = \int_{\Omega} f\phi \quad \forall \phi \in S_0(\Omega)$; then

$$|||u-v|||_{k,\Lambda} < c\{||u-v||_{0,\Omega} + h^{r-k}|||u|||_{r,\Omega}\} \quad 0 < k < r,$$

where c depends on Λ.

<u>PROOF</u>

c will denote a generic constant independent of u and v. Let $\Lambda \subset\subset D \subset\subset C \subset\subset B \subset\subset A \subset\subset \Omega$. By lemma 1, there exists $\tilde{u} \in H^m(\Omega)$ such that $b(\tilde{u},\phi) = \int_{\Omega} f\phi \quad \forall \phi \in H_0^m(A)$ and $||\tilde{u}-v||_{0,A} \leqslant ch^{r}(||v||_{0,\Omega} + ||f||_{r-2m,\Omega})$. Let (H5) $w \in S$ such that $b(\tilde{u}-w,\phi)=0 \quad \forall \phi \in S_0(B)$, $|||\tilde{u}-w|||_{0,C} \leqslant ch^{r}|||\tilde{u}|||_{r,B}$. We apply H4b and lemma 3 to $v-w \in S$ which satisfies the relation $b(v-w,\phi)=0 \quad \forall \phi \in S_0(b)$:

$$|||v-w|||_{k,\Lambda} \leqslant ch^{-k}|||v-w|||_{0,D} \leqslant ch^{-k}||v-w||_{0,C} \leqslant ch^{-k}(||\tilde{u}-v||_{0,C} + ||\tilde{u}-w||_{0,C}) \leqslant$$
$$\leqslant ch^{r-k}(||v||_{0,\Omega} + ||f||_{r-2m,\Omega} + ||u||_{r,B});$$

by lemma 4, $|||\tilde{u}-w|||_{k,\Lambda} \leqslant ch^{r-k}|||\tilde{u}|||_{r,B}$ and consequently

$$|||u-v|||_{k,\Lambda} \leqslant |||u-\tilde{u}|||_{k,\Lambda} + |||\tilde{u}-w|||_{k,\Lambda} + |||w-v|||_{k,\Lambda} \leqslant$$

$$\leqslant |||u-\tilde{u}|||_{k,\Lambda} + ch^{r-k}(|||\tilde{u}|||_{r,B} + ||v||_{0,\Omega} + ||f||_{r-2m,\Omega}) \leqslant$$

$$\leqslant c|||u-\tilde{u}|||_{r,B} + ch^{r-k}(|||u|||_{r,\Omega} + ||v||_{0,\Omega} + ||f||_{r-2m,\Omega});$$

now by lemma 2 one has $|||u-\tilde{u}|||_{r,B} \leqslant c||u-\tilde{u}||_{0,A} \leqslant c\{||u-v||_{0,A} + ||v-\tilde{u}||_{0,A}\}$ $\leqslant c||u-v||_{0,A} + ch^{r}(||v||_{0,\Omega} + ||f||_{r-2m,\Omega})$; finally, remarking that $||f||_{r-2m,\Omega} \leqslant c||u||_{r,\Omega} \leqslant c|||u|||_{r,\Omega}$, one gets

$$|||u-v|||_{k,\Lambda} \leqslant c||u-v||_{0,A} + ch^{r-k}(|||u|||_{r,\Omega} + ||v||_{0,\Omega} + ||f||_{r-2m,\Omega}) \leqslant$$

$$\leqslant c||u-v||_{0,A} + ch^{r-k}|||u|||_{r,\Omega}.$$

From now on, we restrict ourselves to the regular case defined by H6. We recall theorem 5 of [3].

LEMMA 5

One supposes H1, H2, H3, H4 and H6. Let $\Gamma \subset\subset \Lambda \subset \Omega$, $f \in C^{\infty}(\bar{\Omega})$, $u \in H^{m}(\Omega)$ with $b(u,\phi) = \int_{\Lambda} f\phi \quad \forall \phi \in H_0^m(\Lambda)$, $v \in S$ with $b(v,\phi) = \int_{\Lambda} f\phi \quad \forall \phi \in S_0(\Lambda)$, T be a finite difference operator of order M; then for $h<h_0$ one has $||Tu-Tv||_{k,\Gamma} \leqslant c\{||u-v||_{0,\Lambda} + h^{r-k}(||u||_{0,\Omega} + ||f||_{r+M-2m,\Lambda})\}$ where c and $h_0>0$ depends on Γ,Λ and T.

Lemma 6 can be considered as a discrete version of Sobolev's imbeedding theorem; our proof uses the extension theorem of Rayben'kii [9]. In the remaining part of this paper we shall denote by T_M the set of finite difference operators which are the product of at most M operators of the form $v(x) \to h^{-1}(v(x_1,\ldots,x_i+h,\ldots,x_n) - v(x_1,\ldots,x_i,\ldots,x_n))$; T_m contains the identity.

LEMMA 6

Let R_h be a net of step h, $D \subset\subset B \subset\subset A \subset \Omega$, $M>N/2$, and v be defined on $R_h \cap A$; then for $h<h_0$, $|||v|||^2_{R_h\cap D} \leqslant c \sum_{T\in T_M} ||Tv||^2_{R_h\cap B}$, where c and $h_0>0$ are independent of R_h, but depend on D,B.

PROOF

c will denote a generic constant idependent of R_h. Let $D \subset\subset C \subset\subset B$. Following [8], one constructs a piecewise polynomial function $u \in C^M(\bar{C})$ interpolating v on $R_h \cap C$ such that $|D^s u(x)| \leqslant c \max_{y \in R_h, |y-x|<ah} \max_{T \in \mathcal{T}_M} |(Tv)(y)|$ for $x \in C$ and $|s| \leqslant M$; here a is independent of x and R_h. One deduces immediately the inequality $\|u\|^2_{M,C} \leqslant \sum_{T \in \mathcal{T}_M} \|Tv\|^2_{R_h \cap B}$

and the lemma follows by the theorem of Sobolev.

THEOREM 2

One supposes H1,H2,H3,H4,H6. Let M be the smallest integer larger than N/2, $\Lambda \subset\subset \Omega$, U be a finite difference operator of order P, $u \in C^\infty(\bar{\Omega})$ with $b(u,\phi) = \int_\Omega f\phi \quad \forall \phi \in H^m_0(\Omega)$, $v \in S$ with $b(v,\phi) = \int_\Omega f\phi \quad \forall \phi \in S_0(\Omega)$. Then for $h<h_0$

$$|||U(u,v)|||_{k,\Lambda} \leqslant c\ \{ \|u-v\|_{0,\Omega} + h^{r-k}(\|u\|_{0,\Omega} + \|f\|_{r+P+2M-2m,\Omega})\},\ 0 \leqslant k \leqslant r,$$

where $h_0 > 0$ and c depend on U and Λ.

PROOF

Let $\Lambda \subset\subset D \subset\subset C \subset\subset B \subset\subset A \subset\subset \Omega$. c will denote a generic constant independent of u and v. As in the proof of theorem 1, one introduces two auxiliary functions $\tilde{u} \in H^m(\Omega)$ (lemma 1) and $w \in S$ (H6b) such that

1) $b(\tilde{u},\phi) = \int_\Omega f\phi \quad \forall \phi \in H^m_0(A)$, $\|\tilde{u}-v\|_{0,A} \leqslant ch^r\{\|v\|_{0,\Omega} + \|f\|_{r-2m,\Omega}\}$,

2) $\|TU\tilde{u}-TUw\|_{0,C} \leqslant ch^r\ |||\tilde{u}|||_{r+P+M,B} \quad \forall T \in \mathcal{T}_M$.

From (1), lemma 5 and then by (2), one obtains

$$\|TU\tilde{u}-TUv\|_{0,B} \leqslant ch^r(\|v\|_{0,\Omega} + \|\tilde{u}\|_{0,A} + \|f\|_{r+P+M-2m,\Omega})$$

$$(3)\qquad \|TU(v-w)\|_{0,C} \leqslant ch^r\{\|v\|_{0,\Omega} + \|\tilde{u}\|_{0,A} + |||\tilde{u}|||_{r+P+M,B} + \|f\|_{r+P+M-2m,\Omega}\}$$

Let R_h be any net of step h: since (H6a) $TU(v-w) \in S$, one has by H6c $\|TU(v-w)\|_{R_h \cap D} \leqslant c\|TU(v-w)\|_{0,C}$ and then by lemma 6

$||| U(v-w) |||_{R_h \cap \Lambda} \leq c(\sum_{T \in T_M} \| TU(v-w) \|^2_{0,C})^{\frac{1}{2}}$; for h=h(S) fixed one applies the last relations to all possible nets R_h and gets by (3);

$||| U(v-w) |||_{0,\Lambda} \leq ch^r \{ \| v \|_{0,\Omega} + \| u \|_{0,A} + |||\tilde{u}|||_{r+P+M,B} + \| f \|_{r+P+M-2m,\Omega} \}.$

The rest of the proof is similar to that of theorem 1. Note that in order to get rid of the term $|||u|||_{r+P+M,B}$ we use lemma 2 which introduces the term $\| f \|_{r+P+2M-2m,\Omega}$.

REFERENCES

[1] J. BRAMBLE, J. NITSCHE, A. SCHATZ: "Maximum-norm interior estimates for Ritz-Galerkin methods".Mathematics of Computation, vol.29,n. 131, 1975, 677-689.

[2] PH. CLEMENT: "Approximation by finite element functions in Sobolev norms". To appear in Revue Française d'automatisme, informatique et recherche opérationnelle.

[3] J. DESCLOUX: "Interior regularity and local convergence of Galerkin finite element approximations for elliptic equations". Topics in numerical analysis II (J. Miller, ed.), Academic Press 1975.

[4] J. DESCLOUX:"Some properties of approximations by finite elements". Report EPFL, Dept. Math. Lausanne 1969.

[5] J. DESCLOUX: "Two basic properties of finite elements".Report EPFL, Dept. Math. Lausanne 1973.

[6] G. FICHERA: "Linear elliptic differential systems and eigenvalue problems".Lecture Notes in Mathematics 8, 1965, Springer-Verlag.

[7] J. NITSHE: "L_∞-convergence of finite element approximation". 2. Conference on finite elements, Rennes 1975.

[8] J. NITSCHE, A. SCHATZ: "Interior estimates for Ritz-Galerkin methods. Mathematics of Computation, vol. 28, N. 128, 1974, 937-959.

[9] V.S. RAYABEN'KII: "Local splines. Computer Methods in Applied Mechanics and Engineering".5 (1975), 211-225.

[10] R. SCOTT: "Optimal L^∞ estimates for the finite element method on irregular meshes".Report, University of Chicago 1975.

[11] G. STRANG: "Approximation in the finite element method".Num. Math. 19, 1972, 81-98.

[12] V. THOMEE, B. WESTERGREN: "Elliptic difference equations and interior regularity". Num. Math. 11, 1968, 196-210.

H^1-Galerkin Methods for a Nonlinear Dirichlet Problem

Jim Douglas, Jr.
University of Chicago

1. Introduction. Consider the Dirichlet problem

$$(1.1)\qquad \begin{aligned} \nabla\cdot(a(x,u)\nabla u) &= f(x),\ x \in \Omega,\\ u &= g(x),\ x \in \partial\Omega, \end{aligned}$$

for a bounded domain $\Omega \subset \mathbb{R}^2$. The boundary $\partial\Omega$ will be assumed smooth in order that duality can be employed in the convergence proofs. The coefficient $a(x,u)$ is assumed to be such that $a:\bar{\Omega} \times \mathbb{R} \to [\alpha\ \beta]$, where $0 < \alpha \le \beta < \infty$, and $|D^\gamma(x,v)| \le M$, $x \in \bar{\Omega}$, $v \in \mathbb{R}$, $|\gamma| \le 2$ The approximate solution will be shown to converge in $L^\infty(\Omega)$; consequently, the assumption that $a(x,v) \in [\alpha,\beta]$ causes no loss of generality provided only that the solution u of (1.1) is such that $a(x,u(x)) \in [\alpha_o,\beta_o]$ for some positive constants α_o and β_o, since asymptotically only values of $a(x,v)$ in the neighborhood of the set $S = \{(x,u(x)):x \in \bar{\Omega}\}$ need be considered. Similarly the constraint $|D^\gamma a| \le M$ need be valid only in a neighborhood of S.

Let M_h be a finite-dimensional finite element space such that $M_h \subset W_2^2(\Omega) \cap W_\infty^1(\Omega)$ and such that for almost every $x \in \partial\Omega$ there exists a ball $B(x)$ about x as center for which, if $v \in M_h$, $v \in W_2^3(\Omega \cap B(x))$. Thus, for any $v \in M_h$ a trace of v on $\partial\Omega$ can be defined, and this trace belongs to $W_2^2(\partial\Omega)$. Denote the inner products in $L^2(\Omega)$ and $L^2(\partial\Omega)$ by (v,w) and $\langle v,w\rangle$, respectively. Let

$$||v||_k = ||v||_{W_2^k(\Omega)}, \quad |v|_k = ||v||_{W_2^k(\partial\Omega)} .$$

Assume the following approximation property for M_h:

$$\inf_{v\in M_h} \sum_{j=0}^{2} (h^j||z-v||_j + h^{j+1/2}|z-v|_j) \le C||z||_k h^k, \quad \tfrac{5}{2} \le k \le r+1, \tag{1.2}$$

where $r \geq 3$ is a positive integer and $z \in W_2^k(\Omega)$. Examples of admissible spaces are constructed in [2, 3, 5, 7, 13, 16]. Restrictions to Ω of tensor products of C^1-piecewise-polynomial functions (of degree r) are discussed in [7]. Clough-Tocher elements and their generalizations are treated in [5]. See [9] for C^1-piecewise-polynomial elements of degree at least five on triangulations, and see [2, 3] for isoparametric elements. Note that quasi-regularity (or quasi-uniformity) is not required of the elements so long as (1.2) is satisfied.

Let $T_2 : W_2^2(\partial\Omega) \to L^2(\Omega)$ be such that there exist constants c_1 and c_2 such that

$$\begin{aligned} &c_1|z|_2 \le |T_2 z|_o + |z|_o \le c_2|z|_2 , \\ &c_1|z|_2 \le |T_2^* z|_o + |z|_o \le c_2|z|_2 , \\ &|T_2^* T_2 z|_k \le c_2|z|_{k+4}, \quad 0 \le k \le r-3 , \end{aligned} \tag{1.3}$$

for $z \in C^\infty(\bar{\Omega})$. For $\Omega \subset \mathbb{R}^2$ it is sufficient to take $T_2 = -\partial^2/\partial s^2$, where s measures arc length along $\partial\Omega$. When $\Omega \subset \mathbb{R}^n$ for $n > 2$, one admissible T_2 would be given by the Laplace-Beltrami operator on $\partial\Omega$. Set

$$A(v;w,z) = (\nabla\cdot(a(x,v)\nabla w), \Delta z) + h^{-3}\langle w,z\rangle + h\langle T_2 w, T_2 z\rangle \tag{1.4}$$

for v, w, and z in M_h or with $v \in W_2^1(\Omega)$ and w and z in $W_2^{5/2}(\Omega)$.

An H^1-Galerkin procedure can be defined as follows. Find $u_h \varepsilon M_h$ such that

(1.5) $$A(u_h;u_h,v) = (f,\Delta v) + h^{-3}\langle g,v\rangle + h\langle T_2 g,T_2 v\rangle,\ v \varepsilon M_h .$$

It will be shown below that, for h sufficiently small, a solution of u_h of (1.5) exists tending to u as h tends to zero. Moreover, there is only solution inside any ball $\{||u-u_h||_2 \leq K\}$ for small h. The rate of convergence of u_h to u is optimal in $W_2^j(\Omega)$, $0 \leq j \leq 2$.

It is convenient to treat a linear Dirichlet problem before considering the nonlinear one. In the linear case the dimension of Ω will not be restricted.

A second H^1-Galerkin method will also be presented, although the analysis of this method will be omitted. It is based on interpolation of the boundary data and is designed to employ the generalized Clough-Tocher elements [5].

The procedure given by (1.5) is a direct generalization of a method of Dupont, Wheeler, and the author [7] for the classical Dirichlet problem. It is also related to a method of Thomée and Wahlbin [14]. The second method, to be defined below in (4.4), is in some sense an H^1- analogue of the procedure analyzed by Scott [12] for the linear Dirichlet problem. Both methods can be thought of as non-symmetric versions of the least squares Ritz-Galerkin method [1]. See also [4] for another Galerkin method, based on the Nitsche form, for (1.1).

2. The linear Dirichlet problem. Let

$$(2.1)\qquad \begin{aligned} Lu = \nabla\cdot(a\nabla u) + b\cdot\nabla u + cu &= f,\ x \in \Omega,\\ u &= g,\ x \in \partial\Omega, \end{aligned}$$

be a linear Dirichlet problem on $\Omega \subset \mathbb{R}^n$, where $\partial\Omega$ is smooth and the coefficients of L are smooth functions of $x \in \Omega$ with $a:\Omega \to [\alpha,\beta]$. Assume that (2.1) possesses one and only one solution for each pair $f \in L^2(\Omega)$ and $g \in W_2^{3/2}(\partial\Omega)$. Also, assume elliptic regularity for the adjoint problem; i.e., if

$$(2.2)\qquad \begin{aligned} L^*z = \nabla\cdot(a\nabla z - bz) + cz &= \phi,\ x \in \Omega,\\ z &= \psi,\ x \in \partial\Omega, \end{aligned}$$

then, with $0 < c_3 < c_4 < \infty$,

$$(2.3)\qquad c_3||z||_{k+2} \le ||\phi||_k + |\psi|_{k+\frac{3}{2}} \le c_4\,||z||_{k+2},\quad 0 \le k \le r-1.$$

Let the bilinear form corresponding to (1.4) be given by

$$(2.4)\qquad A(w,z) = (Lw,\Delta z) + h^{-3}\langle w,z\rangle + h\langle T_2w,T_2z\rangle.$$

Then the H^1-Galerkin method for (2.1) consists in determining $u_h \in M_h$ such that

$$(2.5)\qquad A(u_h,v) = (f,\Delta v) + h^{-3}\langle g,v\rangle + h\langle T_2g,T_2v\rangle,\quad v \in M_h .$$

The space M_h is assumed to satisfy exactly the same conditions as required above; in particular, (1.2) is assumed to hold.

<u>Lemma 1</u>. (Gårding Inequality) There exist constants $h_o > 0$, $\rho > 0$, and C such that

$$(2.6)\qquad A(v,v) \ge \rho||v||_2^2 - C||v||_1^2,$$

provided that $v \in M_h + W_2^{5/2}(\Omega)$ and that $h \le h_o$.

Proof. The hypothesis that $v \in M_h + W_2^{5/2}(\Omega)$ implies that $v \in W_2^2(\partial\Omega)$. Since $\alpha \le a(x) \le \beta$ and the coefficients are smooth, (1.3) implies that

$$\begin{aligned} A(v,v) &= (Lv,\Delta v) + h^{-3}|v|_o^2 + h|T_2 v|_o^2 \\ &\ge (a\Delta v,\Delta v) - C||v||_1||\Delta v||_o + h^{-3}|v|_o^2 + h\{c_1|v|_2 - |v|_0\}^2 \\ &\ge \frac{\alpha}{2}||\Delta v||_o^2 - C||v||_1^2 + (1-Ch^4)h^{-3}|v|_o^2 + \frac{1}{2}c_1^2 h|v|_2^2 . \end{aligned}$$

Since [8]

$$||v||_2^2 \le C\{||\Delta v||_o^2 + |v|_{3/2}^2\}$$

and

$$|v|_{3/2}^2 \le (h^{-3}|v|_o^2)^{\frac{1}{4}}(h|v|_2^2)^{\frac{3}{4}} \le \frac{1}{4}h^{-3}|v|_o^2 + \frac{3}{4}h|v|_2^2,$$

it follows that (2.6) holds as soon as $1 - Ch^4$ is greater than some positive number. q.e.d.

Let

$$(2.7) \qquad |||v|||^2 = ||z||_2^2 + h^{-3}|z|_o^2 + h|z|_2^2 .$$

A glance at the proof of Lemma 1 shows that the Gårding inequality can be stated in the form

$$(2.8) \qquad A(v,v) \ge \rho|||v|||^2 - C||v||_1^2, \; v \in M_h + W_2^{5/2}(\Omega), \; h \le h_o .$$

It follows directly from (1.2) that

$$(2.9) \qquad \inf_{v\in M_h} |||z-v||| \le C||z||_k h^{k-2}, \; \frac{5}{2} \le k \le r+1,$$

for $z \in W_2^k(\Omega)$.

Lemma 2. Let $z \in M_h + W_2^{5/2}(\Omega)$ satisfy $A(z,v) = 0$, $v \in M_h$. Then there exists a constant C such that

$$(2.10)\quad \begin{aligned} ||z||_{-s} &\le C|||z|||h^{s+2}, \ -2 \le s \le r-3, \\ |z|_{-q} &\le C|||z|||h^{q+\frac{3}{2}}, \ -2 \le q \le r+\tfrac{1}{2}, \end{aligned}$$

where, for $s \ge 0$ and $q \ge 0$,

$$(2.11)\quad ||z||_{-s} = \sup_{\psi \varepsilon W_2^s(\Omega)} \frac{(z,\psi)}{||\psi||_s}, \quad |z|_{-q} = \sup_{\psi \varepsilon W_2^q(\partial\Omega)} \frac{\langle z,\psi\rangle}{|\psi|_q}.$$

Proof. This proof is modelled on the arguments of [7]. The boundary estimate will be demonstrated first. Let $\gamma \varepsilon W_2^q(\partial\Omega)$, $2 \le q \le r + \frac{1}{2}$, and define $\phi \varepsilon W_2^{q+1/2}(\Omega)$ through the boundary problem

$$\begin{aligned} L^*\Delta\phi &= 0, \ x \varepsilon \Omega, \\ \Delta\phi &= 0, \ x \varepsilon \partial\Omega, \\ (a\tfrac{\partial}{\partial n} - b\cdot n)\Delta\phi - h^{-3}\phi &= -\gamma, \ x \varepsilon \partial\Omega. \end{aligned}$$

It follows from (2.3) that $\Delta\phi = 0$ on $\bar{\Omega}$; hence, ϕ satisfies the simple Dirichlet problem $\Delta\phi = 0$ in Ω and $\phi = h^3\gamma$ on $\partial\Omega$. Thus,

$$||\phi||_{q+1/2} \le Ch^3|\gamma|_q,$$

by the regularity of the Dirichlet problem for the Laplace operator, given that $\partial\Omega$ is smooth. Note that

$$0 = (z,L^*\Delta\phi) = (Lz,\Delta\phi) + h^{-3}\langle z,\phi\rangle - \langle z,\gamma\rangle,$$

or

$$\langle z,\gamma\rangle = A(z,\phi) - h\langle T_2 z, T_2\phi\rangle.$$

Since $A(z,\phi) = A(z,\phi-v)$ for $v \varepsilon M_h$, then for $2 \le q \le r + \frac{1}{2}$,

$$|A(z,\phi)| \leq C|||z||| \inf_{v\varepsilon M_h} |||\phi-v|||$$

$$\leq C|||z|||h^{q-3/2}||\phi||_{q+1/2}$$

$$\leq C|||z|||h^{q+3/2}|\gamma|_q .$$

Next, note that

$$|\langle T_2 z, T_2\phi\rangle| = |\langle z, T_2^* T_2\phi\rangle|$$

$$\leq C|z|_{-q+4}|\phi|_q$$

$$\leq C|z|_{-q+4}h^3|\gamma|_q .$$

Thus, it follows that

(2.12) $\quad |z|_{-q} \leq C\{|||z|||h^{q+3/2} + |z|_{-q+4}h^4\},\ 2 \leq q \leq r + \frac{1}{2}.$

Note the trivial inequality

$$h^{-3}|z|_0^2 + h|z|_2^2 \leq |||z|||^2$$

that follows from (2.7). Hence, by interpolation,

(2.13) $\quad |z|_q \leq |||z|||h^{\frac{3}{2}-q},\ 0 \leq q \leq 2.$

Let $q = 2$ in (2.12); then

$$|z|_{-2} \leq C\{|||z|||h^{7/2} + |z|_2 h^4\} \leq C|||z|||h^{7/2} .$$

Consequently,

$$|z|_{-q} \leq C|||z|||h^{q+3/2},\ -2 \leq q \leq 2.$$

Recurse by intervals of length four until $[-2, r+\frac{1}{2}]$ is covered to complete the proof of the boundary estimate.

In order to estimate z in $W_2^s(\Omega)'$, let $\psi \in W_2^s(\Omega)$ for

$0 \le s \le r-3$. It is technically convenient to approximate ψ by a function ψ_h having the following properties:

$$(2.14)\quad \begin{aligned} &||\psi-\psi_h||_{s-j} \le C||\psi||_s h^j, \quad 0 \le j \le 2r-s+1, \\ &||\psi_h||_{s+j} + |\psi_h|_{s+j-1/2} \le C||\psi||_s h^{-j}, \quad 0 \le j \le 2r-s. \end{aligned}$$

The existence of ψ_h is demonstrated in Theorem 7 of [7] by using a particular spline approximation of ψ. Now, define $\phi \in W_2^{s+4}(\Omega)$ by the boundary problem

$$\begin{aligned} &L^*\Delta\phi = \psi_h, \quad x \in \Omega, \\ &\Delta\phi = (a\tfrac{\partial}{\partial n} - b\cdot n)\Delta\phi - h^{-3}\phi = 0, \quad x \in \partial\Omega. \end{aligned}$$

By the regularity assumption (2.3) and the boundary condition $\Delta\phi = 0$,

$$||\Delta\phi||_{k+2} \le C||\psi_h||_k, \quad 0 \le k \le 2r.$$

Since $\phi = h^3(a\frac{\partial}{\partial n} - b\cdot n)\Delta\phi$ on $\partial\Omega$, regularity of the Dirichlet problem for the Laplace operator implies that

$$\begin{aligned} ||\phi||_{s+4} &\le C\{||\Delta\phi||_{s+2} + |\phi|_{s+7/2}\} \\ &= C\{||\Delta\phi||_{s+2} + h^3|(a\tfrac{\partial}{\partial n} + b\cdot n)\Delta\phi|_{s+7/2}\} \\ &\le C\{||\Delta\phi||_{s+2} + h^3||\Delta\phi||_{s+5}\} \\ &\le C\{||\psi_h||_s + h^3||\psi_h||_{s+3}\} \\ &\le C||\psi||_s , \end{aligned}$$

by (2.14). Next, note that the usual integration by parts shows that

$$(z,\psi_h) = A(z,\phi) - h\langle T_2 z, T_2\phi\rangle .$$

Again, since $A(z,\phi) = A(z,\phi-v)$ for $v \in M_h$ and $0 \leq s \leq r-3$,

$$|A(z,\phi)| \leq C|||z|||\inf_{v\in M_h} |||\phi-v||| \leq C|||z|||h^{s+2}||\psi||_s .$$

If the boundary estimate already derived is applied, then it follows that

$$\begin{aligned} |h<z,T_2^*T_2\phi>| &\leq Ch|z|_{-s+\frac{1}{2}}|\phi|_{s+7/2} \\ &\leq Ch|z|_{-s+\frac{1}{2}}||\psi||_s \\ &\leq C|||z|||h^{s+2}||\psi||_s . \end{aligned}$$

Hence,

$$||z||_{-s} \leq C|||z|||h^{s+2}, \quad 0 \leq s \leq r-3.$$

Interpolation between $s = -2$ and $s = 0$ completes the argument. q.e.d.

Lemma 2 can be applied to obtain existence and uniqueness of a solution of (2.5) and to obtain error estimates.

Theorem 3. There exists $h_o > 0$ such that (2.5) has one and only one solution $u_h \in M_h$ for $0 < h \leq h_o$, and $u - u_h$ satisfies the following inequalities:

(2.15)

$$|||u-u_h||| \leq C\inf_{v\in M_h}|||u-v||| \leq C||u||_k h^{k-2}, \quad \frac{5}{2} \leq k \leq r+1,$$

$$||u-u_h||_{-s} \leq C||u||_k h^{k+s}, \quad \frac{5}{2} \leq k \leq r+1, \ -2 \leq s \leq r-3,$$

$$|u-u_h|_{-q} \leq C||u||_k h^{k+q-1/2}, \quad \frac{5}{2} \leq k \leq r+1, \ -2 \leq q \leq r+\frac{1}{2} .$$

Proof. First, consider the existence and uniqueness of u_h; clearly, it is sufficient to show uniqueness. So, assume that $w \in M_h$ and that $A(w,v) = 0$ for $v \in M_h$. By Lemma 2, $||w||_1 \leq Ch|||w|||$.

By (2.8),

$$\rho|||w|||^2 \leq A(w,w) + C||w||_1^2 \leq Ch^2|||w|||^2 .$$

Consequently, $w = 0$ for small h, as was to be shown.

Let $\zeta = u - u_h$. Since $\zeta \in M_h + W_2^{5/2}(\Omega)$ and $A(\zeta,v) = 0$ for $v \in M_h$, Lemma 2 can be applied to ζ; thus, it is sufficient to obtain the estimate for $|||\zeta|||$ in (2.15). Again using the Gårding inequality along with the relation $||\zeta||_1 \leq Ch|||\zeta|||$, it follows that

$$\begin{aligned}(\rho - Ch^2)|||\zeta|||^2 &\leq A(\zeta,\zeta) = A(\zeta,u-v)\\ &\leq C|||\zeta|||\cdot|||u - v|||\end{aligned}$$

for $v \in M_h$, and the inequality holds for small h. q.e.d.

Note that (2.15) implies uniform convergence of u_h to u if $\dim(\Omega) \leq 3$, by the Sobolev embedding theorem. This justifies the assumptions made on the coefficients of (1.1).

3. The nonlinear Dirichlet problem.

Return to the Galerkin procedure given by (1.5); recall that Ω is assumed to be a bounded domain in the plane. A generalization of Lemma 1 will be required in the analysis of (1.5). Let

$$||z||_{k,p} = ||z||_{W_p^k(\Omega)},$$

with the convention that the index p will be omitted if $p = 2$.

Lemma 4. For any $\delta \in (0,1)$ there exist constants $\rho > 0$ and C, independent in particular of z, such that

(3.1) $$A(z;v,v) \geq \rho|||v|||^2 - C\{1+||z||_{1,2(1-\delta)^{-1}}\}^2||v||_{2-\delta}^2$$

for $z \in W^1_{2(1-\delta)^{-1}}(\Omega)$ and $v \in M_h + W_2^{5/2}(\Omega)$.

Proof. The argument is essentially the same as in the linear case, except that the term $((\nabla_x a)(z)\cdot\nabla v + a_u(z)\nabla z\cdot\nabla v,\Delta v)$ must be estimated using the Sobolev embedding theorem. Note that, as a result of the assumptions on $a(x,u)$,

$$|((\nabla_x a)\cdot\nabla v + a_u\nabla z\cdot\nabla v,\Delta v)| \le C[||\nabla v||_o + ||\nabla z||_{o,p}||\nabla v||_{o,q}]||\Delta v||_o$$

for $p^{-1} + q^{-1} \ge \frac{1}{2}$. By Sobolev for $\Omega \subset\subset \mathbb{R}^2$,

$$||\nabla v||_{o,2\delta^{-1}} \le C||v||_{2-\delta},$$

and the corresponding value of p is $2(1-\delta)^{-1}$; the remainder of the argument then follows easily.

Although logical order would seem to indicate that a demonstration of existence of a solution of (1.5) should come at this point, it is more enlightening to defer existence until after a treatment of convergence. Assume that there exist positive constraints h_o and K such that a solution $u_h \in M_h$ of (1.5) exists for $0 < h \le h_o$ for which $|||u_h||| \le K$ for $0 < h \le h_0$. Then, fixing δ and again using Sobolev,

$$||u_h||_{1,2(1-\delta)^{-1}} \le C||u_h||_2 \le CK.$$

Hence, for any $v \in M_h$,

$$\begin{aligned}\rho|||u_h-v|||^2 &\le A(u_h;u_h-v,u_h-v) + C\{1+||u_h||_{1,2(1-\delta)^{-1}}\}^2||u_h-v||^2_{2-\delta}\\ &\le A(u_h;u_h-v,u_h-v) + C'||u_h-v||^2_{2-\delta}\\ &= A(u_h;u-v,u_h-v) + [A(u;u,u_h-v)-A(u_h;u,u_h-v)]\\ &\quad + C'||u_h-v||^2_{2-\delta}\end{aligned}$$

$$\leq C|||u-v|||\cdot|||u_h-v||| + (\nabla\cdot((a(u)-a(u_h))\nabla u),\Delta(u_h-v)) + C'||u_h-v||^2_{2-\delta} .$$

Since

$$|(\nabla\cdot((a(u)-a(u_h))\nabla u),\Delta(u_h-v))| \leq C||u||_{2,\infty}||u-u_h||_1|||u_h-v|||,$$

it follows that

$$(3.2)\qquad |||u-u_h||| \leq C_1(\inf_{v\in M_h}|||u-v||| + ||u-u_h||_{2-\delta}),$$

where C_1 depends on $||u||_{2,\infty}$ and K.

Just as for the linear problem it is necessary to introduce a duality argument and to treat estimates of $\zeta = u - u_h$ in $W_2^s(\Omega)$, $0 \leq s \leq 2$, simultaneously with the estimate of $|||\zeta|||$. Let

$$Mw = \nabla\cdot(a(u)\nabla w + wa_u(u)\nabla u)$$

denote the linear operator obtained by linearizing the operator of (1.1) about the (unique [6, 15]) solution u of (1.1), and let

$$M^*w = \nabla\cdot(a(u)\nabla w) - a_u(u)\nabla u\cdot\nabla w$$

denote its formal adjoint. Standard elliptic regularity theory combined with the uniqueness results of [6] imply that the Dirichlet problem for M has at most one weak solution in $W_2^1(\Omega)$. Hence, for any $\psi \in L^2(\Omega)$ there is a solution $\xi \in W_2^2(\Omega)$ of the Dirichlet problem

$$M^*\xi = \psi,\ x \in \Omega,$$
$$\xi = 0,\ x \in \partial\Omega.$$

Moreover, it follows from [6] that ξ is unique; consequently, it can be seen that $||\xi||_2 \leq C''||\psi||_o$. The constant C'' depends on the function u in a rather complicated way. See also [15] for another discussion of the uniqueness of solutions of (1.1).

Let $\psi \in L^2(\Omega)$ and define $\phi \in W_2^4(\Omega)$ by

$$(3.3)\qquad \begin{aligned} &M^*\Delta\phi = \psi, \ x \in \Omega, \\ &\Delta\phi = a(g)\frac{\partial}{\partial n}\Delta\phi - h^{-3}\phi = 0, \ x \in \partial\Omega. \end{aligned}$$

An argument similar to that given in §2 shows that

$$(3.4)\qquad ||\phi||_4 \leq C||\psi||,$$

where C now depends on u. Integration by parts shows that

$$(v_1, M^*v_2) = (Mv_1, v_2) + \langle v_1, a(g)\frac{\partial v_2}{\partial n}\rangle$$
$$- \langle a(g)\frac{\partial v_1}{\partial n} + a_u(g)\frac{\partial u}{\partial n}v_1, v_2\rangle.$$

Hence,

$$(3.5)\qquad (\zeta,\psi) = (M\zeta,\Delta\phi) + h^{-3}\langle\zeta,\phi\rangle.$$

A direct calculation shows that

$$M\zeta = \nabla\cdot(a(u)\nabla u) - \nabla\cdot(a(u_h)\nabla u_h) - \nabla\cdot(\zeta a_u(u)\nabla\zeta + \bar{a}_{uu}\zeta^2\nabla(u-\zeta)),$$

where

$$\bar{a}_{uu}(x) = \int_o^1 (1-t)a_{uu}(x,u(x)-t\zeta(x))dt.$$

Consequently, for $v \in M_h$,

$$(\zeta,\psi) = A(u;u,\phi-v) - A(u_h;u_h,\phi-v) - h\langle T_2\zeta, T_2\phi\rangle$$
$$- (\nabla\cdot(a_u\zeta\nabla\zeta + \bar{a}_{uu}\zeta^2\nabla(u-\zeta)),\Delta\phi).$$

Since $\Delta\phi = 0$ on $\partial\Omega$ and since $||u-\zeta||_{o,\infty} = ||u_h||_{o,\infty} \leq C||u_h||_2 \leq CK$,

$$|(\zeta,\psi)| \le |||\zeta|||\cdot||\psi||_o h^2 + Ch|\zeta|_{1/2}||\psi||_o$$
$$+ |(a_u\zeta\nabla\zeta+\bar{a}_{uu}\zeta^2\nabla(u-\zeta,\nabla(\Delta\phi))|$$
$$\le C|||\zeta|||\cdot||\psi||_o h^2 + C\{||\zeta||_{o,3}||\nabla\zeta||_o+||\zeta^2||_{o,6/5}\}\cdot$$
$$\cdot||\nabla\Delta\phi||_{o,6}$$
$$\le \{|||\zeta|||h^2 + ||\zeta||_1^2\}||\psi||_o .$$

It results from (2.14), (3.6), and the converse of Hölder's inequality that (with C depending on u)

(3.7) $$||\zeta|| \le C\{|||\zeta|||h^2 + ||\zeta||_1^2\} .$$

<u>Lemma 5</u>. Let $u_h \in M_h$ be a solution of (1.5) such that $|||u_h||| \le K$, and set $\zeta = u - u_h$. If $||\zeta||_o \to 0$ as $h \to 0$, then there exists a constant $C^* = C^*(u)$, u being the solution of (1.1), such that the optimal order estimate

(3.8) $$||\zeta||_o + h^2|||\zeta||| \le C^*||u||_k h^k, \tfrac{5}{2} \le k \le r + 1,$$

holds.

Proof. Since for any $\gamma > 0$

$$||\zeta||_{2-\delta} \le \gamma|||\zeta||| + C_\gamma||\zeta||_o$$

and

$$||\zeta||_1^2 \le |||\zeta|||\cdot||\zeta||_o ,$$

(3.2), (3.7), and the approximability assumptions on M_h imply that

$$|||\zeta||| \le C_1(||u||_k h^{k-2} + ||\zeta||_o), \ \tfrac{5}{2} \le k \le r + 1,$$

and

$$||\zeta||_o \le C_2(||\zeta||_o + h^2)|||\zeta||| .$$

Hence,

$$|||\zeta||| \leq C\{||u||_k h^{k-2} + (||\zeta||_o + h^2)|||\zeta|||\},$$

and the conclusion follows easily from the hypothesis that $||\zeta||_o \to 0$ as $h \to 0$.

Standard interpolation theory shows that when (3.8) holds

(3.9) $$||\zeta||_s \leq C^*||u||_k h^{k-s}, \quad 0 \leq s \leq 2, \ \tfrac{5}{2} \leq k \leq r + 1.$$

Retain for the moment the assumption that a solution u_h of (1.5) exists and consider the question posed by the hypothesis in Lemma 5 that $||\zeta||_o \to 0$ as $h \to 0$.

Lemma 6. Let $u_h \in M_h$ be a solution of (1.5) and let there exist a constant K, independent of h, such that $|||u - u_h||| \leq K$. Assume that $f \in C^\alpha(\Omega)$ and $g \in C^{2+\alpha}(\partial\Omega)$ for some $\alpha \in (0,1)$. Then, $||u - u_h||_o \to 0$ as $h \to 0$.

Proof. As usual, let $\zeta = u - u_h$. Since $|||\zeta||| \leq K$, then $|\zeta|_s \to 0$ for $0 \leq s < \frac{3}{2}$ as h tends to zero; consequently, u_h converges uniformly to g on $\partial\Omega$. Next, extract a sequence $h_k \to 0$ for which u_{h_k} converges weakly in $W_2^2(\Omega)$ to a function $z \in W_2^2(\Omega)$; u_{h_k} converges strongly to z in $W_2^s(\Omega)$ and in $W_2^q(\partial\Omega)$ for $0 \leq s < 2$ and $0 \leq q < \frac{3}{2}$. In particular, $z = g$ on $\partial\Omega$.

Let $\phi \in C_o^\infty(\Omega)$, and let v be the solution of the Dirichlet problem

$$\Delta v = \phi, \quad x \in \Omega,$$

$$v = 0, \quad x \in \partial\Omega.$$

A simple application of Theorem 3 above or of Lemma 2 of [7] shows that, for any $h > 0$, there exists $v_h \in M_h$ such that

$|||v - v_h||| \leq C(\phi)h^{r-1}$. This implies that

$$h^{-3}\langle g,v_h\rangle + h\langle T_2 g,T_2 v_h\rangle \to 0$$

as $h \to 0$, provided that $r \geq 3$ and $g \in W_2^2(\partial\Omega)$. Also, note that

$$\begin{aligned} h^{-3}\langle u_h,v_h\rangle &+ h\langle T_2 u_h,T_2 v_h\rangle \\ &= h^{-3}\langle g,v_h\rangle + h\langle T_2 g,T_2 v_h\rangle \\ &\quad + h^{-3}\langle u_h - g,v_h\rangle + h\langle T_2(u_h - g),T_2 v_h\rangle \end{aligned}$$

tends to zero as h tends to zero, by the assumption that $|||\zeta||| \leq K$. The equation (1.5) for $h = h_k$ and the test function v_{h_k} reads as follows (with the index k suppressed):

$$\begin{aligned} (\nabla\cdot(a(u_h)\nabla u_h),\Delta v_h) &+ h^{-3}\langle u_h,v_h\rangle + h\langle T_2 u_h,T_2 v_h\rangle \\ &= (f,\Delta v_h) + h^{-3}\langle g,v_h\rangle + h\langle T_2 g,T_2 v_h\rangle. \end{aligned}$$

The remarks immediately above plus a simple calculation show that, as $k \to \infty$,

$$(\nabla\cdot(a(z)\nabla z),\phi) = (f,\phi);$$

hence, $z \in W_2^2(\Omega)$ is a weak solution of (1.1). Schauder estimates [10, Theorem 5.6.3] show that $z \in C^{2+\alpha}(\bar{\Omega})$ if $f \in C^{\alpha}(\Omega)$ and $g \in C^{2+\alpha}(\partial\Omega)$; thus, z is a classical solution of (1.1) and must coincide with u [6, 15]. Consequently, the entire family u_h must converge to u in $W_2^s(\Omega)$, $0 \leq s < 2$. See also [4, pp. 693-694] for a similar argument.

Finally, it is necessary to show existence of a solution $u_h \in \mathcal{M}_h$ of (1.5) such that $|||u - u_h|||$ is bounded independently of h. The argument will combine the Brouwer fixed point theorem with a development similar to that leading to (3.2) and (3.7).

Let M and M^* denote the same operators as previously. Set

$$A_M(z,v) = (Mz,\Delta v) + h^{-3}\langle z,v\rangle + h\langle T_2 z, T_2 v\rangle.$$

Then, $w \in M_h$ is a solution of (1.5) if and only if (with $\zeta = u - w$)

$$(3.10) \quad A_M(\zeta,v) = (\nabla\cdot(a_u(u)\zeta\nabla\zeta + \bar{a}_{uu}\zeta^2\nabla(u-\zeta)),\Delta v), \quad v \in M_h.$$

Thus, it is sufficient to seek a fixed point for the map $w \to z \in M_h$ given by the equation (where $\bar{a}_{uu}$ depends on w)

$$(3.11) \quad A_M(u-z,v) = (\nabla\cdot(a_u(u-w)\nabla(u-w) + \bar{a}_{uu}(u-w)^2\nabla w),\Delta v), \quad v \in M_h,$$

which can be solved for a unique z for h sufficiently small. Let $\delta > 0$ and assume that

$$(3.12) \quad |||u - w||| \le \delta.$$

A calculation shows that

$$(3.13) \quad |(\nabla\cdot(a_u(u-w)\nabla(u-w) + \bar{a}_{uu}(u-w)^2\nabla w),\Delta v)| \le C\delta^2||\Delta v||_o .$$

Write the Gårding inequality for A_M in the form

$$\rho|||u-z|||^2 - C||u-z||_o^2 \le A_M(u-z,u-z)$$
$$= A_M(u-z,u-v) + A_M(u-z,v-z), \quad v \in M_h .$$

Use (3.11) with the test function $v - z$, apply (3.13), simplify, and then choose v to give good approximation of u. It follows that

$$(3.14) \quad |||u-z||| \le C\{||u||_k h^{k-2} + \delta^2 + ||u-z||_o\}.$$

Clearly, duality is again called for. Let $\phi \in W_2^4(\Omega)$ satisfy

$$M^*\Delta\phi = u - z,\ x \in \Omega,$$

$$\Delta\phi = a(g)\frac{\partial}{\partial n}\Delta\phi - h^{-3}\phi = 0,\ x \in \partial\Omega.$$

Now, $||\phi||_4 \le C||u-z||_o$, and for $v \in M_h$ properly chosen

$$\begin{aligned} ||u-z||_o^2 &= A_M(u-z,\phi) - h\langle T_2(u-z),T_2\phi\rangle \\ &= A_M(u-z,\phi-v) + (\nabla\cdot(a_u(u-w)\nabla(u-w) + \bar{a}_{uu}(u-w)^2\nabla w),\Delta v) \\ &\qquad -h\langle T_2(u-z),T_2\phi\rangle \\ &\le C(h^2|||u-z||| + \delta^2)||u-z||_o + Ch|u-z|_{1/2}||u-z||_o \\ &\le C(h^2|||u-z||| + \delta^2)||u-z||_o, \end{aligned}$$

since $|u-z|_{1/2} \le C|||u-z|||h$. Thus,

$$(3.15)\qquad ||u-z||_o \le C\{h^2|||u-z||| + \delta^2\}.$$

Substitute (3.15) into (3.14); it follows that for h sufficiently small

$$(3.16)\qquad |||u-z||| \le \{||u||_k h^{k-2} + \delta^2\},$$

and the right-hand side is smaller than δ for h and δ small enough. Consequently, the mapping defined by (3.11) maps a ball into itself for small h. Since the map is obviously continuous on the finite-dimensional space M_h, the Brouwer fixed point theorem implies the existence of a solution $u_h \in M_h$ of (1.5) lying in the ball $|||u-u_h||| \le \delta$. The above results are summarized in the lemma below.

<u>Lemma 7.</u> If the solution u of (1.1) belongs to $W_2^{5/2}(\Omega)$ and if $\delta > 0$, then for h sufficiently small there exists a solution $u_h \in M_h$ of (1.5) lying inside the ball $\{|||u-u_h||| \le \delta\}$.

Consider next the uniqueness of u_h. Let K be an arbitrary positive number, conceivably quite large. The object will be to show that there is exactly one solution $u_h \in M_h \cap \{||u-u_h||_2 \leq K\}$ of (1.5) for small h; while this is not all that could be desired, it is sufficient for practical purposes, since any implementation of the method could easily contain tests to prevent looking for a solution having an enormous $W_2^2(\Omega)$-norm.

Let $u_h^{(1)}$ and $u_h^{(2)}$ be solutions of (1.5), and set $\zeta_i = u - u_h^{(i)}$ and $\xi = \zeta_1 - \zeta_2$. Let $|||\zeta_i||| \leq K$. Then,

$$(3.17) \quad A_M(\xi,v) = (\nabla\cdot q,\Delta v),\ v \in M_h,$$

where

$$(3.18) \quad q = a_u(u)\{\zeta_1\nabla\zeta_1 - \zeta_2\nabla\zeta_2\} + \bar{a}_{uu}^{(1)}\zeta_1^2\nabla u_h^{(1)} - \bar{a}_{uu}^{(2)}\zeta_2^2\nabla u_h^{(2)} .$$

It is easy to show that

$$(3.19) \quad ||\nabla\cdot q||_o \leq C \max_{i=1,2} |||\zeta_i|||\cdot|||\xi|||.$$

Thus,

$$\rho|||\xi|||^2 - C||\xi||_o^2 \leq A_M(\xi,\xi) \leq C \max_i |||\zeta_i|||\cdot|||\xi|||^2.$$

Since Lemmas 5 and 6 imply that $|||\zeta_i||| \to 0$ as $h \to 0$,

$$|||\xi||| \leq C||\xi||_o$$

for h sufficiently small.

Again employ duality. Let

$$M^*\Delta\phi = \xi,\ x \in \Omega,$$

$$\Delta\phi = a(g)\frac{\partial}{\partial n}\Delta\phi - h^{-3}\phi = 0,\ x \in \partial\Omega.$$

Then, for $v \in M_h$ properly selected,

$$||\xi||_o^2 = A_M(\xi,\phi-v) + (\nabla\cdot q,\Delta v) - h<T_2\xi,T_2\phi>$$
$$\leq C|||\xi|||\{h^2 + \max_i|||\zeta_i|||\}||\xi_o||.$$

Hence,

$$||\xi||_o^2 \leq C\{h^2 + \max_i|||\zeta_i|||\}||\xi||_o^2,$$

and $\xi = 0$ for h sufficiently small.

The following theorem has been proved.

Theorem 8. Let $f \in W_2^j(\Omega) \cap C^\alpha(\Omega)$ for some $j \geq \frac{1}{2}$ and some $\alpha > 0$, and let $g \in C^{2+\alpha}(\Omega)$. Let $K > 0$. Then, there exists an $h_o > 0$, depending on the solution u of (1.1) as well as Ω and the coefficients of (1.1), such that one and only one solution $u_h \in M_h$ of (1.5) exists in the ball $\{||u-u_h||_2 \leq K\}$ for $0 < h \leq h_o$. Moreover, there exists a constant C_o, depending on u, such that

(3.20) $$||u-u_h||_s h^s + |||u-u_h|||h^2 \leq C_o||u||_k h^k$$

for $0 \leq s \leq 2$ and $\frac{5}{2} \leq k \leq r + 1$.

Again uniform convergence of u_h to u is assured by (3.20).

4. Boundary values by interpolation. An alternative approach to approximating the solution of (1.1) can be based on finite element spaces designed to allow the boundary values $g(x)$ to be approximated by interpolation. It is slightly easier to describe the special case given by $g \equiv 0$; the general case can be handled using the full spaces introduced in [5] instead of the subspaces having "nearly zero" boundary values. Let N_h be a finite-dimensional

subspace of $W_2^2(\Omega) \cap C^1(\Omega)$ satisfying the following conditions:

There exists a constant C_o such that, if $1 \le p \le 2$

(4.1) and $-\frac{3}{2} \le s \le r - \frac{7}{2}$ and if $v \in N_h$,

$$|v|_{-s} \le C||v||_p h^{p+s+1/2} .$$

There exists C_1 such that, if $v \in W_2^p(\Omega)$, $2 \le p \le r+1$, and $v = 0$ on $\partial\Omega$, then for $0 \le q \le 2$, $2 \le p_1 \le p$,

(4.2) $-\frac{3}{2} \le s \le r - \frac{7}{2}$, and $2 \le p_2 \le p$, there exists $z \in N_h$ such that

$$||v-z||_q \le C||v||_{p_1} h^{p_1-q} ,$$

$$|z|_{-s} \le C||v||_{p_2} h^{p_2+s+1/2} .$$

The integer r should be at least three. Families of elements satisfying (4.1) and (4.2) have been constructed [5] using C^1-piecewise-polynomial spaces over quasi-regular triangulations of Ω and interpolation of the null function at certain Gauss points on $\partial\Omega$.

Define a form by

(4.3) $$B(z;v,w) = (\nabla\cdot(a(z)\nabla v),\Delta w).$$

A second H^1-Galerkin method is given by seeking $u_h \in N_h$ such that

(4.4) $$B(u_h;u_h,v) = (f,\Delta v),\ v \in N_h .$$

An entirely analogous argument can be given to show that there is a unique solution $u_h \in N_h$ of (4.4) lying in the ball $\{||u-u_h||_2 \le K\}$ for h small. Moreover, it can be shown that

(4.5) $||u-u_h||_s \leq C(u)||u||_k h^{k-s}$, $0 \leq s \leq 2$, $2 \leq k \leq r+1$.

The hypotheses required on f and g are the same as with the penalty-set boundary conditions. The condition (4.1) is used repeatedly to control boundary terms so that a Gårding inequality holds and so that other boundary terms can be shifted to interior terms. In the duality arguments the boundary conditions are usually simplified, often to $\phi = \Delta\phi = 0$ on $\partial\Omega$. Overall, the development in this case is slightly simpler than for the penalty-set case.

If the spaces of [5] are used and g is non-trivial, the trial space should be augmented to permit the interpolation of g; the test space remains N_h.

References

1. J. H. Bramble and A. H. Schatz, Rayleigh-Ritz-Galerkin methods for Dirichlet's problem using subspaces without boundary conditions, Comm. Pure Appl. Math 23 (1970) pp. 653-675.
2. P. G. Ciarlet, Numerical Analysis of the Finite Element Method, Séminaire de Mathématiques Superieures, Université de Montréal, 1975
3. ______________ and P.-A. Raviart, The combined effect of curved boundaries and numberical integration in isoparametric finite element methods, The Mathematical Foundations of the Finite Element Method with Applications to Partial Differential Equations, A. K. Aziz (ed.), Academic Press, New York, 1972.
4. J. Douglas, Jr., and T. Dupont, A Galerkin method for a nonlinear Dirichlet problem, Math. Comp. 29 (1975) pp. 689-696.
5. __________, __________, P. Percell, and R. Scott, A family of C^1 finite elements with optimal approximation properties for various Galerkin methods for second and fourth order problems, to appear.

6. _______, _______, and J. Serrin, Uniqueness and comparison theorems for nonlinear elliptic equations in divergence form, Arch. Rational Mech. Anal. 42 (1971) pp. 157-168.

7. _______, _______, and M. F. Wheeler, H^1-Galerkin methods for the Laplace and heat equations, Mathematical Aspects of Finite Elements in Partial Differential Equations, C. de Boor (ed.), Academic Press, New York, 1974.

8. J.-L. Lions and E. Magenes, Problèmes aux limites non homogènes et applications, vol. 1, Dunod, Paris, 1968.

9. J. Morgan and R. Scott, A nodal basis for C^1-piecewise polynomials of degree $n \geq 5$, Math. Comp. 29 (1975) pp. 736-740.

10. C. B. Morrey, Jr., Multiple Integrals in the Calculus of Variations, Springer-Verlag, New York, 1966.

11. J. A. Nitsche, A projection method for Dirichlet-problems using subspaces with nearly zero boundary conditions, same book as in reference 3.

12. R. Scott, Interpolated boundary conditions in the finite element method, SIAM J. Numer. Anal. 12 (1975) pp. 404-427.

13. G. Strang and G. Fix, An Analysis of the Finite Element Method, Prentice-Hall, Englewood Cliffs, 1973.

14. V. Thomée and L. Wahlbin, On Galerkin methods in semilinear parabolic problems, SIAM J. Numer. Anal. 12 (1975) pp. 378-389.

15. N. S. Trudinger, On the comparison principle for quasilinear divergence structure equations, Arch. Rational Mech. Anal. 57 (1974) pp. 128-133.

16. O. C. Zienkiewicz, The Finite Element Method in Engineering Science, McGraw-Hill, New York, 1971.

DISCRETIZATION OF ROTATIONAL EQUILIBRIUM IN THE FINITE ELEMENT METHOD

B.M. Fraeijs de Veubeke

Laboratoire de Techniques Aéronautiques et Spatiales, Université de Liège, Belgique

SUMMARY

The theory of equilibrium elements [1,2] shows that their stiffness matrices may present a singular behavior due to the presence of mechanisms (deformation modes without strain energy). The origin of such difficulties is easily traced to the rigorous requirement of rotational equilibrium (symmetry of the stress tensor) and equivalently, if the discretization is performed on the basis of stress functions, to the C_1 continuity requirement involved. Moreover loss of diffusivity (reciprocity of surface traction distributions at interfaces) is incurred in an isoparametric coordinate transformation to curved boundaries, whenever preservation of C_1 continuity is at stake.

Both difficulties are resolved by enforcing rotational equilibrium only in weak form. First order stress functions are used to preserve rigorous translational equilibrium and diffusivity. They need only be C_0 continuous, a property that remains invariant under isoparametric coordinate transformations.

The theory of discretized rotational equilibrium has been investigated in detail for membrane elements [3]. The paper is devoted to the more difficult case of axisymmetric elements.

1. AXISYMMETRIC EQUILIBRIUM EQUATIONS

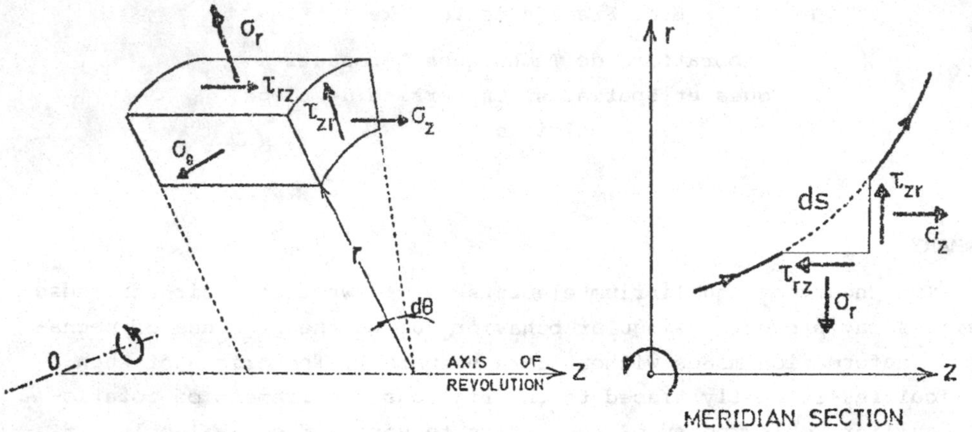

FIgure 1

The translational equilibrium equations of the axisymmetric state of stress are conveniently presented in the following form

axial direction $$\frac{\partial}{\partial r}(r\,\tau_{rz}) + \frac{\partial}{\partial z}(r\,\sigma_z) = 0 \qquad (1)$$

radial direction $$\frac{\partial}{\partial r}(r\,\sigma_r) + \frac{\partial}{\partial z}(r\,\tau_{zr}) = \sigma_\theta \qquad (2)$$

a moment equilibrium of a slice θ_λ about an axis perpendicular to the mean meridian plane requires

$$d\theta\left[\oint r(zt_r - rt_z)\,ds - \iint z\sigma_\theta\,dr\,dz\right] = 0 \qquad (3)$$

where the curvilinear integral is around the boundary of the meridian cross section and the meridian surface tractions are given by

$$t_r\,ds = \tau_{zr}\,dr - \sigma_r\,dz \qquad (4)$$

$$t_z\,ds = \sigma_z\,dr - \tau_{rz}\,dz \qquad (5)$$

The hoopstress σ_θ gives a downward compo,e,t due to the curvature that is responsible for the last term. Substitution of (4) and (5) into (3) and transformation of the curvilinear to a double integral yield

$$d\theta\iint\left[z\left\{\frac{\partial}{\partial r}(r\sigma_r)+\frac{\partial}{\partial z}(r\tau_{zr})-\sigma_\theta\right\}-r\left\{\frac{\partial}{\partial r}(r\tau_{rz})+\frac{\partial}{\partial z}(r\sigma_z)\right\}\right]drdz$$

$$+\ d\theta\iint(\tau_{zr}-\tau_{rz})\,rdrdz = 0$$

Taking (1) and (2) into account this reduces to

$$\iint(\tau_{zr}-\tau_{rz})\,rdrdz = 0 \qquad (6)$$

and, for an elementary surface of the meridian cross section, to the local rotational equilibrium condition

$$\tau_{rz}-\tau_{rz} = 0 \qquad (7)$$

It should be observed that, even if this condition is not fulfilled, the axisymmetric ring of same meridian cross section is, by reason of symmetry, in rotational equilibrium about all axes.

2. A VARIATIONAL PRINCIPLE

We satisfy the axial equilibrium condition (1) by a first order stress function

$$\tau_{rz} = -\frac{1}{r}\frac{\partial\phi}{\partial z} \qquad \sigma_z = \frac{1}{r}\frac{\partial\phi}{\partial r} \qquad (8)$$

that brings the axial component of surface traction to the simple form

$$t_z\,ds = \frac{1}{r}\left(\frac{\partial\phi}{\partial r}dr+\frac{\partial\phi}{\partial z}dz\right) = \frac{1}{r}d\phi \qquad (9)$$

Because we first consider σ_θ to be directly determined through the radial equilibrium condition (2) we do not, at this stage, introduce another stress function. We also consider τ_{zr} separately from τ_{rz} but enforce the equilibrium condition (7) by means of a Lagrangian multiplier ω. The stress energy density is then considered to be a positive definite function of the arguments $(\sigma_r,\ \sigma_z,\ \sigma_\theta,\ \frac{1}{2}(\tau_{zr}+\tau_{rz})$ with the stress-strain properties

$$\varepsilon_r = \frac{\partial\Phi}{\partial\sigma_r} \qquad \varepsilon_z = \frac{\partial\Phi}{\partial\sigma_z} \qquad \varepsilon_\theta = \frac{\partial\Phi}{\partial\sigma_\theta}$$

$$\varepsilon_{rz} = \frac{\partial\Phi}{\partial\tau_{rz}} = \frac{\partial\Phi}{\partial\tau_{zr}} = \varepsilon_{zr} \qquad (10)$$

The fact that ϕ is a symmetrical function with respect to both shearing stresses ensures the symmetry of the corresponding shear strains; moreover translational equilibrium is assumed to hold. Thus the arguments τ_{rz} and σ_z must be expressed in terms of the stress function ϕ as in (8) and the hoopstress is expressed as in (2).
The complementary energy principle then takes the following form

$$\iint\left[\phi+\omega\left(\tau_{zr}+\frac{1}{r}\frac{\partial\phi}{\partial z}\right)\right] rdrdz - \oint \bar{u}r\left(\tau_{zr}\,dr - \sigma_r\,dz\right) + \bar{w}d\phi$$

$$\text{stationary} \tag{11}$$

the displacements being assumed to be given on the boundary of the meridian cross section.

The Euler equations resulting from unconstrained variations on σ_r , τ_{zr} and ϕ are respectively

$$\varepsilon_r = \frac{\partial}{\partial r}(r\,\varepsilon_\theta) \tag{12}$$

$$\omega + \varepsilon_{rz} = \frac{\partial}{\partial z}(r\,\varepsilon_\theta) \tag{13}$$

$$\frac{\partial}{\partial z}(\varepsilon_{rz} - \omega) = \frac{\partial}{\partial r}\,\varepsilon_z \tag{14}$$

Both σ_r and τ_{zr} give the same natural boundary condition

$$\bar{u} = r\,\varepsilon_\theta \tag{15}$$

while for ϕ we obtain

$$d\bar{w} = (\varepsilon_{rz} - \omega)\,dr + \varepsilon_z\,dz \tag{16}$$

3. SOLUTION OF THE VARIATIONAL EQUATIONS

At this stage, in preparation of the imposition of constraints on rotational equilibrium, we consider $\omega(r,z)$ as a given function. Setting

$$u = r\,\varepsilon_\theta \tag{17}$$

The Euler equation (12) becomes

$$\varepsilon_r = \frac{\partial u}{\partial r} \tag{18}$$

and u is recognized to be the radial displacement.
Euler equation (14) is solved by introducing a function w(r,z) such that

$$\varepsilon_z = \frac{\partial w}{\partial z} \tag{19}$$

$$\varepsilon_{rz} - \omega = \frac{\partial w}{\partial r} \tag{20}$$

this new function is thus the axial displacement and, combining (20) with the last Euler equation (13)

$$\varepsilon_{rz} = \frac{1}{2}\left(\frac{\partial u}{\partial z} + \frac{\partial w}{\partial r}\right) \tag{21}$$

$$\omega = \frac{1}{2}\left(\frac{\partial u}{\partial z} - \frac{\partial w}{\partial r}\right) \tag{22}$$

so that the Lagrangian multiplier is, as expected, the material rotation about an axis normal to the meridian plane.

We conclude that for unconstrained variations on σ_r, τ_{zr} and ϕ, the compatibility equations are satisfied and the displacement field obtained, satisfies the given boundary data and has the given $\omega(r,z)$ function as its rotation field. This however requires an obvious global compatibility condition on the data:

$$\oint \bar{u}\, dr + \bar{w}\, dz = 2\iint \omega\, dr\, dz \tag{23}$$

Equation (22) appears in this context as one of the differential equations governing the displacement field.
To obtain the second, we must express the equilibrium equations that are satisfied in terms of displacements.
For simplicity take the isotropic linear stress-strain laws in the form

$\sigma_r = 2G(\varepsilon_r + \eta\varepsilon)$ where G is the shear modulus

$\sigma_\theta = 2G(\varepsilon_\theta + \eta\varepsilon)$ $\qquad \eta = \nu/(1-2\nu)$

$\sigma_z = 2G(\varepsilon_z + \eta\varepsilon)$ $\qquad \varepsilon = \varepsilon_r + \varepsilon_\theta + \varepsilon_z$

$\frac{1}{2}(\tau_{rz} + \tau_{zr}) = 2G\,\varepsilon_{rz}$

The last is the only one generated by the complementary energy density as a symmetrical function. To separate the two shear stresses we introduce a shear strain unbalance function ζ

$$\tau_{rz} = 2G(\varepsilon_{rz} + \zeta) \qquad \tau_{zr} = 2G(\varepsilon_{rz} - \zeta)$$

Replacing the strains in terms of displacements through equations (17), (18) and (21) the equilibrium equations (1) and (2) can now be placed in the form

$$\begin{aligned} &(1-2\nu)\frac{\partial}{\partial r}\{r(\zeta-\omega)\} + (1-\nu)\frac{\partial}{\partial z}(r\,\varepsilon) = 0 \\ &(1-\nu)r\,\frac{\partial \varepsilon}{\partial r} - (1-2\nu)\frac{\partial}{\partial z}\{r(\zeta-\omega)\} = 0 \end{aligned} \tag{24}$$

The elimination of $(\zeta-\omega)$ produce the second differential equation governing with (22), and the boundary conditions $u = \bar{u}$, $v = \bar{v}$, the displacement field

$$\frac{\partial^2}{\partial z^2}(r\varepsilon) + \frac{\partial}{\partial r}\left(r\,\frac{\partial \varepsilon}{\partial r}\right) = 0 \tag{25}$$

$$\varepsilon = \frac{u}{r} + \frac{\partial u}{\partial r} + \frac{\partial w}{\partial z} \tag{26}$$

If, on the contrary, ε is eliminated between equations (24), we obtain a differential equation satisfied by the shear strain unbalance when the rotation ω is imposed

$$\frac{\partial}{\partial r}\left\{\frac{1}{r}\,\frac{\partial}{\partial r}\left[r(\zeta-\omega)\right]\right\} + \frac{\partial^2}{\partial z^2}(\zeta-\omega) = 0 \tag{27}$$

When ω is not given, but an unconstrained Lagrangian multiplier, its variational equation requires the shear unbalance to vanish

$$r\tau_{zr} + \frac{\partial \phi}{\partial z} = r(\tau_{zr} - \tau_{rz}) = -4Gr\zeta = 0$$

and (27) becomes the differential equation governing the distribution of the material rotation

$$\frac{\partial}{\partial r}\left[\frac{1}{r}\,\frac{\partial}{\partial r}(\omega r)\right] + \frac{\partial^2}{\partial z^2}\,\omega = 0 \tag{28}$$

It is worth noticing that this equation is *not* satisfied by $\omega_\lambda = \omega_{\lambda_0}$ a constant but that the simple solutions independent of z are

$$\omega = \frac{\omega_{-1}}{r} + \omega_1 r \tag{29}$$

4. THE ZERO ENERGY STATE

Our separation of the two shear stresses creates a well-defined state of stress for which the complementary energy vanishes. The energy density being a positive definite function, the conditions for zero energy are

$$\sigma_z \equiv 0 \qquad \text{, whence from (1)}$$

$$r\tau_{rz} = f(z)$$

$$\sigma_r \equiv 0 \quad \text{and} \quad \sigma_\theta \equiv 0 \qquad \text{, whence from (2)}$$

$$r\tau = g(r)$$

$$\tau_{rz} + \tau_{zr} \equiv 0 \qquad \text{, whence}$$

$$-f(z) = g(r) = \gamma \qquad \text{a constant}$$

The zero energy stress distribution, in translational but not rotational equilibrium is thus characterized by the shear stresses distribution

$$\tau_{rz} = -\frac{\gamma}{r} \qquad\qquad \tau_{zr} = +\frac{\gamma}{r}$$

Any imposition of a grobal rotational equilibrium condition

$$\iint \omega(\tau_{zr} - \tau_{rz})\, rdzd\omega = 0$$

where ω is one of the simple solutions (29) will eliminate the possibility of such a situation to prevail.

5. STRESS FUNCTIONS DISCRETIZATION

In the discretization we consider the presence of σ_θ in the equilibrium equation (2) as analogous to a body load. We subdivide the stress distribution of σ_r and τ_{zr} in a particular solution taking into account a non zero hoop stress and a general solution without hoop stress. This can be done conveniently by introducing two new first order stress functions as follows:

$$\sigma_r = -\frac{1}{r}\frac{\partial\psi}{\partial z} - \frac{\partial\lambda}{\partial z} \qquad \tau_{zr} = \frac{1}{r}\frac{\partial\psi}{\partial r} + \frac{\partial\lambda}{\partial r} \tag{30}$$

from which we find from (2)

$$\sigma_\theta = -\frac{\partial\lambda}{\partial z} \tag{31}$$

and for the radial surface traction

$$t_r\, ds = \frac{1}{r}\, d\psi + d\lambda \tag{32}$$

C_o continuity of ϕ will thus ensure reciprocity of the axial surface tractions and C_o continuity of ψ and λ, reciprocity of the radial surface tractions.

The stress functions will now be discretized as polynomials in r and z and, to obtain similar surface traction distributions for ψ and λ, the degree of λ will have to be one unit less than that of ψ.

The first model corresponds to

$$\phi = \phi_o + \phi_1 r + \phi_2 z + \phi_3 r^2 + 2\phi_4 rz + \phi_5 z^2$$

$$\psi = \psi_o + \psi_1 r + \psi_2 z + \psi_3 r^2 + 2\psi_4 rz + \psi_5 z^2 \tag{33}$$

$$\lambda = \lambda_o + \lambda_1 r + \lambda_2 z$$

The constants ϕ_o and ψ_o are not producive of stresses, they will only play a role in organizing diffusivity in a finite element of triangular meridian section by expressing, in the usual way, the function ϕ and ψ by interpolation functions related to the local values at vertices and mid-edges. When such local values are taken as nodal values at interfaces, C_o continuity follows. In the case of λ, we see that neither λ_o, nor λ_λ, produce any hoop stress. As the general equilibrium state without hoop stress is already accounted for by ψ, these terms can be dropped. However they are needed again when organizing diffusivity, this time by expressing λ in terms of interpolation functions related to the three vertex values.

We now define the generalized boundary loads associated with the linear distributions of rt_r and rt_z on a slice $d\theta$.

Let $2c_{ij}$ denote the length of side ij of the triangle and the distance s be measured in anticlockwise sense from i to j with origin at the mid point. The non dimensional distance

$$\sigma = \frac{s}{c_{ij}}$$

will vary in the interval [-1, +1]. We then introduce

$$R_{ij} = \int_i^j rt_r ds \quad \text{and} \quad Z_{ij} = \int_i^j rt_z ds \tag{34}$$

the total, respectively radial and axial, loads associated with the surface tractions on a slice per unit angle θ. Correspondingly we introduce the total reduced moments

$$\rho_{ij} = \int_i^j rt_r \sigma ds \quad \text{and} \quad \zeta_{ij} = \int_i^j rt_z \sigma ds \tag{35}$$

We then find easily that along ij

$$\begin{aligned} rt_r &= \frac{1}{2c_{ij}} R_{ij} + \frac{3\sigma}{3c_{ij}} \rho_{ij} \\ rt_z &= \frac{1}{2c_{ij}} Z_{ij} + \frac{3}{2c_{ij}} \zeta_{ij} \end{aligned} \tag{36}$$

and furthermore

$$r = \frac{1}{2} r_i (1-\sigma) + \frac{1}{2} r_j (1+\sigma) \qquad dr = \frac{1}{2}(r_j - r_i) d\sigma$$

$$z = \frac{1}{2} z_i (1-\sigma) + \frac{1}{2} z_j (1+\sigma) \qquad dz = \frac{1}{2}(z_j - z_i) d\sigma$$

From this it becomes possible to compute the matrix S relating the generalized boundary loads to the *active* stress parameters listed in the vector s:

$$s^T = (\phi_1\, \phi_2\, \phi_3\, \phi_4\, \phi_5\, \psi_1\, \psi_2\, \psi_3\, \psi_4\, \psi_5\, \lambda_2)$$

through the relation

$$g = Ss \tag{37}$$

where in g the generalized loads are conventionally sequenced

$$g^T = (R_{12}\, R_{23}\, R_{31}\, Z_{12}\, Z_{23}\, Z_{31}\, \rho_{12}\, \rho_{23}\, \rho_{31}\, \zeta_{12}\, \zeta_{23}\, \zeta_{31}) \tag{38}$$

The first row of S is obtained by replacing in the definition of R_{12}, the expression of $rt_r ds$ in terms of the stress parameters

$$R_{12} = \int_1^2 d\psi + r d\lambda = \psi_1 \int_1^2 dr + 2\psi_3 \int_1^2 r dr + 2\psi_4 \int_1^2 z dr + \psi_2 \int_1^2 dz + 2\psi_4 \int_1^2 r dz + 2\psi_5 \int_1^2 z dz + \lambda_2 \int_1^2 r dz$$

There follows for this first row

$$\left\{ 0\ 0\ 0\ 0\ 0\ \int_1^2 dr \quad \int_1^2 dz \quad \int_1^2 d(r^2)\ 2 \int_1^2 d(rz) \quad \int_1^2 d(z^2) \int_1^2 rdz \right\}$$

The only geometrical integral that is not immediatly expressible in terms of the vertex coordinates is the last

$$\int_1^2 rdz = \frac{1}{2}(z_2 - z_1)\left\{\frac{1}{2}\ r_1 \int_{-1}^{1} (1-\sigma)d\sigma + \frac{1}{2}\ r_2 \int_{-1}^{1} (1+\sigma)d\sigma\right\} = \frac{1}{2}(z_2 - z_1)(r_1 + r_2)$$

The other rows follow by similar procedures.

We now define the (weak) generalized boundary displacements, conjugate to the loads, by expressing the virtual work at each partial boundary in canonical form.

$$\int_1^2 (urt_r + wrt_z)ds = R_{12}\, U_{12} + Z_{12}\, W_{12} + \rho_{12}\, \alpha_{12} + \zeta_{12}\, \beta_{12} \tag{39}$$

Substituting the surface traction distributions in terms of the generalized loads, as in (36) and comparing, there follows

$$U_{12} = \frac{1}{2c_{12}} \int_1^2 uds \qquad W_{12} = \frac{1}{2c_{12}} \int_1^2 ds \tag{40}$$

the ordinary averages of displacements, and

$$\alpha_{12} = \frac{3}{2c_{12}} \int_1^2 u\sigma ds \qquad \beta_{12} = \frac{3}{2c_{12}} \int_1^2 w\sigma ds \tag{41}$$

which are "moments" of the displacement distribution. Similar definitions ensure for the two other partial boundaries.

The generalized boundary displacements are sequenced in the corresponding order as that choosen for g

$$q^T = (U_{12}\ U_{23}\ U_{31}\ W_{12}\ W_{23}\ W_{31}\ \alpha_{12}\ \alpha_{23}\ \alpha_{31}\ \beta_{12}\ \beta_{23}\ \beta_{31}\ ) \tag{42}$$

so that the scalar product $q^T g$ reproduces the complete canonical espansion of virtual work:

$$\oint (urt_r + wrt_z)ds = q^T g = q^T Ss \tag{43}$$

We have now available the discretized form of the last term in the variational principle (11).

For linear homogeneous stress-strain relations, Φ will be a quadratic form of its arguments, and, after discretization of the stress distribution by means of the stress functions, the complementary energy becomes a quadratic form

$$\iint \Phi \; r dr dz = \frac{1}{2} s^T F c \tag{44}$$

in the active stress parameters. This quadratic form is merely non negative, because of the existence of the zero-energy state.

Indeed this state is included in our approximation as corresponding to the choice of parameters:

$$\psi_1 = \phi_2 = \gamma \qquad \text{all other parameters zero.}$$

This means that the stress parameter vector

$$s_o^T = \gamma (0 \; 1 \; 0 \; 0 \; 0 \; 1 \; 0 \; 0 \; 0 \; 0 \; 0) \tag{45}$$

corresponds to $s_o^T F s_o = 0$, and, the flexibility matrix F being non negative, to

$$F s_o = 0 \tag{46}$$

6. ROTATION DISCRETIZATION

There remains to discretize the part corresponding to the Lagrangian multiplier

$$\iint \omega \left(\frac{\partial \psi}{\partial r} + \frac{\partial \phi}{\partial r} + r \frac{\partial \lambda}{\partial r}\right) dr dz = s^T R h \tag{47}$$

a bilinear form in the active stress parameters and whatever coordinates h_i are used in an expansion of ω in interpolation functions.

The linear independence of the columns of the matrix R of the bilinear form will appear later as a necessary condition for a solution to the discretized problem. This impose limitations on the choice of a discretized ω. It easily established that if n+1 is the polynomial degree of ψ and ϕ (and n that of λ), the columns of R are linearly independent when the polynomial degree of ω is not higher than n. The proof can be based on the fact that under the opposite assumption :

columns of R linearly dependent, we reach a contradiction.

If the columns of R are linearly dependent there exists a non zero vector h^* such that

$$Rh^* = 0 \rightarrow s^T Rh^* = 0 \quad \text{for all } s.$$

Thus there would exist a non identically zero polynomial ω^* of degree not higher than n, such that

$$\iint \omega^* \left(\frac{\partial\psi}{\partial r} + \frac{\partial\phi}{\partial z} + r\frac{\partial\lambda}{\partial r}\right) drdz = 0$$

for arbitrary polynomials ψ and ϕ of degree n+1, λ of degree n.

In particular we would have

$$\iint \omega^* \frac{\partial\psi}{\partial r}\, drdz = 0$$

for an arbitrary polynomial ψ of degree n+1. However, as we may choose ψ as a particular integral of

$$\frac{\partial\psi}{\partial r} = \omega^*$$

we reach the contradiction

$$\iint \omega^{*2}\, drdz = 0 \quad \text{for } \omega^* \text{ not identically zero.}$$

On the basis of the discretization (33) of the stress functions, we may thus take

$$\omega = \omega_0 + \omega_1 r + \omega_2 z$$
$$h^T = (\omega_0 \ \ \omega_1 \ \ \omega_2) \tag{48}$$

7. SOLUTION OF THE DISCRETIZED VARIATIONAL EQUATIONS

The discrete form

$$\frac{1}{2}\, s^T Fs + s^T Rh - q^T Ss \quad \text{stationary} \tag{49}$$

of the variational principle (11), where the generalized displacements q are assumed to be given, yields as variational equations

$$
\begin{aligned}
Fs + Rh &= S^T q \\
R^T s &= 0
\end{aligned}
\tag{50}
$$

The first system of equations is generated by variations on s, the second by the variations on h. Although F has been seen to be singular because of (46), the matrix

$$
\begin{pmatrix} F & R \\ R^T & 0 \end{pmatrix}
$$

of the system is not. The proof of this assertion consists in showing that the homogeneous system

$$
Fs + Rh = 0 \qquad R^T s = 0
$$

has only the trivial solution. Premultiply the first equation by s^T and use the second equation in transposed form to obtain

$$
s^T F s = 0 \rightarrow s = \gamma s_\circ
$$

The proof is then achieved if we succed in showing that $R^T s_\circ \neq 0$, because then

$$
\gamma R^T s_\circ = 0 \rightarrow \gamma = 0 \rightarrow s = 0
$$

and then the first equation requires

$$
Rh = 0 \rightarrow h = 0 \quad \text{because R has linearly independent columns.}
$$

Let us examine in succession the influence of the different terms of the polynomial expansion (48) on the condition $R^T s_\circ \neq 0$. For $h_1^T = (1\ 0\ 0)$, that is, using only the constant term of ω, the first column $r_1 = Rh_1$ of R is found to be

$$
r_1^T = \iint drdz\ (0\ 1\ 0\ 2\hat{r}\ 2\hat{z}\ 1\ 0\ 2\hat{r}\ 2\hat{z}\ 0\ 0)
$$

where $\hat{r} = \frac{1}{3}(r_1 + r_2 + r_3) \qquad \hat{z} = \frac{1}{3}(z_1 + z_2 + z_3)$

are the coordinates of the center of area of the meridian section.

Whence, by reference to (45)

$$r_1^T s_\circ = 2\gamma \iint drdz = 1 \qquad (51)$$

if the zero energy vector is "normed" by the condition

$$\gamma = (2 \iint drdz)^{-1}$$

Hence the condition $R^T s_\circ \neq 0$ will be satisfied if the $\omega_\circ$ term is retained.

For $h_2^T = (0\ 1\ 0)$, selecting the term $\omega_1 r$, we find

$$r_2^T = \iint drdz(0\ \ \hat{r}\ \ 0\ \ 2\overline{rr}\ \ 2\overline{rz}\ \ \hat{r}\ \ 0\ \ 2\overline{rr}\ \ 2\overline{rz}\ \ 0\ \ 0)$$

where

$$\overline{rr} \iint drdz = \iint r^2 drdz$$

$$\overline{rz} \iint drdz = \iint rzdrdz$$

and

$$r_2^T s_\circ = \hat{r} > 0 \text{ with the same norm of } s_\circ \qquad (52)$$

Hence again, it is sufficient to retain the $\omega_1 r$ term in ω_λ to satisfy the condition.

However for the last term $\omega_3 z$, $h^T = (0\ 0\ 1)$,

$$r_3^T = \iint drdz(0\ \ \hat{z}\ \ 0\ \ 2\overline{rz}\ \ 2zz\ \ \hat{z}\ \ 0\ \ 2\overline{rz}\ \ 2\overline{zz}\ \ 0\ \ 0)$$

and $\quad r_3^T s_\circ = \hat{z}$

which depends on the origin of axes and can be made to vanish by the r axis pass through the center of area of the element.

In conclusion, the variational equations will be invertible if the discretization of the rotation contains either the $\omega_\circ$ term, or the $\omega_1 r$ term.

The structure of the inverted matrix is

$$\begin{pmatrix} F^* & R^* \\ R^{*T} & G^* \end{pmatrix} \qquad F^* = (F^*)^T \qquad G^* = (G^*)^T$$

Postmultiplying by the original matrix, we find the relations

$$F^*F + R^*R^T = (s/s) \qquad F^*R = 0$$
$$(R^*)^T F + G^* R^T = 0 \qquad R^{*T} R = (h/h) \tag{53}$$

where (s/s) and (h/h) denote identity matrices of respectively the size of s and h. It is seen that F^* typically satisfies a pseudo-inverse relationship with F

$$F^* F F^* = F^*$$

from which it can be concluded that it is also a non negative matrix. In practice the inversion

$$s = F^* S^T q \tag{54}$$

$$h = R^{*T} S^T q \tag{55}$$

is obtained numerically. It gives simultaneously the values of the active stress parameters, thus the state of stress, and the rotation field of the element, when the boundary displacements are given. The stiffness matrix of the element is obtained as a consequence of (37) and (54) in the form

$$g = Kq \qquad K = SF^* S^T \tag{56}$$

The determination of the stiffness matrix allows the use of the same assembling software as in the case of elements based on a discretized displacement field. The modal displacement identification is here replaced by the identification of the weak generalized displacement at the interfaces and insures diffusivity instead of conformity.

8. THE AXIAL RIGID BODY MODE

In principle the stiffness matrix of an axisymmetric element, being representative of a complete "ring", should contain only one rigid body mode, the axial translation mode.

Any radial translation of the meridian section should generate hoop stresses and deformation energy. Likewise, rotation of the meridian section should generate twisting energy. It is easily verified that the axial translation mode is correctly built into the model. If we input

$$w = w_\circ \qquad \text{a constant}$$

into the definition (40) and (41) of the generalized displacements, we find a rigid body mode vector

$$q_\circ^T = w_\circ(0 \quad 0 \quad 0 \quad 1 \quad 1 \quad 1 \quad 0 \quad 0 \quad 0 \quad 0 \quad 0 \quad 0) \qquad (57)$$

that should generate no loads and consequently satisfy

$$Kq_\circ = 0$$

In fact it does so because it already satisfies $S^T q_\circ = 0$. We may prove it by showing that $q_\circ^T Ss = 0$ for all s vectors or, in other terms, by reverting to the discretization (43) of virtual work at the boundary, that

$$\oint (urt_r + wrt_z)ds = \oint u(d\psi + rd\lambda) + wd\phi = 0$$

for any state of discretized stress, when $u \equiv 0$ and $w = w_\circ$. This follows obviously for any discretized model where the stress function ϕ is single-valued

$$\oint d\phi = 0$$

9. SELF STRESSINGS

To see whether this axial rigid body mode is the only solution of problem

$$S^T q = 0 \qquad (58)$$

we can use the algebraic property

$$n(s) + n(r) = n(g) + n(x) \qquad (59)$$

linking the number $n(s)$ of columns of S, $n(r)$ of linearly independent solutions of our problem, $n(g)$ number of rows of S and $n(x)$, number of linearly independent solutions of the homogeneous adjoint problem

$$Sx = 0 \qquad (60)$$

This last problem is that of the so-called self-stressing states of the element, we look for the non zero stress states that produce no boundary loads, that is no surface tractions at all.

In the model proposed under section 5 it is easily shown that no self-stressings exist. For if there are no boundary tractions, we must have by integration of (2)

$$\iint \sigma_\theta drdz = \oint rt_{zr}dr - r\sigma_r dz = \oint rt_r ds = 0 \qquad (61)$$

and, consequently,

$$\sigma_\theta = - \lambda_2 = 0$$

As λ_0 and λ_1 are improductive, we may take $\lambda \equiv 0$. Then, the vanishing of boundary tractions requires

$$d\phi = 0 \qquad d\psi = 0 \qquad \text{on boundary}$$

so that both stress functions must reduce to their improductive constant terms.

Since for the present model n(s) = 11 and n(g) = 12, we have n(r) = 1 and q_0 will be the only non trivial solution to problem (58).

10. MECHANISMS

The other possible solution to the homogeneous problem

$$Kq = SF^*S^Tq = 0$$

may be termed kinematical deformation modes or "mechanisms". They consist in boundary displacements that would normally deform the ring and create strain energy but do, in fact, produce no virtual work because of what may be considered as a deficiency in the model. Since F^* is non negative, such modes are in fact solutions of

$$F^*S^Tq = 0$$

distinct from (58). We must therefore look after solutions of problem

$$F^*m = 0 \qquad (62)$$

and, having found then, look after the solutions of the inhomogeneous problem

$$S^T q = m \qquad (63)$$

From the first of equations (53) in transpose we obtain that if m satisfies (62), it satisfies also

$$R\,R^{*T} m = m$$

so that any solution m is necessarily a linear combination of the columns of R. Furthermore, from the second of equations (53), we see that all columns of R are solutions and, those columns being linearly independent, we have all possible mechanisms by looking after the solutions of

$$S^T q = Rh \qquad h \text{ arbitrary} \qquad (64)$$

The necessary and sufficient condition for the existence of solutions, is that the right-hand side be orthogonal to all the solutions of the homogeneous adjoint problem (60)

$$x^T Rh = 0 \qquad \text{all self-stressings } x \qquad (65)$$

In the present model, there is no self-stressing and equation (64) has a solution, a mechanism, for any choice of h. Thus any weak enforcement of the rotational equilibrium condition (7) will create a mechanism. On the other hand at least one enforcement based on either the constant rotation field ω_o, or on the field $\omega = \omega_1 r$, is necessary to prevent the zero energy state. This is a characteristic weakness of the present model, that has however no counter-part in the simple two-dimensional membrane case. It remains to be seen whether this inconvenience will disappear after assembling at least two elements together.
The following remarks are pertinent to this last aspect :
1. If one uses the complete rotation field (48), the linear function

$$r(\tau_{zr} - \tau_{rz}) = \frac{\partial\psi}{\partial r} + \frac{\partial\phi}{\partial z} + r\,\frac{\partial\lambda}{\partial r} = L(r,z)$$

$$= (\psi_1 + \phi_2) + (2\psi_3 + 2\phi_4 + \lambda_1)\,r + 2(\psi_4 + \phi_5)\,z$$

submitted to the constraints

$$\iint L drdz = 0 \qquad \iint rL drdz = 0 \qquad \iint zL drdz = 0$$

must vanish completely and rotational equilibrium is enforced *exactly*. We thus retrieve a pure equilibrium model with three mechanisms, an interpretation of which can be obtained as follows. Introduce the barycentric coordinates L_i, defined by

$$\begin{aligned} 1 &= L_1 + L_2 + L_3 \\ z &= z_1 L_1 + z_2 L_2 + z_3 L_3 \\ r &= r_1 L_1 + r_2 L_2 + r_3 L_3 \end{aligned} \tag{66}$$

and express the stress functions symmetrically as

$$\begin{aligned} \phi &= \phi_1 L_1^2 + \phi_2 L_2^2 + \phi_3 L_3^3 + 2\phi_{12} L_1 L_2 + 2\phi_{23} L_2 L_3 + 2\phi_{31} L_3 L_1 \\ \psi &= \psi_1 L_1^2 + \psi_2 L_2^2 + \psi_3 L_3^2 + 2\psi_{12} L_1 L_2 + 2\psi_{23} L_2 L_3 + 2\psi_{31} L_3 L_1 \\ \lambda &= \lambda_1 L_1 + \lambda_2 L_2 + \lambda_3 L_3 \end{aligned} \tag{67}$$

(the coefficients ϕ_i, ψ_i, λ_i bear no direct relationship with the preceding ones). If $A = \iint drdz$ denotes the area of the triangle, it is easily found that

$$2A \frac{\partial L_1}{\partial z} = r_2 - r_3 \qquad 2A \frac{\partial L_1}{\partial r} = z_3 - z_2 \tag{68}$$

and the other derivative follow by cyclic permutation.
The quantity

$$\frac{\partial \phi}{\partial z} + \frac{\partial \psi}{\partial r} + r \frac{\partial \lambda}{\partial r} \tag{69}$$

is then easily expressed as a linear homogeneous function of the L_i and its complete vanishing requires the vanishing of the coefficient of L_1

$$\begin{aligned} &\phi_1 (r_2 - r_3) + \phi_{12} (r_3 - r_1) + \phi_{31} (r_1 - r_2) + \psi_1 (z_3 - z_2) + \psi_{12} (z_1 - z_3) + \psi_{31} (z_2 - z_1) \\ &\quad + \frac{r_1}{2} \left\{ \lambda_1 (z_3 - z_2) + \lambda_2 (z_1 - z_3) + \lambda_3 (z_2 - z_1) \right\} = 0 \end{aligned} \tag{70}$$

and those of L_2 and L_3 that follow by cyclic subscript permutations.

Equation (70) is now reinterpreted as a constraint between boundary loads in the vicinity of vertex 1.

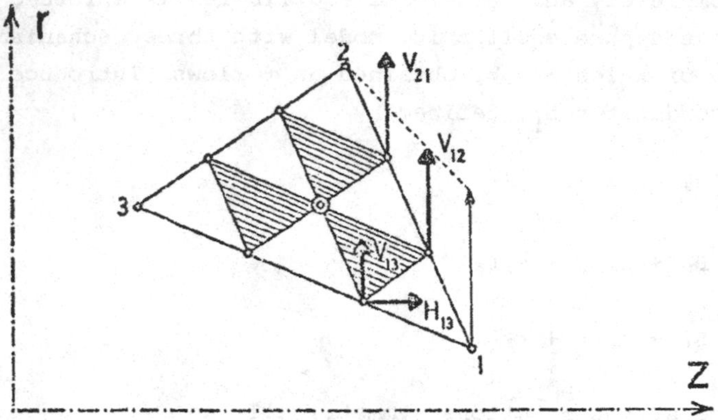

Along boundary 12, where $L_3 = 0$, $dL_1 = -dL_2$, $ds = 2c_{12}\, dL_2$, we have

$$rt_r ds = 2\psi_1 L_1 dL_1 + 2\psi_2 L_2 dL_2 + 2\psi_{12}(L_1 dL_2 + L_2 dL_1) + (r_1 L_1 + r_2 L_2)(\lambda_1 dL_1' + \lambda_2 dL_2)$$

or $$c_{12}\, rt_r = -\psi_1 L_1 + \psi_2 L_2 + \psi_{12}(L_1 - L_2) + \frac{1}{2}(r_1 L_1 + r_2 L_2)(\lambda_2 - \lambda_1)$$

By setting $L_1 = 1$, $L_2 = 0$, in this relation we obtain the resultant load V_{12}, applied at one third of the edge from 1 and due to the linear rt_r distribution sketched on figure 2.

$$V_{12} = \psi_{12} - \psi_1 + \frac{1}{2} r_1 (\lambda_2 - \lambda_1)$$

The complementary distribution has the resultant

$$V_{21} = \psi_2 - \psi_{12} + \frac{1}{2} r_2 (\lambda_2 - \lambda_1)$$

obtained by setting $L_2 = 1$ and $L_1 = 0$. By cyclic permutation of this last result we obtain also

$$V_{13} = \psi_1 - \psi_{13} + \frac{1}{2} r_1 (\lambda_1 - \lambda_3)$$

In a similar fashion we can obtain from the rt_z distribution

$$H_{12} = \phi_{12} - \phi_1 \quad \text{and} \quad H_{13} = \phi_1 - \phi_{13}$$

We can then observe that the condition for the resultant moment of V_{12}, V_{13}, H_{12} and H_{13} with respect to the barycenter of the element to vanish

$$\frac{1}{3}\left\{V_{12}(z_1 - z_3)+V_{13}(z_1 - z_2)-H_{12}(r_1 - r_3)-H_{13}(r_1 - r_2)\right\} = 0$$

turns out to be identical to the requirement (70).

Hence, as sketched on the figure, the element behaves as if made of three parts articulated at the barycenter. The situation is exaxtly similar to that of the pure equilibrium membrane element of same degree. In that case however, the rotation of the element as a whole about its barycenter is a rigid body mode on its own right and the relative rotations of the parts only represent two mechanisms. Here this global rotation is also a mechanism as it represents an energyless torsion of the ring.

The interpretation of the mechanisms yields at the same time the answer to the problem of their inhibition by the composite element technique

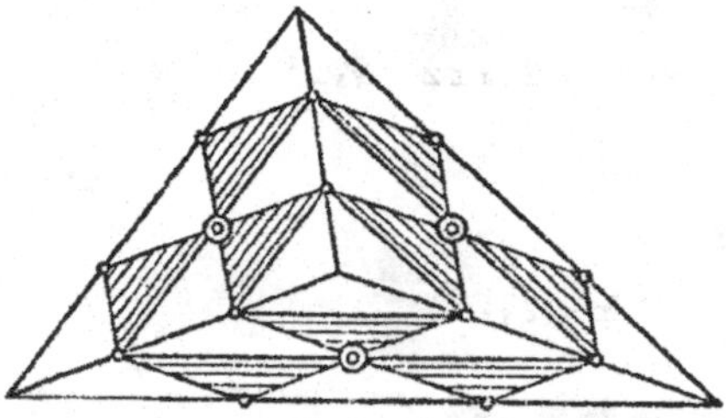

Figure 3

Locking of mechanisms by composite element technique.

2. If we really discretize rotational equilibrium by restricting the rotation field to one of the terms ω_o or $\omega_1 r$, necessary to prevent the zero energy state, the element will present a single mechanism.

In the case of $\omega = \omega_o$, only the average value of expression (69) must vanish.

The corresponding requirement follows by taking the arithmetic mean of the 3 equations of type (70) and is reinterpreted as a rigid body rotation of the meridian section about the barycenter; this represents a pure twisting mechanism of the ring. It is inhibited as soon as we assemble two elements with barycenters of different z coordinate.

11. HIGHER ORDER APPROXIMATIONS. THE LINEAR HOOP STRESS MODEL

The stress functions are of higher degree; complete cubics (in barycentric coordinates)

$$\phi = \phi_1 L_1^3 + \phi_2 L_2^3 + \phi_3 L_3^3$$

$$+ \phi_{12} L_1^2 L_2 + \phi_{21} L_2^2 L_1 + \phi_{23} L_2^2 L_3 + \phi_{32} L_3^2 L_2 + \phi_{31} L_3^2 L_1 + \phi_{13} L_1^2 L_3$$

$$+ \phi_{123} L_1 L_2 L_3$$

$$\psi = \psi_1 L_1^3 + \dots\dots$$

for ϕ and ψ , complete quadratic for λ

$$\lambda = \lambda_1 L_1^2 + \lambda_2 L_2^2 + \lambda_3 L_3^2 + 2\lambda_{12} L_1 L_2 + 2\lambda_{23} L_2 L_3 + 2\lambda_{31} L_3 L_1$$

this corresponds in cartesian coordinates to

$$\lambda = \gamma_0 + \gamma_1 r + \gamma_2 z + \gamma_3 r^2 + 2\gamma_4 rz + \gamma_5 z^2$$

and the hoop stress

$$\sigma_\theta = - \frac{\partial \lambda}{\partial z} = -(\gamma_2 + 2\gamma_4 r + 2\gamma_5 z)$$

can have a linear distribution. The coefficients $(\gamma_0, \gamma_1, \gamma_3)$ are in fact improductive and may be cancelled at will. The number of active stress parameters is thus, discounting one improductive in ϕ, one in ψ , and three in λ, $n(s) = 21$. The parabolic distributions of rt_r and rt_z require a total of 6 generalized loads per side, a total of $n(g) = 18$. Let us now make a count of the independent self-tressings. The absence of loads requires that at the boundary

$$d\phi = 0 \quad \text{and} \quad d\psi + r d\lambda = 0$$

The first condition is equivalent to $\phi = 0$; by adjustement of the improductive additive constant, and is satisfied by the last term

$$\phi = \phi_{123} L_1 L_2 L_3$$

that represents a self-stressing of the axial traction loads t_z alone. For the self-stressings between the radial traction loads, it is pre-

ferable, for reasons of symmetry, to treat ψ and λ together and revert to σ_r and τ_{zr}.
Those are quadratic polynomials, and we can describe them as

$$r\tau_{zr} = \alpha_1 L_1^2 + \alpha_2 L_2^2 + \alpha_3 L_3^2 + 2\alpha_{12} L_1 L_2 + 2\alpha_{23} L_2 L_3 + 2\alpha_{31} L_3 L_1$$

$$r\sigma_r = \beta_1 L_1^2 + \beta_2 L_2^2 + \beta_3 L_3^2 + 2\beta_{12} L_1 L_2 + 2\beta_{23} L_2 L_3 + 2\beta_{31} L_3 L_1$$

Along the boundary $L_3 = 0$ we must have

$$r\tau_{zr} dr = r\sigma_r dz$$

$$(\alpha_1 L_1^2 + \alpha_2 L_2^2 + 2\alpha_{12} L_1 L_2)(r_1 dL_1 + r_2 dL_2) = (\beta_1 L_1^2 + \beta_2 L_2^2 + 2\beta_{12} L_1 L_2)$$

$$(z_1 dL_1 + z_2 dL_2)$$

As $dL_1 = -dL_2$ we obtain, equating the coefficients of L_1^2, L_2^2 and $L_1 L_2$

$$\alpha_1 (r_1 - r_2) = \beta_1 (z_1 - z_2)$$

$$\alpha_2 (r_1 - r_2) = \beta_2 (z_1 - z_2)$$

$$\alpha_{12}(r_1 - r_2) = \beta_{12}(z_1 - z_2)$$

Proceeding in the same manner for the two other sides, we find that we must have

$\alpha_1 = \alpha_2 = \alpha_3 = 0$ $\qquad$ $\beta_1 = \beta_2 = \beta_3 = 0$

but $\alpha_{12} = \sigma_3 (z_1 - z_2)$ $\qquad$ $\beta_{12} = \sigma_3 (r_1 - r_2)$

$\alpha_{23} = \sigma_1 (z_2 - z_3)$ $\qquad$ $\beta_{23} = \sigma_1 (r_2 - r_3)$ $\qquad$ $\sigma_1, \sigma_2, \sigma_3$ arbitrary

$\alpha_{31} = \sigma_2 (z_3 - z_1)$ $\qquad$ $\beta_{31} = \sigma_2 (r_3 - r_1)$

We thus find 3 self-stressing states, each proportional to a σ_i

$$r\tau_{zr} = \sigma_3 (z_1 - z_2) L_1 L_2 + \sigma_1 (z_2 - z_3) L_2 L_3 + \sigma_2 (z_3 - z_1) L_3 L_1$$

$$r\sigma_r = \sigma_1 (r_2 - r_3) L_1 L_2 + \sigma_1 (r_2 - r_3) L_2 L_3 + \sigma_2 (r_3 - r_1) L_3 L_1$$

From this it is easy to compute the hoop stress distribution

$$\sigma_\theta = \sigma_1 (L_2 - L_3) + \sigma_2 (L_3 - L_1) + \sigma_3 (L_1 - L_2)$$

whose average value is zero in accordance with condition (61). From the result n(x) = 4 we deduce that, again,

$$n(r) = n(g) + n(x) - n(s) = 18 + 4 - 21 = 1$$

so that equation (58) has, again, only one solution, the axial rigid body mode. The mechanisms will be the solutions, if they exist, of equation (64). By the same argument as before, a matrix R with linearly independent columns is found by the choice of a complete quadratic for the rotation (one degree less than the stress functions ϕ and ψ):

$$\omega = \omega_1 L_1^2 + \omega_2 L_2^2 + \omega_3 L_3^2 + 2\omega_{12} L_1 L_2 + 2\omega_{23} L_2 L_3 + 2\omega_{31} L_3 L_1 \tag{74}$$

and the weak enforcement of rotational equilibrium of a slice is equivalent to rigorous enforcement (pure equilibrium model), if we keep the 6 parameters. The existence condition for solutions of (64) will be, as before,

$$x^T Rh = \iint \omega (\tau_{zr} - \tau_{rz})\, rdrdz = \iint \omega \left(r\tau_{zr} + \frac{\partial \phi}{\partial z}\right) drdz = 0 \tag{75}$$

where $r\tau_{zr}$ is replaced by (72) and ϕ by (71), the result holding for arbitrary self-stressing intensities $(\sigma_1, \sigma_2, \sigma_3, \phi_{123})$.

To reach conclusions, the bilinear form is presented in the reduced format

$$y^T M h \text{ with } y^T = (\sigma_1(z_2 - z_3) \quad \sigma_2(z_3 - z_1) \quad \sigma_3(z_1 - z_2) \quad \phi_{123})$$

$$h^T = (\omega_1 \quad \omega_2 \quad \omega_3 \quad \omega_{23} \quad \omega_{31} \quad \omega_{12})$$

The integrations required to obtain the matrix M were performed in barycentric coordinates, using formulas like (68) and as surface element

$$\frac{drdz}{A} = dL_i dL_j$$

the pair of subscripts being chosen according to the integrand. We found

$M =$

-5A	9A	9A	16A	-10A	-10A
9A	-5A	9A	-10A	16A	-10A
9A	9A	-5A	-10A	-10A	16A
$7(r_3 - r_2)$	$7(r_1 - r_3)$	$7(r_2 - r_1)$	$13(r_2 - r_3)$	$13(r_3 - r_1)$	$13(r_1 - r_2)$

Mechanisms will be present for a choice of h such that

$$Mh = 0 \rightarrow y^T Mh = 0 \quad \text{for any } y \ .$$

The rank of the matrix is 3 and there will be 3 independent solutions yielding mechanisms. By simple inspection it is found that

$$h^T = (13\alpha \ 13\beta \ 13\gamma \ 7\alpha \ 7\beta \ 7\gamma)$$

is a solution, provided $\alpha + \beta + \gamma = 0$, which gives two independent solutions.

$$h^T = (4 \ 4 \ 4 \ 13 \ 13 \ 13)$$

is a third solution, independent of the others.

However when the rotation field is reduced to its linear part

$$\omega = \hat{\omega}_1 L_1 + \hat{\omega}_2 L_2 + \hat{\omega}_3 L_3$$

in which case we know that the zero energy state is prevented, the existence condition for mechanisms

$$y^T \hat{M} \hat{h} = 0 \qquad \text{for every } y$$

$$h^T = (\hat{\omega}_1 \ \hat{\omega}_2 \ \hat{\omega}_3)$$

has a matrix $\hat{M}$ of maximal rank

-2A	4A	4A
4A	-2A	4A
4A	4A	-2A
$3(r_2 - r_3)$	$3(r_3 - r_1)$	$3(r_1 - r_2)$

and no mechanisms are generated; the stiffness matrix is well behaved.

12. REFERENCES

[1] B.Fraeijs de Veubeke : *Upper and lowers bounds in Matrix Structural Analysis.* AGARDograph 72, Pergamon Press 1964,pp.165-201.

[2] *Displacement and Equilibrium models in the Finite Element Method.*
Chap. 9 in "Stress Analysis" Ed. Zienkiewicz and Molister, John Wiley & Sons, 1965.

[3] *Stress Function Approach.*
World Congress on Finite Element Methods in Structural Mechanics, Bournemouth, 1975, pp. J1-J48.

INTEGRATION TECHNIQUES FOR SOLVING ALGEBRAIC SYSTEMS

Ilio Galligani
Istituto Applicazioni Calcolo C.N.R., Roma

Donato Trigiante
Centro Ricerche IBM & Università di Bari

ABSTRACT

By using results contained in [5] on the caracterization of natural L-spline functions, in this paper we will give a convenient general criterion for smoothing, in the successive approximation method, the solution of badly ill-conditioned linear least squares systems. Besided, for solving a special class of non linear algebraic systems (with properties a), b) and c) stated in § 3)), we shall present an efficient iterative method which has been generated by imposing particular conditions on the discrete dynamical systems associated to the given algebraic systems.

1. INTRODUCTION

The construction of effective iterative methods for the solution of algebraic systems may be considered as the construction of finite-difference methods which reach the equilibrium of a system of ordinary differential equations on infinite intervals.

a) *Linear algebraic systems.* The solution of the linear algebraic system of order n

$$(1) \qquad -A\underline{x} + \underline{b} = 0$$

with det A ≠ 0, may be determined by searching for the steady-state solution $A^{-1}\underline{b}$ of the first order ordinary differential system

$$(2) \qquad \begin{cases} B\dfrac{d\underline{v}(t)}{dt} = -A\underline{v}(t) + \underline{b} \\ \qquad\qquad\qquad\qquad\qquad t > 0 \\ \underline{v}(0) = \underline{g} \end{cases}$$

in which B is any non singular matrix such that the real parts of all the eigenvalues of $Q = -B^{-1}A$, $\mathrm{Re}\lambda_i(Q)$, $(1 \leq i \leq n)$, are negative and $\underline{g}$ is any vector.

For any time-step $\tau > 0$, at each mesh-point $j\tau$ $(j = 0,1,\ldots)$ of the t-axis, the solution of (2) may be written in the form $(\underline{v}(j\tau) \equiv \underline{v}_j)$:

$$
(3) \qquad \begin{cases} \underline{v}_{j+1} = A^{-1}\underline{b} + e^{\tau Q}(\underline{v}_j - A^{-1}\underline{b}) \\ \underline{v}_o = \underline{g} \end{cases}
$$

and

$$
\lim_{j \to \infty} \underline{v}_j = \underline{x} = A^{-1}\underline{b}
$$

In consequence of the impractical determination of the exponential matrix of Q, the system (3) is approximated by $(j = 0,1,\ldots)$

$$
(4) \qquad \begin{cases} \underline{W}_{j+1} = A^{-1}\underline{b} + T(Q\tau)(\underline{W}_j - A^{-1}\underline{b}) \\ \underline{W}_o = \underline{g} \end{cases}
$$

where $T(Q\tau)$ is a matrix approximation to $e^{\tau Q}$. Therefore $\lim_{j \to \infty} \underline{W}_j = \tilde{\underline{x}}$ is the computed solution of the system (1).

It is known that any A-acceptable approximation $T(Q\tau)$ to $e^{\tau Q}$ gives rise to a convergent iterative method (4), whose solution is the steady-state solution $A^{-1}\underline{b}$ of (2). Besides when the matrix A is ill-conditioned, in order to obtain an accurate solution of (1), it is required to use in the iterative method (4) a L-acceptable approximation to $e^{\tau Q}$.

Many iterative methods based on different splittings of the matrix A have been developed; in particular, the explicit and implicit alternating direction methods are very interesting, and the convergence of these methods has been analyzed for many classes of matrices. In [6] a special explicit alternating direction method has been developed which is effective for the numerical solution of the time-dependent multigroup neutron diffusion equations in bidimensional domains [9]. Besides in [6] we have proved that the explicit and implicit alternating direction methods are *optimal* schemes for a large class of matrices whose elements depend on a generic parameter ξ. For example, these matrices may be generated by discretizing a boundary-value problem for partial differential equations of elliptic type.

Other iterative methods (4) may be generated by considering the (μ,ν) rational or Padè approximations to $e^{\tau Q}$. In particular, when

$A = M^T M$ with the rank of M equal to n and $\underline{b} = M^T f$, the *implicit Euler* L-acceptable approximation to $e^{\tau Q}$, $T(Q\tau) = (I-\tau Q)^{-1} = (I+\tau B^{-1}A)^{-1}$, generates a classical iterative scheme for solving the linear least squares system:

(1') $$-M^T M\underline{x} + M^T f = 0$$

Generally B is the identity matrix. For those classes of linear systems where we know, from the origin of the problem, that there is a considerable degree of regularity and smoothness in the solution, the matrix B is determined by using special regularization functionals. In this paper we will describe a general procedure for determining the matrix B for these classes of linear least squares systems.

b) *Non linear algebraic systems.* It is known that the solutions, if they exist, of the non linear algebraic system

(5) $$\underline{f}(\underline{x}) = 0 \qquad \underline{f},\ \underline{x} \in \mathbb{R}^n$$

may be obtained by searching for the steady-state solutions of the first order ordinary differential system

(6) $$\begin{cases} H \dfrac{d\underline{v}(t)}{dt} = \underline{f}(\underline{v}) & t > 0 \\ \underline{v}(0) = \underline{g} \end{cases}$$

where H is a non singular matrix of order n.

We shall assume that $\underline{f}(\underline{x})$ has continuous first partial derivatives with respect to all components of $\underline{x}$.

Among the one-step methods for solving the system (6), the Newton-like and the A-stable methods provide suitable means to determine good approximations to the solutions of (5). The Newton-like methods are obtained by integrating with the Euler method the differential system (6) with $H = \pm J(\underline{v})$, where $J(\underline{v})$ denotes the Jacobian matrix of $\underline{f}(\underline{v})$. In [6] the behaviour of the solutions of the system (6) in the neighbourhood of the points in which $\underline{f}(\underline{x}) = 0$, $\det J(\underline{x}) = \infty$ and $\det J(\underline{x}) = 0$ has been studied. This analysis on the singular points allows, in particular, to find further solutions of $f(\underline{x}) = 0$ in addition to those found during earlier calculations without using deflation techniques. In [2] the Newton method has been applied to solve (5) when $\underline{f}(\underline{x}) = A\underline{x} + \underline{\Phi}(\underline{x})$ (A is a $n \times n$ non singular matrix and Φ is a mapping from $\mathbb{R}^n$ into $\mathbb{R}^n$), has been generated by a consistent discretization on a compact Ω of the boundary value problem

$$\begin{cases} L[x(P)] = \varphi(P,x(P)) & P \in \Omega \\ R[x(P)] = \phi(P) & P \in \partial\Omega \end{cases}$$

with (L,R) operators of monotone kind.

Since the Newton's method for solving (5) may be considered as the Euler method applied to the corresponding differential equation (6), it is natural to attempt to find the "best" way to integrate this differential equation. The A-stable methods for integrating (6) have certain desiderable properties to justify this class of methods as the primary integration schemes. Indeed, it is known the following statement: let be given in $(0,\infty)$ the differential equation $\frac{dv(t)}{dt} = f(v(t))$ with $v(0) = g$ and let x^* be a root of $f(x)$ with $f'(x^*) \neq 0$. If the equation is asymptotically stable at $v = x^*$, then there exists a neighbourhood of x^* in which each A-stable method is an attractor.

In this paper, by imposing some conditions on the discrete dynamical systems generated by integrating (6) with one-step discretization schemes, we shall present some A-stable methods of the first and second order which are very efficient for solving non linear algebraic systems with properties a), b) and c) stated in § 3.

2. SUCCESSIVE APPROXIMATION METHOD WITH SMOOTHING

When the $m \times n$ matrix M of rank n is ill-conditioned, the solution of the linear least squares system

$$(7) \qquad -M^T M \underline{x} + M^T \underline{f} = 0$$

may be obtained by the following Successive Approximation Method with Smoothing [1, pg.143]:

$$(8) \qquad \begin{cases} (A + \tau^{-1} B)\underline{W}_{j+1} = \tau^{-1} B\underline{W}_j + \underline{b} & (j = 0,1,\dots) \\ \underline{W}_o = 0 \end{cases}$$

where $A = M^T M$, $\underline{b} = M^T \underline{f}$ and B is a non-singular matrix of order n with $\mathrm{Re}\lambda_i(-B^{-1}A) < 0 \quad (i = 1,2,\dots,n)$.

The method (8) is generated by using in (4) the implicit Euler L-acceptable approximation to $e^{\tau Q}$:

$$T(\tau Q) = (I - \tau Q)^{-1} = (I + \tau B^{-1} A)^{-1}$$

In order to have a *regular solution* $\underset{\sim}{\underline{x}} = \lim_{j \to \infty} W_j$ of (8), we minimize, at each jth time-step, the smoothing functional ($\tau > 0$) :

$$\text{(9)} \qquad J_\tau(\underline{e}_j) = \| M \underline{e}_j - \underline{r}_j \|^2 + 1/\tau \int_a^b \left| L[e_j(x)] \right|^2 dx$$

where

$$\text{(10)} \qquad L[c(x)] = \sum_{\ell=0}^{p} a_\ell(x) \frac{d^\ell e(x)}{dx^\ell}$$

with $a_\ell(x) \in C^p[a,b]$, $\ell = 0,1,\ldots,p$ and $a_p(x) = 1$.

The vector $\underline{e}$ is a restriction of the function $e(x)$ on the decomposition D of $[a,b]$:

$$D \equiv \left\{ x_i \middle| a < x_1 < x_2 < \ldots x_n < b \right\} \qquad n \geq 2p$$

and $\underline{e}_j = \underline{W}_{j+1} - \underline{W}_j$, $\underline{r}_j = \underline{f} - M\underline{W}_j$.

If L^* is the formal adjoint operator of L, we suppose that the null space $N\{L^*L\}$ is spanned by a Tschebycheff system.

In these hypotheses there exists [5] an unique natural L-spline function $e_j(x)$ which minimizes the functional (9). The restriction $\underline{e}_j$ on D of this natural L-spline function $e_j(x)$ is obtained by solving the following linear algebraic system:

$$(M^T M + \tau^{-1} H_1 H_2^{-1})\, \underline{e}_j = M^T \underline{r}_j$$

or

$$\text{(11)} \qquad (A + \tau^{-1} H_1 H_2^{-1})\, \underline{W}_{j+1} = \tau^{-1} H_1 H_2^{-1} \underline{W}_j + \underline{b}$$

where

$$H_1 = \left\{ (-1)^{p-1} \left(\varphi_h^{(2p-1)}(x_i-) - \varphi_h^{(2p-1)}(x_i+) \right) \right\}$$

$$H_2 = \left\{ \varphi_h(x_i) \right\} \qquad i,h = 1,2,\ldots,n$$

and $\left\{ \varphi_h(x) \right\}$ is a basis for the linear n-dimensional space of the natural L-spline functions related to the operator L and to the decomposition D of $[a,b]$.

In [7] a procedure has been described to represent these functions in terms of a local basis. In [4] the explicit representation of $H_1 H_2^{-1}$ has been given when L is the particular operator $L = \frac{d^2}{dx^2}$. In [4] we have also considered the smoothing functional of Whittaker; in this case the (9) must be replaced by

$$J_\tau(\underline{e}_j) = \|M\underline{e}_j - \underline{r}_j\|^2 + 1/\tau \sum_{t=0}^{p} \mu_t \Big(\sum_{\ell=1}^{n-t} \nu_{t\ell} (\Delta^t e_j(x_\ell))^2 \Big)$$

where $\Delta^t e_j(x_\ell)$ is the t-order forward difference operator applied to the function $e_j(x)$ at the point $x = x_\ell$.

Now, if we compare the equation (11) with (8), we obtain a general criterion for choosing the matrix B in the Successive Approximation Method with Smoothing when the coefficients of the operator L and the elements of the matrix M have values such that $Re\lambda_i((-H_1 \cdot H_2^{-1})^{-1} M^T M) < 0$, $i = 1,2,\dots,n$. In [4] we have proved that the matrix $H_1 H_2^{-1}$ is symmetric and positive definite when $L[e(x)] = \frac{d^2 e(x)}{dx^2}$; therefore in this case $Re\lambda_i(-(H_1 H_2^{-1})^{-1} A) < 0$ for $i = 1,2,\dots,n$ and $H_1 H_2^{-1}$ is a convenient choice for the matrix B. With this choice of B, it is possible to solve the linear least squares system (7) also in the hypothesis that the rank of M is less than n: indeed we can prove that the iterative procedure (8) converges to the solution of (7) which minimizes $\underline{x}^T \cdot B\underline{x}$. For the general differential operator L of the form (10), it is difficult to give a-priori conditions on the coefficients $a_j(x)$ in order to have $Re\lambda_i(-(H_1 H_2^{-1})^{-1} A) < 0$ for all $i = 1,2,\dots,n$. Sufficient and useful conditions to test $Re\lambda_i(-(H_1 H_2^{-1})^{-1} A) < 0$, $i = 1,2,\dots,n$, are to analyse whether the matrix $-(H_1 H_2^{-1})^{-1} A$ is dissipative or whether the matrix $(A + \rho H_1 H_2^{-1})^{-1}(A - \rho H_1 H_2^{-1})$ is convergent for any $\rho > 0$, or whether the matrix $(H_1 H_2^{-1})^{-1} + ((H_1 H_2^{-1}))^{-1} - M^T M$ is positive definite. In [4] we have proved that the matrix B which appears in (8) is symmetric and positive definite when we consider the smoothing functional of Whittaker with $\mu_t \geq 0$ and $\nu_{t\ell} \geq 0$. Therefore in this case we have a convenient general criterion for choosing a matrix B which gives a regular solution $\tilde{\underline{x}}$ of (8).

Furthermore the relationship between the Successive Approximation Method with Smoothing and the implicit Euler Method for solving stiff ordinary differential equations gives a criterion for choosing the parameter τ in (8). Since the implicit Euler approximations $(1-\tau\lambda_i)^{-1}$ to $e^{\tau\lambda_i}$ $(\lambda_i \equiv \lambda_i(-B^{-1}A) = \lambda_i(Q))$ improve as $\tau\lambda_i \to 0$, the poorer approximations will be associated with the eigenvalues with larger real parts in absolute value. Therefore, in order to maintain a good accuracy of the overall approximation to $\exp(\tau Q)$, it is required to choose a value of τ for which $\tau \cdot \mathrm{Re}\lambda_{max}(-B^{-1}A)$ is "small".

3. A-STABLE INTEGRATION TECHNIQUES FOR THE SOLUTION OF NONLINEAR ALGEBRAIC SYSTEMS

Let be given in $\mathbb{R}^n$ the non linear algebraic system

$$\underline{f}(\underline{x}) = 0 \tag{12}$$

with the following properties:

a) $\underline{f}(\underline{x})$ has continuous first partial derivatives with respect to all components of $\underline{x}$;

b) there exists a unique solution $\underline{x}^*$ of (12);

c) the matrix $J(x) + J^T(\underline{x})$ is symmetrix and negative definite for all $\underline{x}$, where $J(\underline{x})$ denotes the Jacobian matrix of $\underline{f}(\underline{x})$.

To the algebraic system (12) may be associated the continuous dynamical autonomous system

$$\begin{cases} \dfrac{d\underline{v}(t)}{dt} = \underline{f}(\underline{v}(t)) & t \in (0,\infty) \\ \underline{v}(0) = \underline{g} & \underline{g} \in \mathbb{R}^n \end{cases} \tag{13}$$

which, for the Krasovskii's theorem, is asymptotically stable at x^*. Therefore the solution $\underline{x}^*$ of (12) may be obtained by searching for the steady-state solution of (13).

Now we associate to (13) a discrete dynamical system of the form $(j = 0,1,2,\ldots)$:

$$\begin{cases} \underline{W}_{j+1} = \underline{W}_j + \tau T_j \underline{f}_j \\ \underline{W}_o = \underline{g} \end{cases} \tag{14}$$

where $\underline{f}_j = \underline{f}(\underline{W}_j)$ and T_j is a $n \times n$ non singular matrix, which satisfies the following conditions [8]:

1) the discrete dynamical system is consistent at order $p \geq 1$ with the continuous dynamical system;

2) the discrete dynamical system is asymptotically stable at $\underline{x}^*$ for "small" $\tau > 0$ (+);

(+) A discrete dynamical system (14), i.e. a discrete scheme of the form (14) which satisfies the condition 2), is *absolutely stable*.

3) T_j is continuous with respect to τ.

To this end, we consider the Lyapunov function $V(\underline{w}) = \underline{f}^T(\underline{w}) \cdot \underline{f}(\underline{w})$, which is zero for $\underline{w} = \underline{x}^*$ and positive for any $\underline{w} \neq \underline{x}^*$. Besides

$$\Delta V(\underline{\tilde{W}}_j) = V(\underline{W}_{j+1}) - V(\underline{W}_j) = \underline{f}^T(\underline{W}_j + \tau T_j \underline{f}_j) \cdot \underline{f}(\underline{W}_j + \tau T_j \underline{f}_j) - \underline{f}^T(\underline{W}_j) \cdot \underline{f}(\underline{W}_j) =$$

$$= (\tau J(\underline{\tilde{W}}_j) T_j \underline{f}_j)^T \underline{f}_j + \underline{f}_j^T (\tau J(\underline{\tilde{W}}_j) T_j \underline{f}_j) + (\tau J(\underline{\tilde{W}}_j) T_j \underline{f}_j)^T (\tau J(\underline{W}_j) T_j \underline{f}_j)$$

where $\underline{\tilde{W}}_j \in [\underline{W}_j, \underline{W}_{j+1}]$. If we put $J_j = J(\underline{W}_j)$ and $\underline{\varphi}_j = T_j \underline{f}_j$, since the elements of $J(\underline{\tilde{W}}_j)$ and T_j are continuous with respect to τ, for small τ the sign of the above $\Delta V(\underline{\tilde{W}}_j)$ is given by the sign of the expression:

$$\Delta V(\underline{W}_j) = \underline{\varphi}_j^T (\tau J_j)^T \cdot \underline{f}_j + \underline{f}_j^T (\tau J_j) \underline{\varphi}_j + \underline{\varphi}_j^T (\tau J_j)^T \cdot (\tau J_j) \cdot \underline{\varphi}_j$$

With the choice $T_j = (I - \alpha\tau J_j)^{-1}$, $\frac{1}{2} \leq \alpha \leq 1$, (which satisfies the condition 3)) by the using the hypothesis c) on $J(\underline{v})$, we have $\Delta V(\underline{W}_j) < 0$. Therefore the condition 2) for the discrete dynamical system (14) is satisfied. With the same choice of T_j also the condition 1) is satisfied and the asymptotic error not only remains bounded but tends to zero [8]; in particular, for $\alpha = 1$ we have $p = 1$ and for $\alpha = \frac{1}{2}$, $p = 2$.

In conclusion the choice $T_j = (I - \alpha\tau J_i)^{-1}$, with $\alpha = 1$ or $\alpha = \frac{1}{2}$, in the formula (14) allows to obtain a discrete scheme which is A-stable and rapidly convergent to the solution of the system (12). It is possible also to prove that, when $T_j = (I - \tau J_j)^{-1}$, the scheme (14) becomes a L-stable method. Numerical experiments, carried out on some common stiff test problems, have confirmed the efficiency of these methods [8].

It is interesting to observe that the above choice of T_j generates many *variable metric methods* for minimizing the function $F(\underline{x})$ for which $\underline{f}(\underline{x}) = \text{grad } F(\underline{x})$.

REFERENCES

[1] R.Bellman, R.E.Kalaba, J.A.Lockett : *Numerical Inversion of the Laplace Transform.* American Elsevier, New York (1966).

[2] E.Böhl : *On Finite Difference Methods as Applied to Boundary Value Problems.* Pubbl. I.A.C. n. 100 (1975), Roma.

[3] E.Böhl : *Iterative procedures in the study of discrete analogues for nonlinear boundary value problems.* Pubbl. I.A.C. n.107 (1975) Roma.

[4] I.Galligani : *Sulla "regolarizzazione" dei dati sperimentali.* Calcolo 8 (1971), 359-376.

[5] I.Galligani, S.Seatzu : *Sulle funzioni interpolanti e regolarizzanti in una e due dimensioni.* In Applicazioni del Calcolo (I.Galligani ed.)Editrice Veschi, Roma (1975).

[6] I.Galligani, D.Trigiante : *Numerical Methods for Solving Large Algebraic Systems.* Intern. Symp. on Discrete Methods in Engineering (Sept.1974), CISE, Milano. Pubbl. I.A.C. n.98 (1974), Roma.

[7] S.Seatzu, F.Testa : *Interpolazione e "smoothing" mono e bidimensionali relativi ad operatori differenziali lineari.* Calcolo (to appear).

[8] D.Trigiante : *Stabilità asintotica e discretizzazione su un intervallo infinito.* Pubbl. I.A.C. n.105 (1975), Roma.

[9] V.Valente : *Un metodo alle direzioni alternate esplicito per la risoluzione numerica delle equazioni della diffusione.* Calcolo 4 (1974), 435-452.

ON THE APPLICATION OF THE MINIMUM DEGREE ALGORITHM TO FINITE ELEMENT SYSTEMS

Alan George
and
David R. McIntyre

Department of Computer Science
University of Waterloo
Waterloo, Ontario, Canada

ABSTRACT

We describe an efficient implementation of the so-called minimum degree algorithm, which experience has shown to produce efficient orderings for sparse positive definite systems. Our algorithm is a modification of the original, tailored to finite element problems, and is shown to induce a partitioning in a natural way. The partitioning is then refined so as to significantly reduce the number of non-null off-diagonal blocks. This refinement is important in practical terms because it reduces storage overhead in our linear equation solver, which utilizes the ordering and partitioning produced by our algorithm. Finally, we provide some numerical experiments comparing our ordering/solver package to more conventional band-oriented packages.

1. Introduction

In this paper we consider the problem of directly solving the linear equations

(1.1) $Ax = b,$

where A is a sparse N by N positive definite matrix arising in certain finite element applications. We solve (1.1) using Cholesky's method or symmetric Gaussian elimination by first factoring A into the product LL^T, where L is lower triangular, and then solving the triangular systems $Ly = b$ and $L^Tx = y$.

It is well known that when a sparse matrix is factored using Cholesky's method, the matrix normally suffers fill; that is, the triangular factor L will typically have nonzeros in some of the positions which are zero in A. Thus, with the usual assumption that exact numerical cancellation does not occur, $L+L^T$ is usually fuller than A.

For any N by N permutation matrix P, the matrix PAP^T remains sparse and positive definite, so Cholesky's method still applies. Thus, we could instead solve

(1.2) $(PAP^T)(Px) = Pb.$

In general, the permuted matrix PAP^T fills in differently, and a judicious choice of P can often drastically reduce fill. If zeros are exploited, this can in turn imply a reduction in storage requirements and/or arithmetic requirements for the linear equation solver. The permutation P may also be chosen to ease data management, simplify coding etc. In general, these varying desiderata conflict, and various compromises must be made.

A heuristic algorithm which has been found to be very effective in finding efficient orderings (in the low-fill and low-arithmetic sense) is the so-called minimum degree algorithm [10]. The ordering algorithm we propose in this paper is a modification of this algorithm, changed in a number of ways to improve its performance for finite element matrix problems, which we now characterize.

Let M be any mesh formed by subdividing a planar region R with boundary ∂R by a number of lines, all of which terminate on a line or on ∂R. The mesh so formed consists of a set of regions enclosed by lines, which we call elements. The mesh M has a node at each vertex (a point of intersection of lines and/or ∂R), and may also have nodes on the lines, on ∂R, and in the interiors of the elements. An example of such a finite element mesh is given in Figure 1.1.

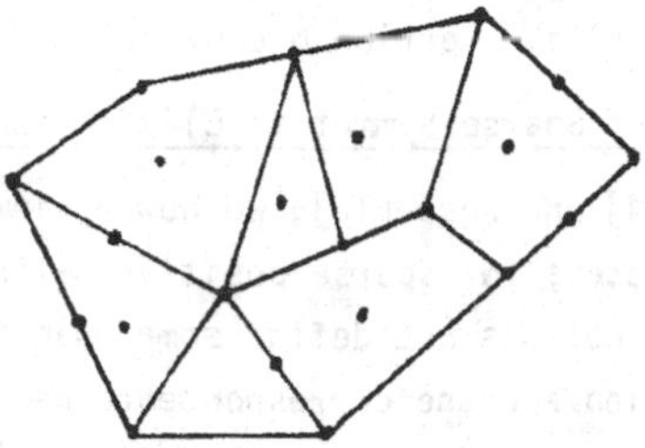

Figure 1.1 An example of a finite element mesh

Now let M have N nodes, labelled from 1 to N, and associate a variable x_i with the i-th node.

Definition 1.1 [4]

A finite element system of equations associated with the finite element mesh M is any N by N symmetric positive definite system $Ax = b$ having the property that $A_{ij} \neq 0$ if and only if x_i and x_j are associated with nodes of the same element. □

In one respect this definition is more general than required to characterize matrix problems arising in actual finite element applications in two dimensions. Usually M is restricted to consist of triangles or quadrilaterals, with adjacent elements having a common side. However, just as in [4], we intend to associate such meshes with matrices which arise when Gaussian elimination is applied to A, and these matrices require meshes having a less restricted topology in order that the correspondence be correct in the sense of Definition 1.1.

In a second respect, the above definition is not quite general enough to cover many matrix problems which arise in finite element applications because more than one variable is often associated with each node. However, the extension of our ideas to this situation is immediate, so to simplify the presentation we assume only one variable is associated with each node. (The code which produced the numerical results of section 5 works for this more general case with no changes). Thus, in this paper we make no distinction between nodes and variables.

An outline of the paper is as follows. In section 2 we review two closely related models for symmetric Gaussian elimination, and describe the basic minimum degree algorithm. In section 3 we show how the special structure of finite element matrix problems can be exploited in the implementation of the minimum degree algorithm. We also show how the algorithm induces a natural partitioning of the matrix. In section 4 we describe a refinement of the partitioning produced by our version of the minimum degree algorithm which normally leads to a considerable reduction in the number of non-null off-diagonal blocks of the partitioning. This refinement is important since it reduces storage overhead when the matrix is processed as a block matrix, with only the non-null blocks being stored. Section 5 contains a brief description of the method of computer implementation of the ideas of sections 3 and 4, along with some numerical results. Section 6 contains our concluding remarks.

2. Models for the Analysis of Sparse Symmetric Elimination

Following George [4] and Rose [10], we now review for completeness some basics of the elimination process for sparse positive definite matrices. We begin with some basic graph theory notions and define some quantities we need in subsequent sections. Much of the notation and the correspondence between symmetric Gaussian elimination and graph theory is due to the work of Parter [9] and Rose [10].

An undirected graph $G = (X,E)$ consists of a finite nonempty set X of nodes together with a set E of edges, which are unordered pairs of distinct nodes of X. A graph $G' = (X',E')$ is a subgraph of $G = (X,E)$ if $X' \subset X$ and $E' \subset E$. The nodes x and y of G are adjacent (connected) if $(x,y) \in E$. For $Y \subset X$, the adjacent set of Y, denoted by $\mathrm{Adj}(Y)$, is

$$\mathrm{Adj}(Y) = \{x \mid x \in X \backslash Y \text{ and } \exists y \in Y \ni (x,y) \in E\}.$$

When Y is a single node y, we write $\mathrm{Adj}(y)$ rather than $\mathrm{Adj}(\{y\})$. The degree of a node x, denoted by $\deg(x)$, is the number $|\mathrm{Adj}(x)|$, where $|S|$ denotes the cardinality of the finite set S. The incidence set of Y, $Y \subset X$, is denoted by $\mathrm{Inc}(Y)$ and defined by

$$\mathrm{Inc}(Y) = \{(x,y) \mid y \in Y \text{ and } x \in \mathrm{Adj}(Y)\}.$$

For a graph $G = (X,E)$ with $|X| = N$, an ordering (numbering, labelling) of G is a bijective mapping $\alpha:\{1,2,\ldots,N\} \to X$. We denote the labelled graph and node set by G^α and X^α respectively.

A path in G is a sequence of distinct edges $\{x_0,x_1\},\{x_1,x_2\}\ldots\{x_{r-1},x_r\}$ where all nodes except possibly x_0 and x_r are distinct. If $x_0 = x_r$, then the path is a cycle. The distance between two nodes is the length of the shortest path joining them. A graph G is connected if every pair of nodes is connected by at least one path. If G is disconnected, it consists of two or more maximal connected components.

We now establish a correspondence between graphs and matrices. Let A be an N by N symmetric matrix. The labelled undirected graph corresponding to A is denoted by $G^A = (X^A,E^A)$, and is one for which X^A is labelled as the rows of A, and $(x_i,x_j) \in E^A \iff A_{ij} \neq 0$, $i \neq j$. The unlabelled graph corresponding to A is simply

G^A with its labels removed. Obviously, for any N by N permutation matrix $P \neq I$, the unlabelled graphs of A and PAP^T are identical but the associated labellings differ. Thus, finding a good ordering of A can be viewed as finding a good labelling for the graph associated with A.

A symmetrix matrix A is reducible if there exists a permutation matrix P such that PAP^T is block diagonal, with more than one diagonal block. This implies G^A is disconnected. In terms of solving linear equations, this means that solving $Ax = b$ can be reduced to that of solving two or more smaller problems. Thus, in this paper we assume A is irreducible, which means that the graph associated with A is connected.

Following Rose [10], we now make the connection between symmetric Gaussian elimination applied to A, and the corresponding graph transformations on G^A. Symmetric Gaussian elimination applied to A can be described by the following equations.

$$A = A_0 = H_0 = \begin{bmatrix} d_1 & v_1^T \\ v_1 & \bar{H}_1 \end{bmatrix} = \begin{bmatrix} \sqrt{d_1} & 0 \\ \frac{v_1}{\sqrt{d_1}} & I_{N-1} \end{bmatrix} \begin{bmatrix} 1 & 0 \\ 0 & H_1 \end{bmatrix} \begin{bmatrix} \sqrt{d_1} & v_1^T/\sqrt{d_1} \\ 0 & I_{N-1} \end{bmatrix}$$

$$= L_1 A_1 L_1^T,$$

$$(2.1) \qquad A_1 = \begin{bmatrix} 1 & 0 & \\ & d_2 & v_2^T \\ 0 & v_2 & \bar{H}_2 \end{bmatrix} = \begin{bmatrix} 1 & & \\ & \sqrt{d_2} & \\ 0 & \frac{v_2}{\sqrt{d_2}} & I_{N-2} \end{bmatrix} \begin{bmatrix} 1 & 0 & \\ & 1 & 0 \\ 0 & 0 & H_2 \end{bmatrix} \begin{bmatrix} 1 & 0 & \\ & \sqrt{d_2} & v_2^T/\sqrt{d_2} \\ & & I_{N-2} \end{bmatrix}$$

$$= L_2 A_2 L_2^T$$

$$\vdots$$

$$A_N = I$$

It is easy to verify that $A = LL^T$, where

$$(2.2) \qquad L = \left(\sum_{i=1}^{N} L_i\right) - (N-1)I.$$

Consider now the labelled graph G^A, with the labelling denoted by the mapping α. The deficiency Def(x) is the set of all pairs of distinct nodes in Adj(x) which are not themselves adjacent. Thus,

$$\mathrm{Def}(x) = \{\{y,z\} \mid y,z \in \mathrm{Adj}(x),\ y \neq z,\ y \notin \mathrm{Adj}(z)\}.$$

For a graph $G = (X,E)$ and a subset $C \subseteq X$, the section graph G(C) is the subgraph $G = (C,E(C))$, where

$$E(C) = \{(x,y) \mid (x,y) \in E,\ x \in C,\ y \in C\}.$$

Given a vertex y of a graph G, define the y-elimination graph G_y by

$G_y = \{X\backslash\{y\}, E\backslash Inc(y) \cup Def(y)\}$.

The sequence of <u>elimination graphs</u> $G_1, G_2, \ldots, G_{N-1}$ is then defined by $G_1 = G_{x_1}$ and $G_i = (G_{i-1})_{x_i}$, $i = 2,3,\ldots,N-1$.

The elimination graph G_i, $0 < i < N$, is simply <u>the graph associated with the matrix H_i</u>; i.e. $G_i = G^{H_i}$. We define $G_0 = G^A$, and note that G_{N-1} consists of a single node. The recipe for obtaining G_i from G_{i-1}, which is to delete x_i and its incident edges, and to then add edges so that $Adj(x_i)$ is a <u>clique</u>, is due to Parter [9]. A clique is a set of nodes all of which are adjacent to each other.

This graph model has many advantages for describing and analyzing sparse matrix computations. However, except for rather small examples, it is not easy to visualize; although G_0 may sometimes be planar, the G_i rapidly become nonplanar with increasing i and become difficult to draw and interpret. For our class of matrix problems, which are associated with planar mesh problems, we can define a sequence of finite element meshes $M = M_0, M_1, \ldots, M_N$ such that G_i can easily be constructed from M_i, $i = 0,1,\ldots,N-1$.

Formally, a mesh $M = (X,S)$ is an ordered pair of sets, with X a (possibly empty) set of nodes and S a set of mesh lines, where each mesh line either joins two nodes, is incident to only one node (forming a loop), or else forms a nodeless loop. Let $M_0 = (X_0, S_0)$ be the original finite element mesh M, with nodes of the mesh forming the set X_0, and the lines joining the nodes comprising the set S_0. A <u>boundary mesh line</u> is a member of S shared by only one element.

Starting with M_0, the mesh $M_i = (X_i, S_i)$, $i = 1,2,\ldots,N$ is obtained from $M_{i-1} = (X_{i-1}, S_{i-1})$ by

a) Deleting node x_i and its non-boundary incident mesh lines.

b) Repeatedly deleting mesh lines incident to a node having degree equal to one.

Here the degree of a node y in a mesh is the number of times mesh lines are incident to y. When x_i is eliminated from M_{i-1} and x_i has incident boundary lines, these boundary lines are simply "fused" to form a new line of S_i, as depicted in the transformations $M_5 \rightarrow M_6$ and $M_{12} \rightarrow M_{13}$ in Figure 2.1. The application of step b) is illustrated in the transformation $M_6 \rightarrow M_7$ and $M_{11} \rightarrow M_{12}$.

We now describe how to obtain G_i from M_i. Since the node sets are identical, we need only describe how to construct E_i. Since the sequence M_i is generated by removing nodes and/or mesh lines, the meshes of the sequence are all planar, having faces (elements) with nodes on their periphery and/or in their interior as shown in Figure 2.1. Also recall that by definition 1.1, G_0 is a graph such that each set of nodes $C_0^{\ell} \subseteq X_0$ associated with an element (interior to and/or on the periphery of a face of M_0) forms a clique. This construction is illustrated in Figure 2.2.

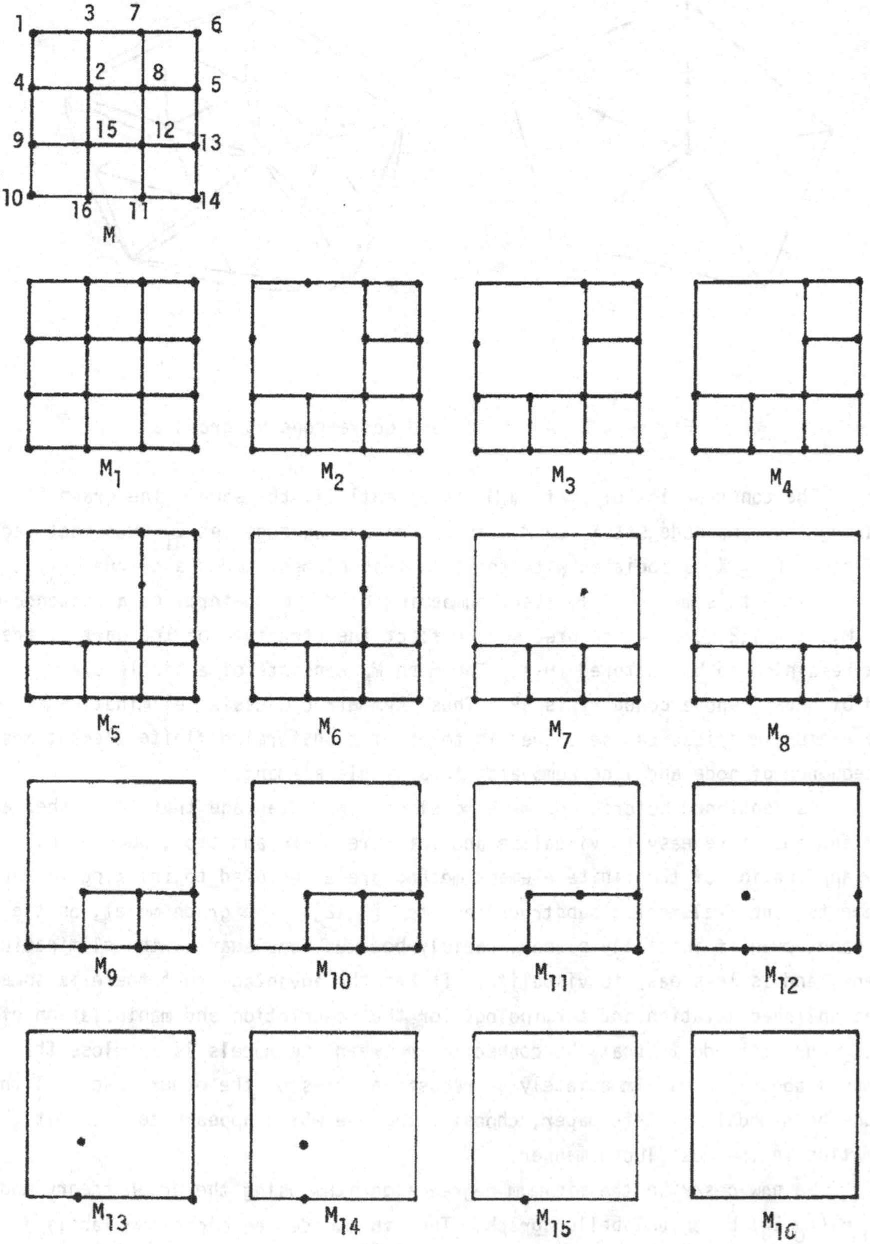

Figure 2.1 Meshes M_1 through M_{16}

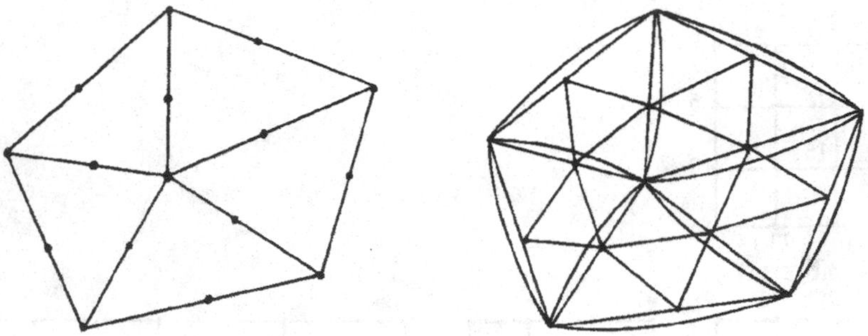

Figure 2.2 A mesh M_0 and corresponding graph G_0

The construction of G_i from M_i is essentially the same. The graph G_i is one having the same node set X_i as M_i, and which has an edge set E_i such that each set of nodes $C_i^\ell \subseteq X_i$ associated with the same mesh element forms a clique.

Using this mesh model, every numbering of $M = M_0$ determines a sequence of meshes M_i, $i = 1,2,\ldots,N$ which precisely reflect the structure of the part of the matrix remaining to be factored (H_i). The mesh M_N consists of a single element, devoid of nodes, whose boundary is ∂R. Thus, symmetric Gaussian elimination on finite element matrices can be viewed in terms of transforming finite element meshes, by a sequence of node and line removals, to a single element.

As mentioned before, our mesh model has the advantage that the meshes are planar and therefore easy to visualize and interpret. In addition, people involved in the application of the finite element method are accustomed to thinking in terms of elements, super-elements, substructures etc. [1,12]. The graph model, on the other hand, even if initially planar, rapidly becomes nonplanar as the elimination proceeds, and is less easy to visualize. It has the advantage that there is some well established notation and terminology for the description and manipulation of graphs. Our attitude is that the connection between the models is so close that statements about one can immediately be recast in terms of the other. We will therefore use both models in this paper, choosing the one which appears to transmit the information in the most lucid manner.

We now describe the minimum degree algorithm using the graph theory model. Let $G_0 = (X_0,E_0)$ be an unlabelled graph. The minimum degree algorithm labels X (determines the mapping α) according to the following pseudo-Algol program:

for $i = 1,2,\ldots,N$ do

1) Find $y \in X_{i-1}$ such that $\deg(y) \leq \deg(x)$ for all $x \in X_{i-1}$.

2) Set $\alpha^{-1}(y) = i$ and if $i < N$ form G_i from G_{i-1}.

Various strategies for breaking ties have been proposed, but in our application we found they made little difference to the quality of the ordering produced. Thus, we break ties arbitrarily.

Notice that the ordering algorithm requires knowledge about the current state of the factorization; that is, the choice of the i-th variable is a function of the structure of the partially factored matrix. Thus, one could argue that the ordering algorithm (i.e., the pivot selection) should be imbedded in the actual numerical factorization code. (Of course, this is more or less essential when the pivot selection is partially determined by numerical stability considerations.)

However, in our situation we have the option of isolating the ordering and the factorization in separate modules, and we prefer to do so for the following reasons:

1. If the ordering is done a-priori, the factorization code can utilize a static data structure, since the positions where fill will occur can be determined during the ordering. If the ordering is imbedded in the factorization, the data structure must be adaptive to accommodate fill as it occurs. By isolating the ordering and factorization, data structures can be tailored specifically for each function.
2. In some engineering design applications, many problems with the same matrix A, or with coefficient matrices having the same structure, must be solved. In these situations, it makes sense to find an efficient ordering and set up the appropriate data structures only once.

3. The Minimum Degree Algorithm

From the graph model developed in section 2, we see that symmetric Gaussian elimination can be viewed as transforming G^A by a sequence of graph transformation rules to one having a single node and no edges. Using the mesh model, we see that the algorithm can also be viewed as transforming the original mesh according to some precise node and line removal rules so that the final mesh M_N consists of a single nodeless element. We also established a correspondence between the two models, so that G_i can be constructed from M_i, $0 \leq i < N$. In this section we show that for our finite element matrix problems, the local behavior of the minimum degree algorithm can be precisely characterized. In addition, we show that the algorithm induces a natural partitioning of the node set X, or equivalently, of the reordered matrix A. In the next section we show how this ordering can be refined so that the triangular factor L of A can be efficiently stored.

We begin with some definitions. For a graph $G = (X,E)$, let $C \subseteq X$ and let $G(C) = (C,E(C))$ be the corresponding section graph of G determined by C (see section 2). The node x is an interior node of C if $x \in C$ and $\mathrm{Adj}(x) \subseteq C$. If $x \in C$ but $\mathrm{Adj}(x) \nsubseteq C$, then x is a boundary node of C.

As we have described before, nodes associated with a mesh element correspond to a clique in the corresponding graph. Generally, interior nodes of a clique in G_i correspond to interior nodes of an element in M_i. However, there are exceptions.

The nodes on the periphery of an element which forms part of ∂R are interior nodes of the corresponding graph clique. Also, when all the nodes form a clique, in the element model they could be interior nodes and/or boundary nodes. Thus although the element model is very helpful in visualizing the changing structure of the matrix during the decomposition, its meaning in terms of the corresponding graph must be interpreted carefully near ∂R and for the last clique of nodes.

In what follows, the sequence G_i, $i = 0,1,\ldots,N-1$ will refer to the graph sequence generated by the minimum degree algorithm.

<u>Lemma 3.1</u>

Let C be a clique in $G(X,E)$, let x be an interior node of C, and let y be a boundary node of C. Then $\deg(x) < \deg(y)$.

The proof is trivial and is omitted.

<u>Lemma 3.2</u>

Let C be a clique in $G_i = (X_i,E_i)$, $0 \le i < N$, and let x be an interior node of C. Then if x is eliminated, the degree of all nodes in C is reduced by one, and the degrees of all other nodes in G_i remain the same.

<u>Proof</u> Since $\text{Adj}(x) \subset C$, the elimination of x cannot affect nodes $y \notin C$. Thus $\deg(y)$, $y \in X_{i+1}\backslash C$, is the same as it was in G_i. On the other hand, since $x \in C$ and C is a clique, the elimination of x causes no fill. Thus, E_{i+1} is obtained from E_i by deleting $\text{Inc}(x)$ from E_i, thereby reducing the degree of all nodes in $C\backslash\{x\}$ by one. □

<u>Theorem 3.3</u>

Let C be a clique in $G_i = (X_i,E_i)$, $0 \le i < N$, and let Q_i be the set of <u>all interior nodes</u> of C. Then if the minimum degree algorithm chooses $x \in Q_i$, at the i-th step, then it numbers the remaining $|Q_i|-1$ nodes of Q_i next.

<u>Proof</u> Let $\deg(x) = d$, and note that $\deg(y) \ge d$, $y \notin Q_i$, with $\deg(y) > d$ if $y \in C\backslash Q_i$. By Lemma 3.4, after x is eliminated, the degree of all nodes remaining in C will be reduced by one, and nodes not in C will have their degrees unchanged. Thus, if $Q_{i+1} = Q_i\backslash\{x\} \neq \phi$, then the node of minimum degree in G_{i+1} is in Q_{i+1}. Repeatedly using Lemma 3.2, we conclude that the minimum degree algorithm will choose nodes in Q_i until it is exhausted. □

<u>Corollary 3.4</u>

In terms of the mesh model, Lemma 3.2 and Theorem 3.3 imply that nodes in the interior of an element, or those on the boundary of an element which forms part of ∂R, will all be eliminated before other nodes associated with that element.

<u>Theorem 3.5</u>

Let C_1 and C_2 be two cliques in G_i, $0 \le i < N$, with $K_i = C_1 \cap C_2 \neq \phi$, $C_1 \not\subset C_2$, and $C_2 \not\subset C_1$. Let

$$\bar{K}_i = \{y|y \in K_i \text{ and } Adj(y) \subset C_1 \cup C_2\}.$$

Then if the minimum degree algorithm chooses $x \in \bar{K}_1$ at the i-th step, then it numbers the remaining nodes of $\bar{K}_{i+1} = \bar{K}_i\backslash\{x\}$ next, in arbitrary order.

<u>Proof</u> First, note that C_1 and C_2 cannot have any interior nodes, since otherwise the minimum degree algorithm would not choose a node from $\bar{K}_i$. Thus, for all $y \in \bar{K}_i$, $deg(y) = d = |C_1 \cup C_2|-1$. Also, note that if $y \in K_i\backslash\bar{K}_i$, then $deg(y) > d$, because it is connected to all other nodes in $C_1 \cup C_2$, and at least one other node. (Otherwise, it would be in $\bar{K}_i$.) Finally, if $y \in X_i\backslash K_i$, then $deg(y) \geq d$, because otherwise the minimum degree algorithm would not choose x. We now want to show that after x is eliminated, yielding $G_{i+1} = (X_{i+1},E_{i+1})$, that $deg(y) = d-1$ if $y \in \bar{K}_{i+1}$, and $deg(y) \geq d$ for $y \in X_{i+1}\backslash\bar{K}_{i+1}$.

First, if $\bar{K}_{i+1} = \phi$, there is nothing to prove, so suppose $\bar{K}_{i+1} \neq \phi$. Now the elimination of x renders $(C_1 \cup C_2)\backslash\{x\}$ a clique, but since $Adj(x) \subset C_1 \cup C_2$, nodes $y \in X_i\backslash(C_1 \cup C_2)$ are not affected by the elimination of x, so $deg(y) \geq d$ as before. Suppose $y \in C_1\backslash C_2$, and let its degree in G_i be $p \geq d$. Then after elimination it is $p + |C_2\backslash C_1|-1 \geq d$, since $|C_2\backslash C_1| \neq \phi$. Similarly, after elimination of x, $deg(y) \geq d$ for $y \in C_2\backslash C_1$. Now consider $y \in K_{i+1} = K_i\backslash\{x\}$. Before elimination of x, y is connected to all nodes in $(C_1 \cup C_2)\backslash\{y\}$, since C_1 and C_2 are cliques. Since the elimination of x only involves nodes in $C_1 \cup C_2$, y cannot be connected to any new variables. Moreover, after elimination of x, y is no longer connected to x, so deg(y) is reduced by one for $y \in K_{i+1}$. Thus, in G_{i+1}, $deg(y) = d-1$ for $y \in \bar{K}_{i+1}$ and $deg(y) \geq d$ for $y \in K_{i+1}\backslash\bar{K}_{i+1}$.

The minimum degree algorithm will now choose a node in $\bar{K}_{i+1}$, and by Theorem 3.3, it will continue to choose nodes from $\bar{K}_{i+1}$ until it is exhausted. □

<u>Corollary 3.6</u>

Let C_k, $k = 1,2,\ldots,r$ be cliques in G_i, $i \in \{0,1,\ldots,N-1\}$, $K_i = \bigcap_{k=1}^{r} C_k \neq \phi$, and $C_\ell \cap (X_i\backslash(\bigcup_{k\neq\ell} C_k)) \neq \phi$. Let

$$\bar{K}_i = \{y|y \in K_i \text{ and } Adj(y) \subset (\bigcup_{k=1}^{r} C_k)\}.$$

Then if the minimum degree algorithm chooses $x \in \bar{K}_i$ at the i-th step, then it numbers the remaining nodes of $\bar{K}_{i+1} = \bar{K}_i\backslash\{x\}$ next, in arbitrary order. The proof is similar to that of Theorem 3.5 and is omitted.

Theorems 3.3 and 3.5, and Corollary 3.6 have important practical implications. It is easy to recognize when their hypotheses apply, and they allow us to immediately number <u>sets</u> of nodes by doing only <u>one</u> minimum degree search.

Thus, after each minimum degree search, a <u>set</u> of $r \geq 1$ nodes will be numbered, inducing a <u>partitioning</u> of X. We will denote this partitioning by $P = \{P_1,P_2,P_3,\ldots,P_p\}$, where $\bigcup_{i=1}^{p} P_i = X$. Figures 3.1a,b contain an L-shaped triangular

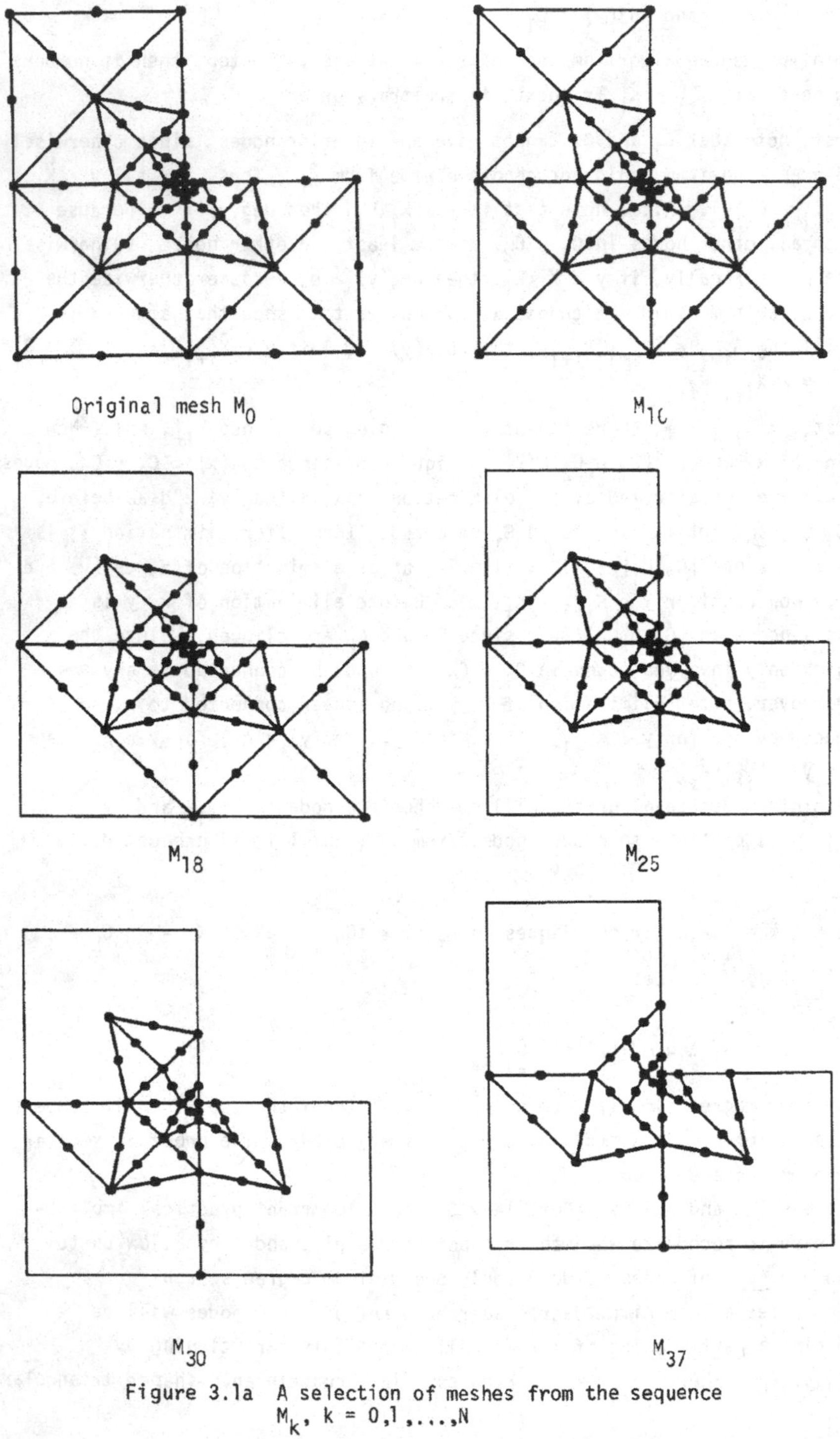

Figure 3.1a A selection of meshes from the sequence M_k, $k = 0,1,\ldots,N$

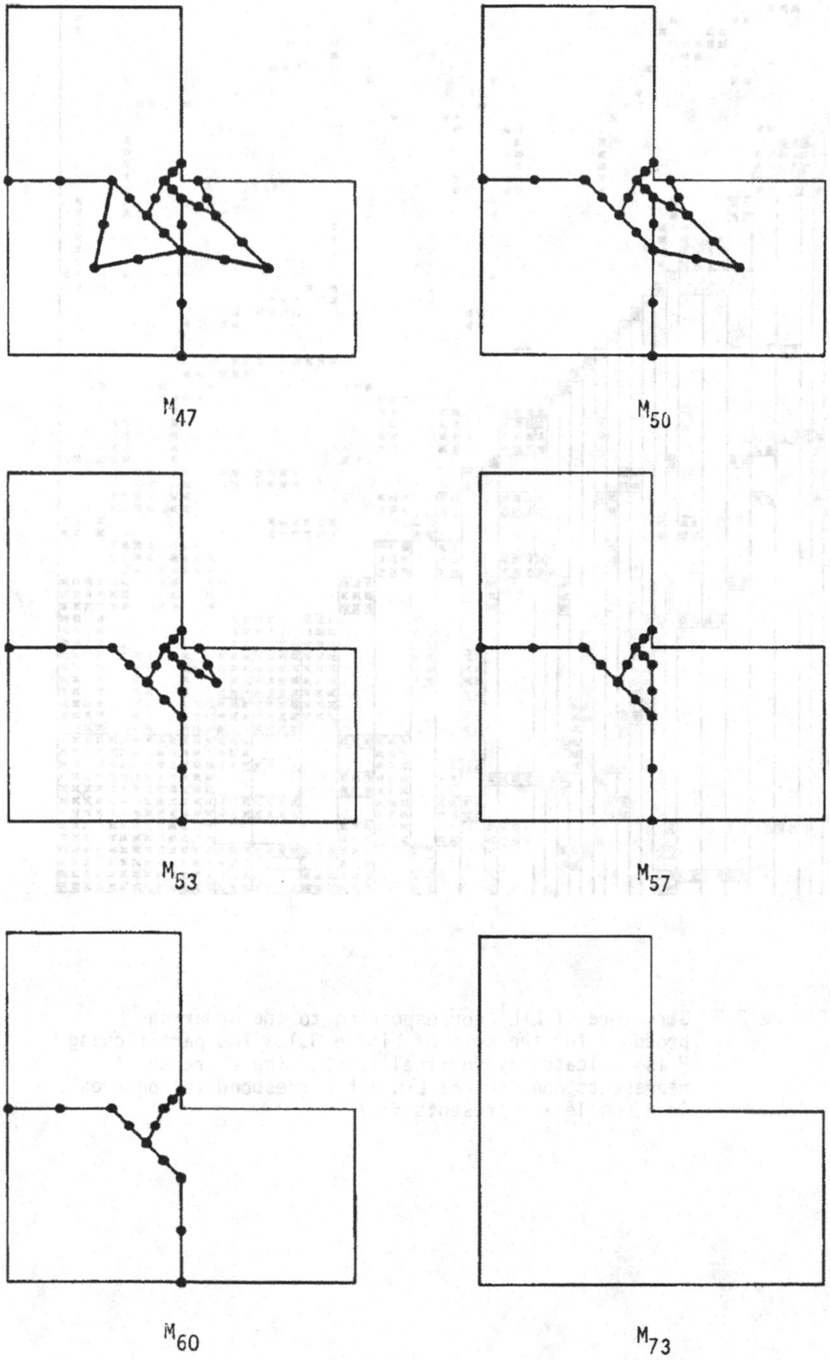

Figure 3.1b A selection of meshes from the sequence M_k, $k = 0,1,\ldots,N$

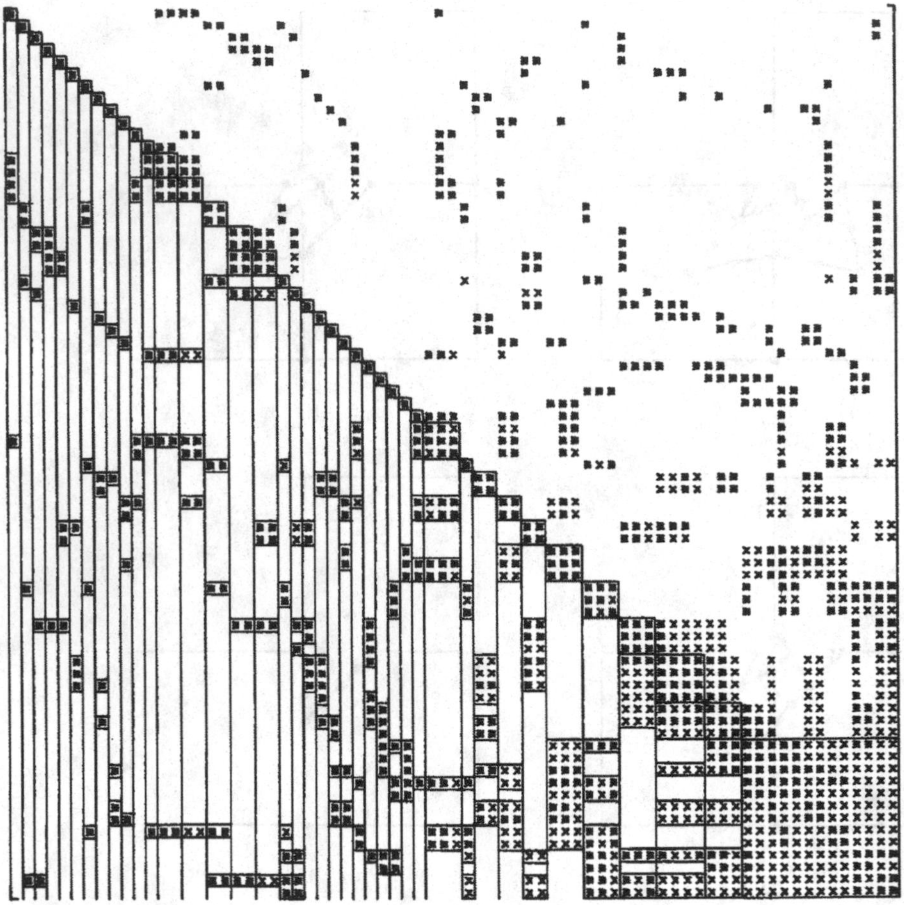

Figure 3.2 Structure of $L+L^T$ corresponding to the ordering produced for the mesh of Figure 3.1. The partitioning P is indicated by vertical lines. The character * represents nonzeros of L which correspond to nonzeros in A, while x represents fill

mesh M_0 together with a few of the meshes M_i, $0 \le i \le N$ generated by the minimum degree algorithm. Figure 3.2 shows the matrix structure of $L+L^T$ corresponding to the ordering, with the column partitioning P indicated by the vertical lines.

4. A Refinement of the Minimum Degree Ordering

In the last section we saw how the minimum degree algorithm naturally induces a partitioning of the matrix A (and therefore also of L). Consider each "block column" of L, determined by the members of the partitioning P. Within each such block column the rows are either empty or full, and we can partition the rows according to contiguous sets of non-null rows, as indicated in Figure 3.2.

Basically, the storage scheme we use in our linear equation solver is one which stores the dense submatrices determined by this row-within-(block) column partitioning. Now each non-null block incurs a certain amount of storage overhead, so we would like the number of these blocks to be as small as possible. The purpose of this section is to describe a way of reordering the members of each partition member P_i so as to reduce the number of non-null blocks in this row-within-column partitioning.

Our reordering scheme is most easily motivated using the mesh model of elimination we introduced in section 2. First, note that most of the partition members (except for some of the initial ones) will correspond to nodes lying on a side of an element in some mesh M_k, $0 \le k \le N$. It is clear that the order in which these nodes are numbered is irrelevant as far as fill or operation counts are concerned, and their relative order is not specified by the minimum degree algorithm. Thus, we are free to choose the ordering within the partitions to reduce the number of individual blocks of L we must store.

How do we achieve this? Consider the schematic drawing in Figure 4.1, indicating a subsequence of meshes taken from the sequence M_k, $k = 0,1,\ldots,N$.

From what we have established about the behavior of the minimum degree algorithm through Theorems 3.3 and 3.5, it is clear that the mesh line segments (element sides) denoted by ① through ⑦ correspond to dense diagonal blocks of L. Now consider the block column of L corresponding to ①. Obviously, block ① is connected to part of block ⑦, but in general the connection will not be reflected as a dense off-diagonal block of L, unless the nodes of ⑦ which are connected to ① are numbered consecutively. Similarly, the connections of block ② to block ⑦ will correspond to a dense block of L only if nodes of ⑦ which are connected to ② are numbered consecutively.

This motivates our reordering algorithm for each partition P_i. We reorder the nodes of each P_i in a way which corresponds to numbering nodes on an element side consecutively, beginning at one end. Figure 4.2 shows the structure $L+L^T$ corresponding to the reordering obtained by applying our reordering scheme to the problem which produced the Figures 3.1 and 3.2. The reduction in the number of off-diagonal blocks is apparent, but it is not particularly impressive because the problem is quite small.

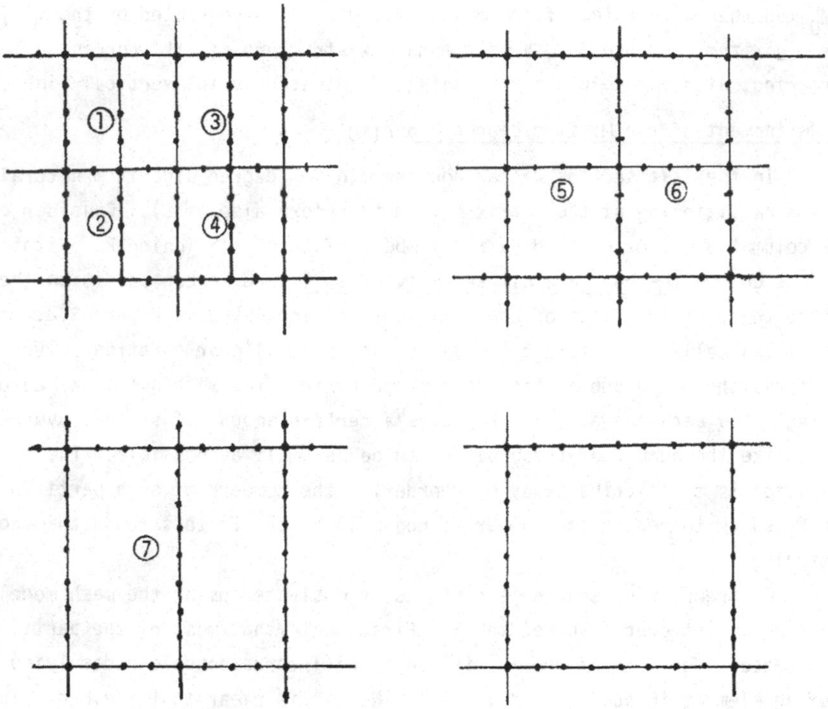

Figure 4.1 A subsequence of meshes from the sequence M_k, $k = 0,1,\ldots,N$

For larger problems, however, the reduction is usually very substantial.

Now that we have established <u>what</u> we want done, how do we achieve it? Obviously, if $|P_i| \leq 2$, there is nothing to do. For $|P_i| > 2$, the nodes in P_i will typically all lie on an element side in some mesh M_k, $0 \leq k < N$, such as indicated in Figure 4.3. (For P_p, the last partition, the situation is somewhat more complicated, since often three element sides are involved, as in M_{60}, Figure 3.1-b. We consider this problem later.)

Let $G = (X,E)$ be the unlabelled graph corresponding to A, and let $\widetilde{G} = (X,\widetilde{E})$ be the subgraph of G obtained by interpreting the mesh M as a graph. Let G_{P_i} be the section graph $G(P_i)$ (see section 2). Now the element model makes it abundantly clear that in general G_{P_i} consists of <u>a single node</u>, or <u>a simple chain</u>, usually the latter. The graph G_{P_p} is a special case, typically consisting of three chains connected by virtue of a small shared clique or two chains connected by a cycle.

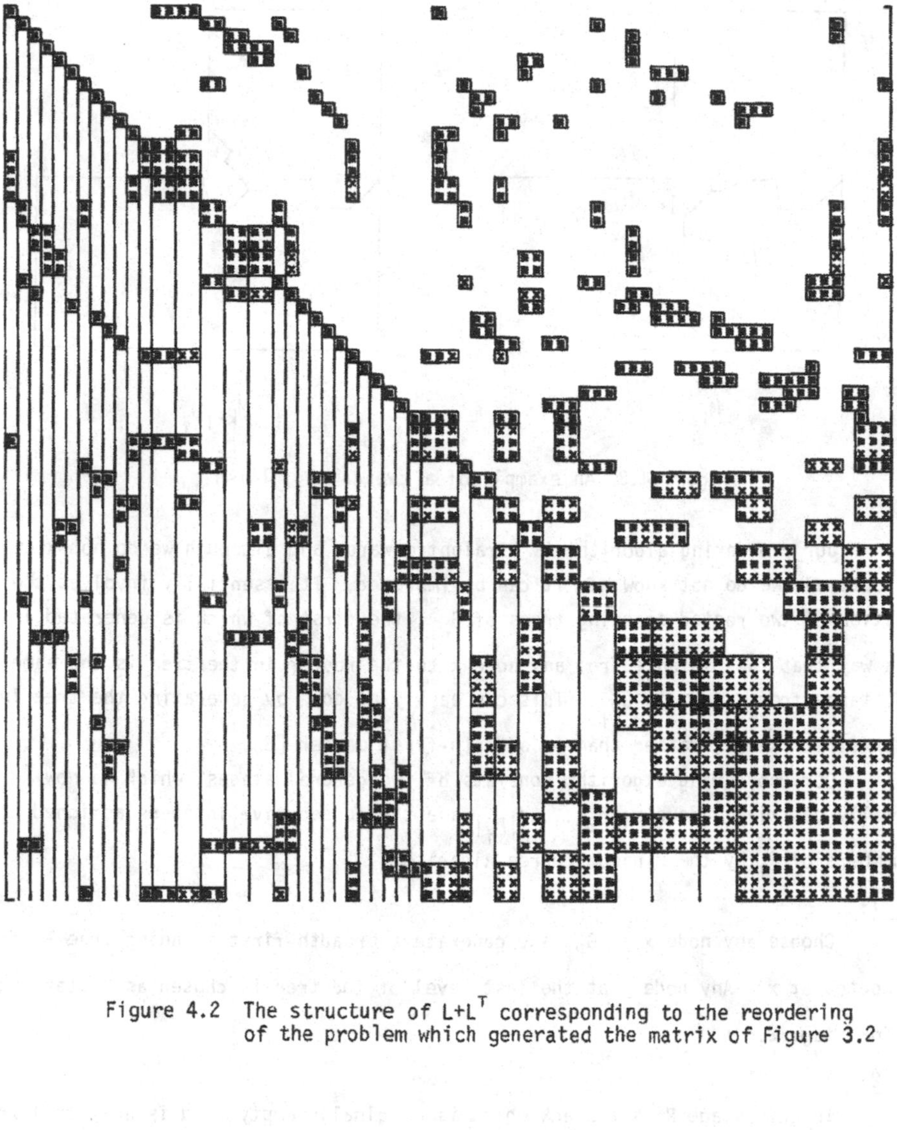

Figure 4.2 The structure of $L+L^T$ corresponding to the reordering of the problem which generated the matrix of Figure 3.2

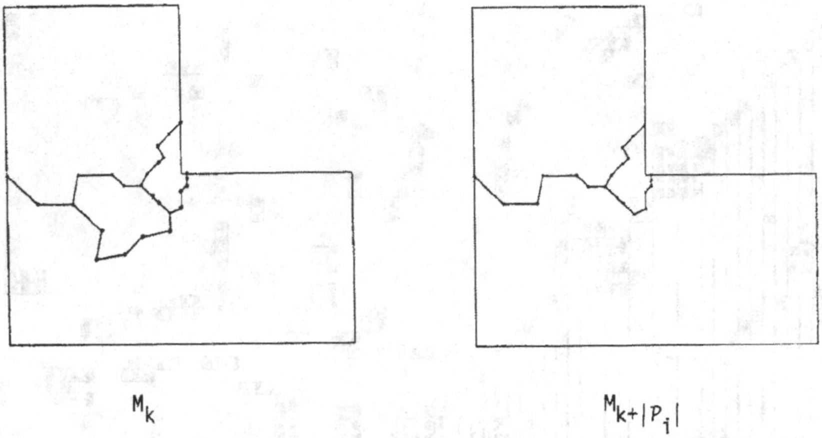

Figure 4.3 An example of a typical P_i, $1 \le i \le p$

Our reordering algorithm is straightforward, and although we do not know if it is optimal, we do not know how it can be improved. It essentially involves the generation of two rooted spanning trees of G_{P_i}, the first of which is generated in such a way that the distance from any node x to the root r in the tree is the same as the distance from x to r in G_{P_i}. This can easily be done by generating the tree in a breadth-first manner, rather than in a depth-first manner [8].

Our reordering algorithm consists of two general stages, which we now informally describe. Here $r_1, r_2, \ldots, r_{|P_i|}$ are the consecutive integers assigned to the members of P_i by the minimum degree algorithm.

Stage 1:

Choose any node x in G_{P_i} and generate a breadth-first spanning tree T_1 for G_{P_i}, rooted at x. Any node y at the last level of the tree is chosen as a starting node for stage 2.

Stage 2:

In this stage R is a stack which is originally empty, and is only utilized if G_{P_i} is not a chain.

1) Label the node y provided by stage 1 as r_1.
2) For each $i = 2,3,\ldots,|P_i|$ do the following:
   a) If x_{r_i-1}, the last labelled node, has only one unlabelled node y adjacent to it, then label it r_i.
   b) If $\mathrm{Adj}(x_{r_i-1})$ has more than one unlabelled node, of those not already in R, label one r_i and place the remainder on the stack R. If all nodes in

Adj(x_{r_i-1}) are also in R, choose one of those and label it r_i.

c) If the members of Adj(x_{r_i-1}) are all numbered, pop the stack R until an unlabelled node y is popped, and label it r_i.

Figure 4.4 illustrates the reordering algorithm. Phase 1 generates the tree T_1 rooted at x, and chooses the node y as the starting node for phase 2. Step a) of phase 2 is executed until node g is labelled. At the next step node h is placed in R and c is labelled. At the next step, the unlabelled nodes of Adj(c) are {h,x}, but since h is already in the stack, node x is labelled, and then step a) of phase 2 operates until node a is labelled. Since Adj(a) is all labelled, node h is obtained from R and labelled, followed by nodes i,j and k via steps a) and c) of phase 2.

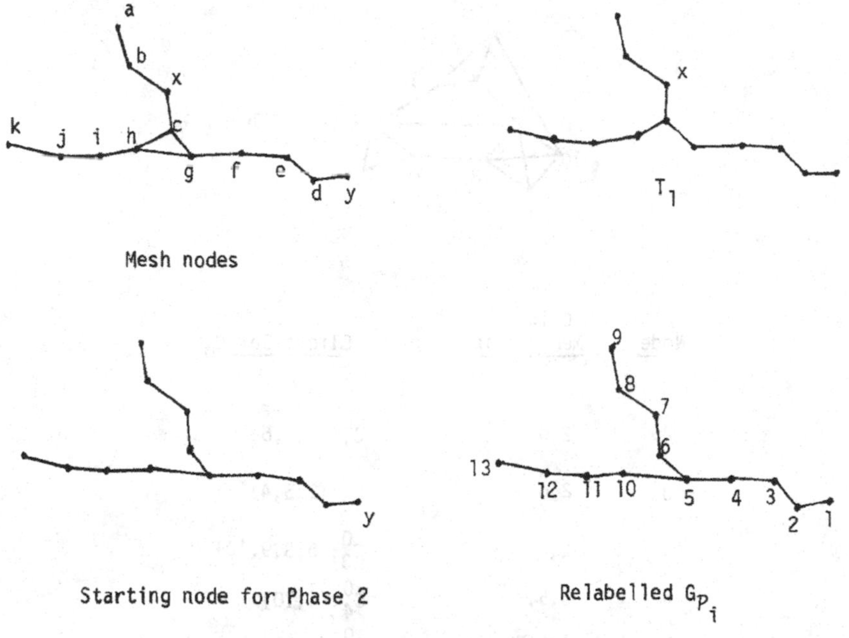

Figure 4.4 Relabelling of G_{P_i}

5. Remarks on Implementation and Some Numerical Experiments

We saw in section 2 how cliques naturally arise during symmetric Gaussian elimination. In matrix problems associated with the use of the finite element method, cliques of size larger than one exist in G^A, and persist for some time during the elimination, typically growing in size by merging with other cliques before finally disappearing through elimination. Moreover, Theorem 3.5 operates for a considerable proportion of the total node numberings.

These observations make it natural to represent the elimination graph sequence through its clique structure, since elimination of variables typically leads to merging of two or more cliques into a new clique. Our approach then, is as follows. The graph $G_i = (X_i, E_i)$ is represented by the set of its cliques $C_i = \{C_\ell^i\}$ along with a <u>clique membership list</u> for each node. An example appears in Figure 5.1.

Now our actual implementation does not represent the entire sequence of graphs G_i, $i = 0,1,2,\ldots,N-1$, during its execution. Only those graphs which would be obtained after each P_i is determined are actually created. That is, we repeatedly apply Theorems 3.3 and 3.5. The general step of our algorithm, described below, is executed p times, where $p = |P|$ and $P = \{P_1, P_2, \ldots, P_p\}$.

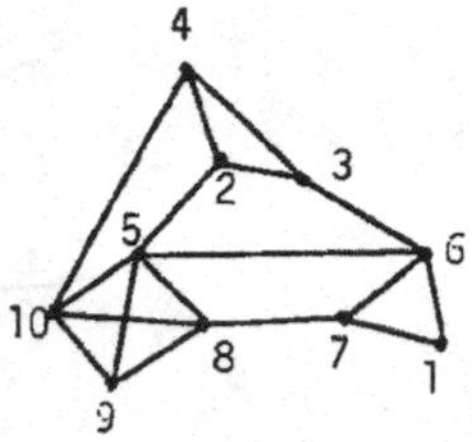

Graph G_0

Node	Clique Membership	Clique Set C_0
1	1	
2	2,5	C_1^0:{1,7,6}
3	2,6	C_2^0:{2,3,4}
4	2,4	C_3^0:{5,8,9,10}
5	3,5,7	C_4^0:{4,10}
6	1,6,7	C_5^0:{2,5}
7	1,8	C_6^0:{3,6}
8	3,8	C_7^0:{5,6}
9	3	C_8^0:{7,8}
10	3,4	

Figure 5.1 The graph G_0 represented by its clique set C_0 and clique membership list

General Step r, r = 1,2,...,p

1) Find an unnumbered node x of minimum degree. If all nodes are numbered, stop.

2) Let $Q_r = \{\ell | x \in C_\ell^{r-1}\}$, and determine the set of nodes P_r in $\tilde{C}_r = \bigcup_{\ell \in Q_r} C_\ell^{r-1}$ which are connected only to nodes in $\tilde{C}_r$.

3) Set $C_r = (C_{r-1} \setminus \bigcup_{\ell \in Q_r} \{C_\ell^{r-1}\}) \cup (\tilde{C}_r \setminus P_r)$.

4) Update the degrees of the nodes in $\tilde{C}_r \setminus P_r$ (the new clique), and their clique membership lists.

5) Increment r and go to step 1).

Our code consists of two phases, the first is simply the minimum degree algorithm, modified to exploit what we know about the behavior of the algorithm, as described by Theorems 3.3 and 3.5. The second phase performs the reordering of each partition member P_i as described in section 4. Although this splitting into two phases is not necessary (since each P_i could be reordered as it is generated), it was done to keep the code modular, and to ease maintenance and subsequent possible enhancements.

Our code accepts as initial input a collection of node sets corresponding to the elements (cliques) of the finite element mesh. This mesh changes as the algorithm proceeds, so its representation must be such that merging cliques (elements) is reasonably efficient and convenient. The data structure we used to represent the graphs is depicted in Figure 5.2. At any stage of the algorithm, the nodes of each clique along with some storage management information are stored in consecutive locations in a storage pool (POOL). A pointer array HDR of length ≤ NCLQS (the initial number of elements) is used to point to the locations of the elements in POOL. Finally, a rectangular array C is used to store the clique membership lists; row i of C contains pointers into HDR corresponding to cliques which have node i as a member.

Step 3) of the algorithm above obviously implies an updating operation of the arrays C, HDR and POOL to reflect the new clique structure of the graph which has seen some of its cliques coalesce into a single new one, along with the removal of some nodes. In general, the node-sets corresponding to each clique to be merged will be scattered throughout POOL, and none of them may occupy enough space so that the new clique to be created could overwrite them. To avoid excessive shuffling of data, we simply allocate space for the new clique from the last-used position in POOL, and mark the space occupied by the coalesced cliques as free. When space for a new element can no longer be found in POOL, a storage compaction is performed. See [8, pp.435-451] for a description of these standard storage management techniques.

Our first objective is to study the behaviour of our ordering algorithm. We ran our code on N by N finite element matrix problems arising from n×n right-triangular meshes of the form shown in Figure 5.3. We ran our code for n = 5(5)35 to study the behavior of various quantities as a function of $N = (n+1)^2$.

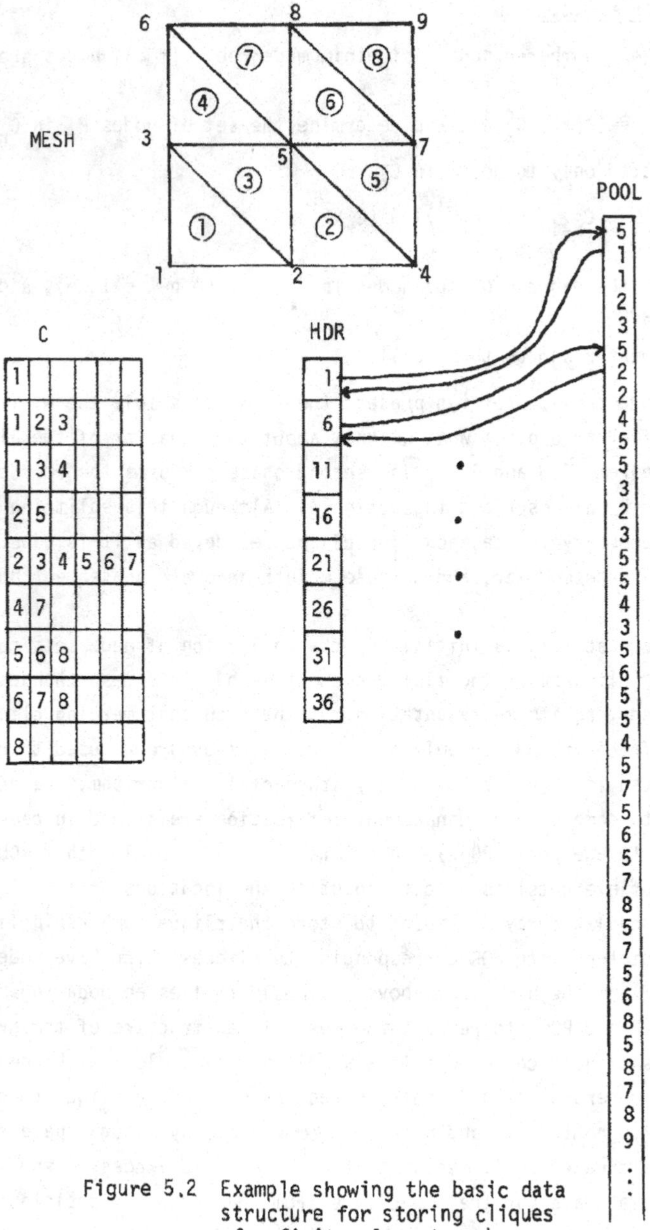

Figure 5.2 Example showing the basic data structure for storing cliques of a finite element mesh

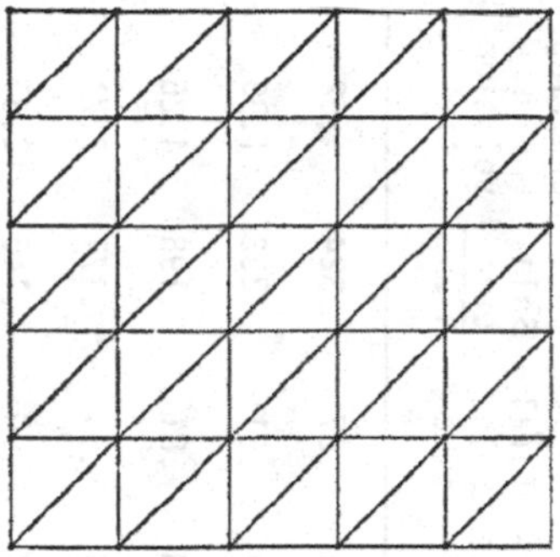

Figure 5.3 A 5 by 5 right-triangular mesh yielding N = 36

The results of our runs are summarized in Tables 5.1-5.3. The "overhead" column in Table 5.1 refers to the number of pointers etc. used by our data structure for L. In our implementation on an IBM 360/75, we used a 32 bit word for both pointers and data. On many machines with a larger wordlength, it would make sense to pack two or perhaps more pointers per word. Thus, in other implementations the overhead for our data structure compared to the storage required for the actual components of L would be much less than appears in Table 5.1.

The overhead and primary column entries in Table 5.1 do not quite add up to the corresponding entry in the total column because we included various other auxiliary vectors and space for the right side b in the total storage count.

TABLE 5.1
Storage statistics for the ordering produced by the minimum degree algorithm followed by the improvement described in section 4

n	N	Overhead	Primary	Total	$\frac{\text{Overhead}}{\text{Total}}$	$\frac{\text{Overhead}}{N}$	$\frac{\text{Total}}{N \log N}$
5	36	316	185	537	.59	8.78	4.16
10	121	1174	1039	2334	.50	9.70	4.02
15	256	2457	2899	5612	.44	9.60	3.95
20	441	4195	5959	10595	.40	9.51	3.95
25	676	6501	10092	17269	.38	9.62	3.92
30	961	9153	17190	27304	.34	9.52	4.14
35	1296	12425	24252	37973	.33	9.59	4.09

TABLE 5.2

Execution Time in Seconds on an IBM 360/75 for the ordering algorithm described in Sections 3 and 4

n	N	Time for Phase 1	Time for Phase 2	Total time	$\frac{\text{Time}}{N \log N}$	Fact. mult.	Fact. time	Back-solve mult.	Back-solve time	Total Soln. time	Soln.† $\frac{\text{time}}{N\sqrt{N}}$	Ordering plus solution time
5	36	.35	.20	.55	.0043	578	.04	370	.03	.07	.324	.62
10	121	1.15	.50	1.65	.0028	5739	.22	2078	.09	.31	.233	1.96
15	256	2.37	1.08	3.45	.0024	21919	.61	5798	.20	.81	.198	4.26
20	441	4.23	1.84	6.07	.0023	56501	1.30	11918	.34	1.64	.177	7.71
25	676	6.57	2.84	9.41	.0021	107474	2.36	20184	.54	2.90	.165	12.31
30	961	9.69	4.01	13.70	.0021	242548	4.36	34380	.77	5.13	.172	18.83
35	1296	13.23	5.39	18.62	.0020	360937	6.14	48504	1.06	7.20	.154	25.82

† Scaled by 10^{-3}

TABLE 5.3

Statistics on P as a function of N for the ordering produced by the algorithm described in Sections 3 and 4

n	N	No. of off-diagonal blocks	$\lvert P\rvert$	$\frac{\text{off-diagonal blocks}}{\lvert P\rvert}$	$\frac{\lvert P\rvert}{N}$
5	36	59	28	2.1	.777
10	121	242	82	3.0	.678
15	256	510	156	3.3	.609
20	441	871	256	3.4	.580
25	676	1362	384	3.5	.568
30	961	1912	531	3.6	.553
35	1296	2608	710	3.7	.548

The following observations are apparent from the data in Tables 5.1-5.3.

1) The overhead storage appears to grow linearly with N, and the total storage requirement for all data associated with solving the matrix problem grows as N log N. This has two important practical implications. First, it implies that (overhead storage)/(total storage) $\to 0$ as $N \to \infty$, in contrast to most sparse matrix solvers for which this ratio is some constant α, usually with $\alpha > 1$. The second implication is perhaps even more important. It is well known that for this problem, the use of bandmatrix methods (i.e., a banded ordering) implies that total storage requirements grow as $N^{3/2}$. Indeed, the best ordering known to the authors (the so-called diagonal dissection ordering [3]) would imply a storage requirement of O(N log N).

2) The entries in Table 5.2 suggest rather strongly that the execution time of our ordering code for this problem grows no faster than N log N. Similar experiments with other mesh problems demonstrate the same behavior.

3) Table 5.3 contains some interesting statistics about P, the partitioning induced by the repeated application of Theorems 3.3 and 3.5 in our ordering algorithm. It appears that $\lvert P\rvert$ is approaching a "limit" near N/2, and that the number of off-diagonal blocks in each "block column" is approaching about 4. Again, similar experiments with other mesh problems indicate that this behavior is not unique to our test problem.

We now turn to a comparison of our ordering algorithm with an alternative. For comparison, we used the recently developed ordering algorithm due to Gibbs et al.

[5], along with a solver which exploits the variation in the bandwidth of the matrix, as suggested by Jennings [7]. In our tables, we denote results for this ordering-solver combination by BAND, as opposed to the results of our ordering algorithm/linear equation solver package, which we denote by BMD (block-minimum-degree).

From Table 5.4 the total storage for the solution of the test problem, using the band ordering, appears to grow as $O(N\sqrt{N})$, as expected for these meshes. The storage overhead is only $\approx N$. However, in contrast, the total storage used for the solver which uses the BMD ordering appears to grow only as $O(N \ell n N)$, despite the larger overhead. Extrapolating the results of the tables suggests that the storage for the BMD ordering will be less than band storage for $N \gtrsim 2000$, with the saving reaching 50 percent by the time N is around 15000.

TABLE 5.4
Storage Statistics for Band Ordering

n	N	Overhead	Primary	Total	$\frac{\text{Overhead}}{\text{Total}}$	$\frac{\text{Overhead}}{N}$	$\frac{\text{Total}}{N\sqrt{N}}$
5	36	39	191	267	.15	1.08	1.236
10	121	124	1056	1302	.10	1.02	.978
15	256	259	3096	3612	.07	1.01	.882
20	441	444	6811	7697	.06	1.01	.831
25	676	679	12701	14057	.05	1.00	.800
30	961	964	21266	23192	.04	1.00	.778
35	1296	1299	33006	35602	.04	1.00	.763

It should be noted here that the BMD ordering algorithm is implemented in $\approx 30N$ storage (i.e., linear in N). This is important because for $N \gtrsim 1000$ the ordering can be done in the space used later for the factorization.

The entries in Table 5.5 suggest that the band ordering time is $O(N^\rho)$, for $\rho \approx 1.05$, and the solution time is $O(N^2)$. A look at the operations for the factorization time for the BMD and band orderings in Tables 5.2 and 5.5 confirm that the apparent differences in factorization times are indeed due to differences in operation counts and not to program complexity.

Least squares approximations to the total execution times were found for the BMD and band algorithms using as basis functions the orders suggested in Tables 5.2 and 5.5. The results suggest that the BMD algorithm will execute faster than the band algorithm for $N \gtrsim 20{,}000$. For $N \approx 60{,}000$ the results imply that the BMD algorithm is twice as fast. Thus, our ordering/solution package is unlikely to be attractive as a one shot scheme.

TABLE 5.5

Execution Time in Seconds on an IBM 360/75 for Band Ordering

n	N	Ordering time	Ordering $\frac{\text{time}}{N^{1.05}}$	Fact. mult.	Fact. time	Back-solve mult.	Back-solve time	Total soln. time	Soln. $\frac{\text{time}}{N^2}$ †	Ordering plus solution time
5	36	.03	.697	610	.02	382	.01	.03	.0231	.06
10	121	.11	.715	5445	.09	2112	.03	.12	.082	.23
15	256	.23	.681	21880	.29	6192	.06	.35	.053	.58
20	441	.41	.686	61040	.75	13622	.13	.88	.045	1.29
25	676	.66	.705	137800	1.48	25402	.22	1.70	.037	2.36
30	961	.93	.686	270785	2.73	42532	.38	3.11	.034	4.04
35	1296	1.31	.706	482370	4.64	66012	.54	5.18	.031	6.49

† Scaled by 10^{-3}

However, in many situations involving mildly nonlinear and/or time dependent problems, many matrix problems having the same structure, or even the same coefficient matrix, must be solved. In these situations it makes sense to ignore ordering time and compare the methods with respect to factorization time or solution time. If we do this we see from Table 5.6 that the cross-over point for factorization time is at $N \approx 1500$, and for solution time the cross-over point is about $N \approx 2200$.

TABLE 5.6

Ratio of BMD/BAND for Various Quantities

n	N	Total time	Total Store	Fact. time	Soln. time
5	36	10.33	2.01	2.00	3.00
10	121	8.52	1.79	2.44	3.00
15	256	7.34	1.55	2.10	3.33
20	441	5.98	1.38	1.73	2.62
25	676	5.22	1.23	1.59	2.45
30	961	4.66	1.18	1.60	2.03
35	1296	3.98	1.07	1.32	1.96

6. Concluding Remarks

In terms of execution time, our numerical experiments suggest that our ordering algorithm/solution package is attractive for "one-of" problems only if N is extremely large. However, in terms of storage requirements, and if only factorization and solution time is considered, our scheme looks attractive compared to band oriented schemes if N is larger than a few thousand.

Our experiments suggest that for our class of finite element problems, the ordering code executes in $O(N \log N)$ time, and the ordering produced for this problem yields storage and operation counts of $O(N \log N)$ and $O(N^{3/2})$ respectively. For the square mesh problem, these counts are known to be optimal, in the order of magnitude sense [4]. It is interesting to observe that the partitioning produced by the minimum degree algorithm prescribes dissecting sets similar in flavor to those for dissection orderings [3,4]. This leads us to speculate whether the minimum degree algorithm generates asymptotically optimal orderings for general finite element matrix problems. Further research in this area seems appropriate.

7. References

[1] Araldsen, P.O., "The application of the superelement method in analysis and design of ship structures and machinery components", National Symposium on Computerized Structural Analysis and Design, George Washington University, March 1972.

[2] C. Berge, *The Theory of Graphs and its Applications*, John Wiley & Sons Inc., New York, 1962.

[3] Garrett Birkhoff and Alan George, "Elimination by nested dissection" in *Complexity of Sequential and Parallel Algorithms* (J.F. Traub, editor), Academic Press, New York, 1973, pp.221-269.

[4] Alan George, "Nested dissection of a regular finite element mesh", SIAM J. Numer. Anal., 10 (1973), pp.345-363.

[5] N.E. Gibbs, W.G. Poole and P.K. Stockmeyer, "An algorithm for reducing the bandwidth and profile of a sparse matrix", SIAM J. Numer. Anal., to appear.

[6] M.J.L. Hussey, R.W. Thatcher and M.J.M. Bernal, "Construction and use of finite elements", J. Inst. Math. Appl., 6(1970), pp.262-283.

[7] A. Jennings, "A compact storage scheme for the solution of symmetric linear simultaneous equations", Computer J.,9(1966), pp.281-285.

[8] D.E. Knuth, *The Art of Computer Programming*, vol.I (Fundamental Algorithms), Addison Wesley, 1968.

[9] S.V. Parter, "The use of linear graphs in Gauss elimination", SIAM Rev., 3(1961), pp.364-369.

[10] D.J. Rose, "A graph-theoretic study of the numerical solution of sparse positive definite systems of linear equations", in *Graph Theory and Computing*, edited by R.C. Read, Academic Press, New York, 1972.

[11] B. Speelpenning, "The generalized element method", unpublished manuscript.

[12] K.L. Stewart and J. Baty, "Dissection of structures", J. Struct. Div., ASCE, Proc. paper No.4665, (1966), pp.75-88.

[13] James H. Wilkinson, *The Algebraic Eigenvalue Problem*, Clarendon Press, Oxford, 1965.

[14] O.C. Zienkiewicz, *The Finite Element Method in Engineering Science*, McGraw Hill, London, 1970.

METHODES D'ELEMENTS FINIS EN VISCOELASTICITE PERIODIQUE

G. GEYMONAT
Ist. Matematico, Politecnico
Torino

M. RAOUS
C.N.R.S. - LMA
Marseille

INTRODUCTION

Les methodes d'éléments finis ont été appliquées avec succès aux problèmes de Cauchy quasi-statiques viscoélastiques nécessairement isotropes (v. par example, Bazant-Wu [2], Carpenter [6], Zienkiewicz [14],....).

Nous présentons ici une estimation d'erreur pour un problème de viscoélasticité périodique (modèle de type Maxwell) pour un corps vieillissant, pas forcément isotrope; l'étude numérique du problème considéré ici est préliminaire au traitement numérique de l'inéquation variationnelle introduite en [4]. Le problème continu donne lieu à une équation différentielle opérationnelle avec un opérateur borné, ce qui nous permet de trouver une condition de stabilité (pour une méthode explicite) indépendante de la discrétisation spatiale. Les estimations d'erreur sont obtenues en adaptant les résultats de Crouzeix [7] .

L'application de la théorie developpée ici au comportement d'une aube de turbine, dont les paramètres caracteristiques (module de Young, temps de relaxation, coefficient de dilatation) sont fonctions de x,t par intermédiaire du champs de temperature, sera donnée ailleurs.

1. Le problème continu

1.1. Soit Ω un ouvert borné régulier de $\mathbb{R}^n$ (dans la pratique on prendra $n = 2$ ou $n = 3$) de frontière Γ assez régulière ; soit Γ_o un sous-ensemble ouvert de Γ de mesure > 0 et soit $\Gamma_1 = \Gamma \setminus \Gamma_o$.

$(H^1(\Omega))^n$ est l'espace des vecteurs de déplacement $u = (u_1, \ldots, u_n)$ avec $u_i \in H^1(\Omega)$ pour $i = 1, \ldots, n$; on pose

$$U = \{ v \in (H^1(\Omega))^n \; ; \; v_i / \Gamma_o = 0 \text{ pour } i = 1, \ldots, n \}$$

muni de la structure hilbertienne induite par $(H^1(\Omega))^n$.
Soit E l'espace des tenseurs des déformations, i.e. des matrices $e = (\!(e_{ij})\!)_{i,j=1,\ldots,n}$ symétriques, à éléments $e_{ij} \in L^2(\Omega)$ et soit S l'espace des tenseurs des contraintes, i.e. des matrices $s = (\!(s_{ij})\!)_{i,j=1,\ldots,n}$ symétriques, à éléments $s_{ij} \in L^2(\Omega)$.

Les espaces E et S sont mis en dualité séparante par la forme bilinéaire

$$(1.1) \qquad < e \, , s > \; = \sum_{i,j=1}^{n} \int_\Omega e_{ij}(x) \, s_{ij}(x) \, dx$$

qui représente, d'un point de vue mécanique, l'opposé du travail de la contrainte s dans la déformation e ; d'un point de vue mathématique, les espaces E et S peuvent aussi être identifiés et alors (1.1) est le produit scalaire.

Soit Φ l'espace des charges, dual de U par rapport à la forme bilinéaire séparante $\ll u \, , \varphi \gg$ qui représente le travail de la charge φ par rapport au déplacement u ; si φ_Ω et φ_{Γ_1} sont des charges volumiques dans Ω et surfaciques sur Γ_1 assez régulières, alors

$$\ll u \, , \varphi \gg \; = \int_\Omega u(x) \, \varphi_\Omega(x) \, dx + \int_{\Gamma_1} u(x) \, \varphi_{\Gamma_1}(x) \, d\Gamma$$

il est aisé de vérifier qu'une telle formule est valable par exemple pour $\varphi_\Omega \in (L^2(\Omega))^n$ et $\varphi_{\Gamma_1} \in (L^2(\Gamma_1))^n$; mais on peut lui donner un sens dans une situation plus générale.

On note par D l'opérateur gradient symétrique :

$$(1.2) \qquad Du = \left(\left(\frac{1}{2} \left(\frac{\partial u_i}{\partial x_j} + \frac{\partial v_j}{\partial x_i} \right) \right) \right)_{i,j=1,\ldots,n}$$

il s'agit d'un opérateur linéaire et continu de U dans E et on suppose que D soit <u>injectif</u> i. e. qu'il n'existe pas de champs de déplacements de solide. Son transposé ${}^tD : S \longrightarrow \Phi$ défini par

$$(1.3) \qquad \langle Dv, s \rangle = \langle\langle v, {}^tDs \rangle\rangle \qquad \forall v \in U \quad , \forall s \in S$$

est linéaire, continu et <u>surjectif.</u>

L'équation d'équilibre d'un milieu continu en petites déformations écrite formellement ${}^tDs = \varphi$, signifie alors

$$\sum_{j=1}^{n} \frac{\partial}{\partial x_j} s_{ij} = \varphi_{i\Omega} \qquad \text{dans } \Omega$$

$$\sum_{j=1}^{n} s_{ij} \cos(\vec{n}, x_j) = \varphi_{i\Gamma_1} \qquad \text{sur } \Gamma_1$$

où $\vec{n}$ désigne la normale à Γ_1 extérieure à Ω et cette interprétation peut être rendue rigoureuse à l'aide des méthodes de Lions-Magenes [11].

$J = \ker({}^tD) = \{ s \in S ; \langle Dv, s \rangle = 0 \ \forall v \in U \}$ est l'ensemble des champs d'autocontraintes.

$I = \mathrm{Im}(D) = \{ e \in E ; \text{ il existe } v \in U \text{ avec } Dv = e \}$ est l'ensemble des champs de déformations cinématiquement admissibles.

Il est aisé de vérifier que I et J sont des sous-espaces fermés mutuellement polaire pour la dualité (1. 1) (ou bien orthogonaux pour le produit scalaire (1. 1) si on identifie E et S et dans ce cas leur somme directe hilbertienne coïncide avec tout l'espace).

1. 2. Les matériaux viscoélastiques linéaires sont des matériaux avec mémoire au sens que le tenseur des contraintes s est lié au tenseur de déformation e par des lois qui s'expriment à l'aide d'équations différentielles linéaires du type (*)

...

(*) D'un point de vue thermodynamique, il est parfois plus intéressant d'introduire les paramètres cachés (voir par exemple Germain [9], chap. VIII, §7).

(1.4) $$\sum_{k=0}^{m} R_k \frac{\partial^k s}{\partial t^k} = \sum_{\ell=0}^{m'} A_\ell \frac{\partial^\ell e}{\partial t^\ell}$$

En général, les matrices R_k et A_ℓ dépendent de $x \in \Omega$ et du temps t et cette dépendance du temps est essentielle quand l'état du matériau dépend du vieillissement ou de la température.

Parmi les cas plus étudiés et utilisés dans la pratique, on a :

a) le modèle de Kelvin-Voigt du matériau à mémoire courte où $m = 0$ et $m' = 1$;

b) le modèle de Maxwell du matériau à mémoire longue où $m = m' = 1$ et $A_0 = 0$.

La loi de comportement d'un matériau viscoélastique peut être aussi donnée en intégrant l'équation différentielle (1.4) en obtenant ainsi une relation intégrale entre les tenseurs s et e ; pour l'étude mathématique de ces problèmes, on peut voir, par exemple, Babuška-Hlaváček [1], Duvaut-Lions [8], Bouc-Geymonat-Jean-Nayroles [3].

1.3. Dans la suite, nous considérerons, pour le modèle de Maxwell, le problème quasi-statique de Cauchy et le problème périodique.

Soit donc $T > 0$ fixé (avec éventuellement $T = +\infty$). Soient donnés $e_0(t) \in H^1(0,T ; E)$, $\varphi(t) \in H^1(0,T ; \Phi)$, $u_0 \in U$, $s_0 \in S$ vérifiant la condition de compatibilité ${}^tDs_0 = \varphi(0)$.

Le problème de Cauchy quasi-statique est alors le suivant :

Problème (P1) : Trouver $u(t) \in H^1(0,T ; U)$, $s(t) \in H^1(0,T ; S)$ tels que

(1.5) $$e(t) = Du(t) + e_0(t)$$

(1.6) $$\dot{s}(t) + R(t)\, s(t) = A(t)\, \dot{e}(t) \quad (*)$$

(1.7) $${}^tDs(t) = \varphi(t)$$

(1.8) $$s(0) = s_0$$

(1.9) $$u(0) = u_0$$

...

(*) Ici et dans la suite $\dot{\psi} = \frac{d\psi}{dt}$.

Ici $A(t) = ((a_{ijkh}(x,t)))_{i,j,k,h=1,\dots,n}$, (il s'agit de la matrice de raideur) et $R(t) = ((r_{ijkh}(x,t)))_{i,j,k,h=1,\dots,n}$ (où interviennent simultanément des termes de raideur et de viscosité) sont des matrices à coefficients réguliers (i. e. au moins mesurables et bornés) en $(x,t) \in \Omega \times]0,T[$ et vérifiant les conditions

(1.10) $\quad a_{ijkh} = a_{jihk} = a_{khij}$ (symétrie)

(1.11) $\quad$ il existe $\alpha > 0$ tel que p.p. en $(x,t) \in \Omega \times]0,T[$

$$\sum_{i,j,k,h=1}^{n} a_{ijkh}(x,t)\,\xi_{ij}\,\xi_{kh} \geq \alpha \sum_{i,j=1}^{n} |\xi_{ij}|^2$$

(1.12) $\quad r_{ijkh} = r_{jikh}$

Dans le cas où $e_o(t)$, $\varphi(t)$, $R(t)$, $A(t)$ sont T-périodiques, nous considérerons aussi le problème périodique quasi-statique suivant

Problème $(P_{\#}1)$: Trouver $u(t) = at + p(t)$, avec $a \in U$ et $p(t) \in H^1_{\#}(0,T\,;U) = \{ q \in H^1(0,T\,;U)\,;\, q(0) = q(T) \}$, et $s(t) \in H^1_{\#}(0,T\,;S)$ tels que (1.5), (1.6) et (1.7) soient vraies.

Remarquons de suite que $(P_{\#}1)$ n'admet pas une solution unique en u mais seulement en $\dot{u}$; $p(t)$ étant donné à une constante près, on peut choisir $p(t)$ de valeur moyenne nulle.

1.4. On peut transformer les équations (1.5), (1.6) et (1.7) à l'aide des résultats suivants.

Lemme 1.1 - Sous les hypothèses (1.10) et (1.11), l'opérateur $H(t) = {}^tD\,A(t)\,D : U \longrightarrow \Phi$ est un isomorphisme algébrique et topologique, dont l'inverse sera noté par $H^{-1}(t)$, et il vérifie aussi

(1.13) $\quad \langle\langle u\,,Hv \rangle\rangle = \langle\langle v\,,Hu\rangle\rangle \qquad \forall u,v \in U$

Il s'agit du résultat classique d'élasticité linéaire statique (voir, par exemple, Campanato [5], Duvaut-Lions [8]).

...

Lemme 1.2- Soient vérifiées les hypothèses (1.10) et (1.11). L'opérateur intégral

(1.14) $O(t) = A(t)\, DH^{-1}(t)\, {}^tD : S \longrightarrow S$

est linéaire et continu et on a

(1.14') $O(t)\, A(t)\, Du = A(t)\, Du \qquad \forall u \in U$

(1.14'') ${}^tD\, O(t) s = {}^tDs \qquad \forall s \in S$

(1.15) $O(t)\, O(t) s = O(t)\, s \qquad \forall s \in S$

(1.15') $O(t) s = 0 \qquad \forall s \in J$

Il est alors aisé de vérifier le résultat suivant.

Lemme 1.3- Soient vérifiées les conditions (1.10) et (1.11). Les équations (1.5), (1.6) et (1.7) sont équivalentes à

(1.16) $e = Du + e_o$

(1.17) $H(t)\, \dot{u}(t) = \dot{\varphi}(t) - {}^tDA(t)\, \dot{e}_o(t) + {}^tDR(t)\, s(t)$

(1.18) $\dot{s}(t) + (\mathbb{1} - O(t))\, R(t)\, s(t) = (\mathbb{1} - O(t)) A(t)\, \dot{e}_o(t) + A(t) DH^{-1}(t)\, \dot{\varphi}(t)$

(1.19) ${}^tDs(0) = \varphi(0)$

où $\mathbb{1}$ est l'identité dans S.

Remarquons que la condition (1.19) est automatiquement vérifiée seulement dans le cas du problème de Cauchy (à cause de la condition de compatibilité entre s_o et $\varphi(0)$).

Les problèmes de Cauchy et périodique sont donc ramenés aux problèmes suivants dans la seule inconnue $s(t)$.

(P2) Problème de Cauchy. Trouver $s(t) \in H^1(0,T\,;S)$ qui vérifie (1.18) et (1.8).

($P_{\#}$2) Problème périodique. Trouver $s(t) \in H^1_{\#}(0,T\,;S)$ qui vérifie (1.18) et (1.19).

Les opérateurs $(\mathbb{1}-O(t))\,R(t) : S \longrightarrow S$ étant linéaires et continus, l'équation (1.18) est une équation différentielle ordinaire dans l'espace de Hilbert S. On en déduit donc que le problème de Cauchy admet toujours une solution unique et on connaît des conditions nécessaires et suffisantes "abstraites" pour que le problème périodique admette une solution unique et pour que l'on ait stabilité asymptotique de la solution du problème de Cauchy sur la solution du problème périodique. Toutefois, ces conditions ne sont pas aisément exploitables et interprétables du point de vue de la Mécanique. Pour cela, dans la suite, nous utiliserons plutôt le point de vue des équations différentielles variationnelles de Lions [10].

1.5. En tenant compte de (1.14') et (1.14''), il paraît naturel de décomposer l'équation (1.18). Pour cela, soit $J^{\perp}$ l'orthogonal de J dans S, de telle façon que

$$s = s_J + s_{J^{\perp}} \qquad \forall s \in S \qquad (*)$$

On en déduit aisément

$$\frac{d}{dt}\, s(t) = \frac{d}{dt}\,(s_J(t) + s_{J^{\perp}}(t)) = \left(\frac{ds}{dt}\right)_J + \left(\frac{ds}{dt}\right)_{J^{\perp}}$$

et alors les équations (1.18) et (1.19) entraînent, puisque O(t) est un projecteur sur $J^{\perp}$ de noyau J (lemme 1.2) :

(1.20) $\quad \dot{s}_{J^{\perp}}(t) = A(t)\, DH^{-1}(t)\, \dot{\varphi}(t)$

(1.21) $\quad \dot{s}_J(t) + (\mathbb{1}-O(t))\, R(t)\, s_J(t) = (\mathbb{1}-O(t))\, A(t)\, \dot{e}_o(t) - (\mathbb{1}-O(t))\, R(t)\, s_{J^{\perp}}(t)$

(1.22) $\quad {}^tD\, s_{J^{\perp}}(0) = \varphi(0)$

Puisque ${}^tD$ est injectif sur $J^{\perp}$ l'équation (1.20) équivaut à

$$\frac{d}{dt}\left({}^tD s_{J^{\perp}}(t)\right) = \dot{\varphi}\,(t)$$

et donc, à l'aide de (1.22), on trouve

(1.23) $\quad {}^tD\, s_{J^{\perp}}(t) = \varphi(t)$

...

(*) Si on identifie S avec E alors on a $J^{\perp} = I$ par un résultat classique d'algèbre linéaire.

d'où on calcule explicitement

(1. 24) $$s_{J^\perp}(t) = A(t)\, DH^{-1}(t)\, \varphi(t)$$

On est ainsi ramené à l'étude des problèmes de Cauchy et périodique seulement pour $s_J(t)$, problèmes qui s'énoncent de la façon suivante.

(P3) PROBLEME DE CAUCHY. Trouver $s_J(t) \in H^1(0, T\,;\, J)$ tel que

(1. 25) $$\dot{s}_J(t) + (\mathbb{1} - O)\, R(t)\, s_J(t) = g(t)$$

(1. 26) $$s_J(0) = s_{oJ}$$

où $g(t) \in L^2(0, T\,;\, J)$ est donnée par

(1. 27) $$g(t) = (\mathbb{1} - O(t))\, A(t)\, \dot{e}_o(t) - (\mathbb{1} - O(t))\, R(t)\, A(t)\, DH^{-1}(t)\, \varphi(t)$$

($P_{\#}3$) PROBLEME PERIODIQUE. Trouver $s_J(t) \in H^1_{\#}(0, T\,;\, J)$ telle que (1. 25) est satisfaite avec $g(t) \in L^2_{\#}(0, T\,;\, J)$ donnée par (1. 27).

1.6. Pour appliquer aux problèmes de Cauchy et périodique les méthodes variationnelles, il est utile de remarquer

(1. 28) pour $\eta \in J$ et $s \in S$ on a $\langle O(t)\, R(t) s\, ,\, A^{-1}(t)\, \eta \rangle = 0$

A l'aide de cette remarque, on obtient le résultat suivant.

Théorème 1. 1 - Soient vérifiées les conditions (1. 10), (1. 11) et (1.12) et, si on pose $\dot{A}(t) = ((\dot{a}_{ijkh}(x, t)))_{i, j, k, h=1, \dots, n}$, soit vérifiée la condition

(1. 29)
L'opérateur $A^{-1}R + \frac{1}{2} A^{-1} \dot{A} A^{-1} = B = ((b_{ijkh}(x, t)))_{i, j, k, h}$ est coercif sur l'espace des matrices symétriques, i. e. il existe $\beta > 0$ tel que

$$\sum_{i, j, k, h=1}^{n} b_{ijkh}(x, t)\, \xi_{ij}\, \xi_{kh} \geq \beta \sum_{i, j=1} |\xi_{ij}|^2 \quad \text{p. p. en } (x, t)$$

pour toute $((\xi_{ij}))_{i, j=1, \dots, n}$ matrice symétrique.

Alors le problème de Cauchy (P3) et le problème périodique ($P_{\#}3$) admettent une solution unique.

De plus, dans le cas où A, R, e_o, φ sont périodiques, si $w \in H^1([0,+\infty[\ ;J)$ est la solution du problème de Cauchy avec donnée initiale $w_o \in J$ et s_J la solution du problème périodique, on a pour tout $t \geqslant 0$

$$(1.30) \quad \|w(t) - s_J(t)\|_S \leq c_1 \exp(-\frac{\beta}{\alpha^*} t) \{ \|g\|_{H^1_{\#}(0,T,S)} + \|w_o\|_S \}$$

où α^* est tel que $< \sigma, A^{-1}\sigma > \geqslant \alpha^* \|\sigma\|_S^2$, $\forall \sigma \in S$.

La démonstration du théorème se fait en appliquant la méthode de Lions [10] après avoir obtenu la formulation faible de (1. 25) en multipliant les deux membres par $A^{-1}(t)\,\eta(t)$ avec $\eta(t)$ prenant ses valeurs dans J et en utilisant (1. 28).

Remarque 1 .1. Le théorème 1. 1 et la formule (1. 30) nous disent que les problèmes (P3) et ($P_{\#}3$) admettent une solution unique et la solution du problème de Cauchy est asymptotiquement stable à la solution du problème périodique.

Remarque 1.2 Si $R(t) = 0$ et $\dot{A}(t) = 0$ (cas élastique à coefficients constants), alors B est nul et (1. 29) n'est plus vérifiée. Dans ce cas, le problème($P_{\#}3$) admet quand même une solution unique, mais la solution du problème de Cauchy n'est plus asymptotiquement stable à la solution du problème périodique.

2. Semi-discrétisation en t par une méthode explicite

2.1. Pour résoudre numériquement les problèmes (P1) ou ($P_{\#}1$), il faut, en particulier, résoudre l'équation différentielle (1.6) ; pour cela, on choisit une méthode explicite de Runge-Kutta et, après en avoir étudié la stabilité, on applique pour les estimations d'erreur les résultats de la thèse de Crouzeix [7].

Pour ne pas alourdir l'exposé, nous allons considérer seulement la méthode de Euler, même si tous les résultats sont valables en général, appliquée à l'équation différentielle (1.18) que l'on écrira :

(2.1) $$\dot{s}(t) + \mathcal{R}(t)\, s(t) = f(t)$$

avec

(2.2) $$\mathcal{R}(t) = (\mathbb{1} - O(t))\, R(t)$$

(2.3) $$f(t) = (\mathbb{1} - O(t))\, A(t)\, \dot{e}_o(t) + A(t)\, D\, H^{-1}(t)\, \dot{\varphi}(t)$$

Soit donc $[0, T]$ l'intervalle de temps où nous étudions (2.1) ; soit $M \geqslant 2$ fixé, $\Delta t = \frac{T}{M}$ et $t_i = i\,\Delta t$. On considère le système :

(2.4) $$s_{n+1} - s_n + \Delta t\, \mathcal{R}_n\, s_n = \Delta t\, f_n \qquad n = 0, \ldots, M-1$$

(2.5) $$\begin{cases} s_o \text{ donné} & \text{(Pb de Cauchy)} \\ s_o = s_M \text{ avec } {}^tDs_o = \varphi(0) & \text{(Pb périodique)} \end{cases}$$

où $\mathcal{R}_n = \mathcal{R}(t_n)$ (ou bien en est une bonne approximation)

et $f_n = f(t_n)$.

Comme il a été fait dans le cas du problème continu, on peut projeter (2.4) sur J et sur $J^{\perp}$ et on obtient alors les équations

(2.4 $J^{\perp}$) $$s_{n+1,J^{\perp}} - s_{n,J^{\perp}} = \Delta t\, f_{n,J^{\perp}} = \Delta t\, A(t_n)\, DH^{-1}(t_n)\, \dot{\varphi}(t_n)$$

(2.4 J) $$s_{n+1,J} - s_{n,J} + \Delta t\, \mathcal{R}_n\, s_{n,J} = \Delta t\, f_{n,J} = \Delta t(\mathbb{1} - O(t_n))A(t_n)\dot{e}_o(t_n)$$

L'équation (2.4 J) peut être écrite de façon faible en utilisant (1.28) (en posant $A_n = A(t_n)$ et $R_n = R(t_n)$)

...

$$(2.6)\quad \langle A_n^{-1}\sigma\ , s_{n+1,J}-s_{n,J}\rangle + \Delta t\langle A_n^{-1}\sigma, R_n s_{n,J}\rangle =$$

$$= \langle A_n^{-1}\sigma\ , f_{n,J}\rangle \text{ pour tout } \sigma \in J.$$

Il faut aussi projecter la condition (2.5); remarquons que l'on obtient la condition initiale ${}^t D S_{0,J^\perp} = \phi(0)$ dans le cas du problème périodique.

2.2. Stabilité. Pour étudier la stabilité de (2.6) dans le cas du problème périodique, on choisit $\sigma = s_{n,J}$ et en sommant on obtient:

$$\sum_{k=0}^{M-1} \langle (A_k^{-1}-A_{k+1}^{-1}) s_{k+1,J},\ s_{k+1,J}\rangle +$$

$$-\sum_{k=0}^{M-1} \langle A_k^{-1}(s_{k+1,J}-s_{k,J}),\ s_{k+1,J}-s_{k,J}\rangle +$$

$$+2\Delta t \sum_{k=0}^{M-1} \langle A_k^{-1} s_{kJ}, R_k s_{kJ}\rangle = 2\Delta t \sum_{k=0}^{M-1} \langle A_k^{-1} s_{k,J}\ , f_{k,J}\rangle$$

En remarquant que

$$A_k^{-1}-A_{k+1}^{-1} = A_{k+1}^{-1}\ (A_{k+1}-A_k)A_k^{-1} = \Delta t\ A_{k+1}^{-1}\ \dot{A}(\xi_k)A_k^{-1}$$

avec $\xi_k \in [t_k, t_{k+1}]$ et en faisant les estimations habituelles dans ce problème, on aboutit au résultat suivant de stabilité.

Théorème 2.1 - Sous les hipothèses du théorème 1.1, il existe $\Delta t_o > 0$ et $L>0$ (dépendant seulement de β, α^* et $\| R(t) \|$) tels que pour $\Delta t < \Delta t_o$, on a

$$(2.7)\quad \Delta t \sum_{k=0}^{M-1} \| s_{kJ} \|^2 \leq L\, \| f_J \|^2_{H^1_\#(0,T;S)}.$$

On a un résultat analogue de stabilité pour le problème de Cauchy.

2.3 L"équation (2.4 $J^\perp$) est faiblement stable puisqu'on a l'estimation

$$(2.8)\quad \Delta t \sum_{k=1}^{M} \| s_{k,J^\perp} \|^2 \leq T^2 L_1 \| f_{J^\perp} \|^2_{H^1_\#(0,T;S)} + L_2\, T\, \| \phi(0) \|^2$$

et une estimation analogue pour le problème de Cauchy.

2.4. Estimation d'erreur locale. On pose pour k=0,1,...,M-1

$$E_k = \frac{1}{\Delta t}(s(t_{k+1})-s(t_k))+\mathcal{R}_k s(t_k)-f_k$$

et on obtient sous des hypothèses naturelles de régularité pour $\mathcal{R}(t)$ et $f(t)$ (par ex. $f \in H^2(O,T,S),\ldots$)

$$(2.9)\qquad \|E_k\|_S \leq \frac{\Delta t}{2}\left\|\frac{d^2s}{dt^2}\right\|_{L^\infty(t_k,t_{k+1};S)} \leq C\Delta t\ |||f|||$$

avec C indépendante de k=0,..., M-1 et $|||f|||$ désigne une norme convenable de f (par la norme dans $H^2(O,T,S),\ldots$).

Remarquons aussi que dans le cas periodique on a $E_o = E_M$.

2.5 Estimation globale d'erreur. Elle s'obtient à l'aide de l'estimation locale (2.9) et de la stabilité; on obtient par example dans le cas du problème periodique:

$$\Delta t \sum_{k=0}^{M-1} \|s(t_k)-s_k\|_S^2 \leq C_1(\Delta t)^2 T\ \{\ |||f_J|||^2 + C_2 T^2\ |||f_{J^\perp}|||^2\}$$

Remarque 2.1. Dans le cas precisé à la remarque 1.2 on a une estimation de stabilité du type (2.8) sur S_{kJ} également.

Remarque 2.2. Si on utilise la méthode implicite

$$s_{n+1}-s_n+\Delta t\ \mathcal{R}_{n+1}s_{n+1} = \Delta t f_{n+1} \qquad n = 0,\ldots,M-1$$

par projection sut $J^\perp$ on trouve

$$s_{n+1J^\perp} - s_{nJ^\perp} = \Delta t f_{n+1J^\perp}$$

et donc on a faible stabilité au sens de l'estimation (2.8).

3. Discrétisation totale

3.1. Pour approcher la solution $(\dot{u}, s)$ des problèmes (P1) et $(P_{\#}1)$ pour chaque entier N, assez grand et destiné à tendre $\to +\infty$, on se donne :

- un sous-espace U_N de U de dimension finie ;
- un sous-espace Φ_N de Φ de dimension finie ;
- un sous-espace E_N de E de dimension finie ;
- un sous-espace S_N de S de dimension finie .

On fera l'hypothèse suivante de compatibilité entre les données :

(3.1) $$D(U_N) \subset E_N \quad \text{et} \quad D(U_N) \neq E_N$$

et puisque D est <u>injectif</u>, l'opérateur transposé ${}^tD_N : S_N \to \Phi_N$ défini par

(3.2) $$\langle\langle v, {}^tD_N s \rangle\rangle = \langle Dv , s \rangle \qquad \forall v \in U_N , \forall s \in S_N$$

est surjectif.

...

Si les espaces E_N et U_N sont obtenus par une méthode d'éléments finis droits, l'hypothèse (3.1) signifie que sur chaque simplexe K de la triangulation $\mathcal{C}_N$ de Ω, on a intérêt à prendre pour E_N des polynômes de degré inférieur de 1 à celui des polynômes de U_N.

De même, on définit les opérateurs $A_N(t) : E_N \longrightarrow S_N$ et $R_N(t) : S_N \longrightarrow S_N$ approchant A et R et l'on suppose que A_N soit <u>uniformément elliptique</u> au sens suivant :

(3.3) <u>il existe</u> $\tilde{\alpha} > 0$, <u>pour tout</u> N, $\forall e \in E_N : \langle e, A_N e \rangle \geq \tilde{\alpha} \| e \|_E^2$

Si on néglige les erreurs dues à l'intégration numérique, on peut définir $A_N(t)$ et $R_N(t)$ par les formules suivantes :

(3.4) $\langle f, A_N(t)\, e_N \rangle = \langle f, A(t)\, e \rangle \qquad \forall f, e \in E_N$

(3.5) $\langle e, R_N(t) s \rangle = \langle e, R(t)\, s \rangle \qquad \forall e \in E_N, \forall s \in S_N$

et alors (3.3) est automatiquement vérifiée avec $\tilde{\alpha} = \alpha$ de (1.11).

On peut alors approcher le système (1.5), (1.6) et (1.7) par :

(3.6) $e_N(t) = D\, u_N(t) + e_{oN}(t)$

(3.7) $\dot{s}_N(t) + R_N(t)\, s_N(t) = A_N(t)\, \dot{e}_N(t)$

(3.8) ${}^tD_N\, s_N = \varphi_N$

où $e_{oN}(t) \in E_N$, pour tout t, et $\varphi_N(t) \in \Phi_N$, pour tout t, sont des approximations de e_o et φ.

Reprenant le raisonnement développé au n. 1.4, on introduit

$$H_N(t) = {}^tD_N\, A_N(t)\, D : U_N \longrightarrow \Phi_N$$

$$O_N(t) = A_N(t)\, D H_N^{-1}(t)\, {}^tD_N : S_N \longrightarrow S_N$$

et on voit que (3.6), (3.7) et (3.8) sont équivalentes à (3.6) et

(3.9) $H_N(t)\, \dot{u}_N(t) = \dot{\varphi}_N(t) - {}^tD_N\, A_N\, \dot{e}_{oN} + {}^tD_N\, R_N\, s_N$

(3.10) $\dot{s}_N(t) + (\mathbb{1} - O_N)\, R_N\, s_N = (\mathbb{1} - O_N)\, A_N\, \dot{e}_{oN} + A_N\, D\, H_N^{-1}(t)\, \dot{\varphi}_N(t)$

(3.11) $$ {}^tD_N\, s_N(0) = \varphi_N(0) $$

où, évidemment, $\mathbb{1}$ est l'opérateur identité dans S_N.

On voit que (3.9) permet de déterminer $\dot{u}_N$ une fois trouvé $s_N(t)$ et les résultats classiques d'estimation d'erreur (voir par exemple Strang-Fix [13], Raviart [12]), nous donnent pour tout t

(3.12) $$ \| \dot{u}(t) - u_N(t) \|_U \leqslant c_1 \inf_{v \in U_N} \| \dot{u}(t) - v \| + c_2 \| s(t) - s_N(t) \| $$

et donc on est ramené à estimer l'erreur de l'approximation de $s(t)$ par $s_N(t)$.

3.2. On définit maintenant $J_N = \{ s \in S_N \,;\, {}^tD_N s = 0 \} = \{ s \in s_N \,; < s, Dv > = 0 \ \forall v \in U_N \}$ il s'agit d'un sous-espace de S_N non réduit à $\{0\}$ grâce à la condition (3.1). Soit $J_N^\perp$ son orthogonal dans S_N ; on peut, comme en 1.5, décomposer $s_N = (s_N)_J + (s_N)_{J^\perp}$; par commodité on écrira $s_N^{(1)} = (s_N)_J$ et $s_N^{(2)} = (s_N)_{J^\perp}$. alors $s_N^{(2)}(t)$ est déterminé par :

(3.13) $$ {}^tD_N\, s_N^{(2)}(t) = \varphi_N(t) $$

i.e. par

(3.13') $$ < s_N^{(2)}(t)\,,\, D\,w > = \ll \varphi_N, w \gg \qquad \forall w \in U_N $$

et $s_N^{(1)}$ est solution de l'équation différentielle

(3.14) $$ \dot{s}_N^{(1)} + (\mathbb{1} - O_N)\, R_N\, s_N^{(1)} = g_N(t) $$

où

(3.15) $$ g_N(t) = (\mathbb{1} - O_N)\, A_N\, \dot{e}_{oN} - (\mathbb{1} - O_N)\, R_N\, s_N^{(2)} $$

En ce qui concerne (3.14), remarquons, avant tout, que sous l'hypothèse :

(3.16) $\exists\, \tilde{\beta} > 0$ <u>pour tout</u> N, $\forall s \in S_N : < \{ A_N^{-1} R_N + \frac{1}{2} A_N^{-1} \dot{A}_N A_N^{-1} \} s, s > \geqslant \tilde{\beta} \|s\|_S^2$

les problèmes périodiques et de Cauchy associés à (3.14) admettent une solution unique. De plus, la solution du problème de Cauchy est asymptotiquement convergente à la solution du problème périodique, au sens qu'on a une estimation analogue à (1.30):

(3.17) $\|w_N^{(1)}(t)-s_N^{(1)}(t)\| \leq c_2 \exp(-\frac{\tilde{\beta}}{\alpha^*}t)\{\|g_N\|_{H^1_{\#}(0,T;S)} + \|w_{o,N}\|\}$

où c_2 ne dépend pas de N.

3.3. On résout numériquement (3.10) à l'aide d'une methode explicite de Runge-Kutta; pour simplifier nous considérons seulement la méthode d'Euler, même si les réultats restent valables dans le cas général que on peut développer en suivant Croizeix [7].

Par application de la méthode d'Euler à (3.10) et par projection sur J_N et $J_N^{\perp}$ on obtient, comme en 2.1, les equations pour n=0,...,M-1:

(3.18) $s^{(2)}_{N,n+1} - s^{(2)}_{N,n} = \Delta t\, f^{(2)}_{N,n}$

(3.19) $s^{(1)}_{N,n+1} - s^{(1)}_{N,n} + \Delta t\, \mathcal{R}_{N,n}\, s^{(1)}_{N,n} = \Delta t\, f^{(1)}_{N,n}$

auxquelles il faut ajouter les conditions initiales et/ou de periodicité.

Pour étudier la stabilité de (3.19) dans le cas du problème periodique on multiplie scalairement par $A^{-1}_{N,n}\, s^{(1)}_{N,n}$ et en sommant on obtient:

$$2\Delta t \sum_{k=0}^{M-1} \langle (A^{-1}_{N,k} R_{N,k} + \frac{1}{2} A^{-1}_{N,k} \dot{A}(\xi_{k-1}) A^{-1}_{N,k-1})\, s^{(1)}_{N,k}, s^{(1)}_{N,k} \rangle$$

$$- \sum_{k=0}^{M-1} \langle A^{-1}_{N,k}(s^{(1)}_{N,k+1} - s^{(1)}_{N,k}), s^{(1)}_{N,k+1} - s^{(1)}_{N,k} \rangle =$$

$$= 2\Delta t \sum_{k=0}^{M-1} \langle A^{-1}_{N,k}\, s^{(1)}_{N,k}, f^{(1)}_{N,k} \rangle$$

Puisque $R(t): S \longrightarrow S$ est uniformément borné on en déduit que aussi la famille $R_N(t)$ est bornée uniformément en t et N et donc le théorème 2.1 est encore valable et de plus <u>la constante</u> Δt_o (assurant la stabilitè) <u>ne dépend pas de l'approximation spatiale.</u>

3.4 - Les estimations d'erreur s'obtiennent aisément puisque

$$\|s(t_k) - s_{N,k}\|_S \leq \|s(t_k) - s_N(t_k)\|_S + \|s_N(t_k) - s_{N,k}\|_S$$

et donc l'erreur est la somme de l'erreur liée à la méthode d'éléments finis spatials et de l'erreur liée à la méthode Runge-Kutta.

L'estimation de l'erreur dans les méthodes d'éléments finis est maintenant bien connue (voir par ex. Raviart [12] et Strang-Fix [13]) et l'erreur liée à la méthode de Runge-Kutta a été evaluée au N. 2.

BIBLIOGRAPHIE

[1] I. BABUŠKA, I. HLAVÁČEK - On the existence and uniqueness of solution in the theory of viscoelasticity. Archivium Mechaniki Stosowanej, 18, (1966), 47-84.

[2] Z.P. BAŽANT, S.T. WU - Rate-type creep law for aging concrete based on Maxwell chain. Rilem Materiaux et Constructions, 7, (1974), 45-60.

[3] R. BOUC, G. GEYMONAT, M. JEAN, B. NAYROLES - Solution périodique du problème quasi-statique d'un solide viscoélastique à coefficients périodiques. Journal de Mécanique, 14, (1975), 609-637.

[4] R. BOUC, G. GEYMONAT, M. JEAN, B. NAYROLES - Cauchy and periodic unilateral problems for aging linear viscoelastic materials. A paraître dans : Journal Math. Anal. and Appl.

[5] S. CAMPANANTO - Sui problemi al contorno per sistemi di equazioni differenziali lineari del tipo dell'elasticità. Ann. sc. Norm. Sup. Pisa, 13 (1959), 223-258 et 275-302.

[6] W.C. CARPENTER - Viscoelastic Stress Analysis. International Journal Num. Meth. Engng., 4, (1972), 357-366.

[7] M. CROUZEIX - Sur l'approximation des équations différentielles opérationnelles linéaires par des méthodes de Runge-Kutta. Thèse, Paris, mars 1975.

[8] G. DUVAUT, J.L. LIONS - Les inéquations en mécanique et en physique. Dunod, Paris, 1972.

[9] P. GERMAIN - Cours de mécanique des milieux continus. Masson, Paris, 1973.

[10] J.L. LIONS - Equations différentielles opérationnelles. Springer-Verlag, Berlin, 1961.

[11] J.L. LIONS, E. MAGENES - Problèmes aux limites non homogènes. Dunod Paris, 1968.

[12] P.A. RAVIART - Méthode des Eléments Finis. Cours 3e Cycle, Paris VI 1971-72.

[13] G. STRANG, J. FIX - An analysis of the finite element method. Prentice Hall, New York, 1973.

[14] O.C. ZIENKIEWICZ - The Finite Element Method in Engineering Science. McGraw-Hill, New York, 1971.

ON SOLVING A MIXED FINITE ELEMENT APPROXIMATION OF THE DIRICHLET PROBLEM FOR THE BIHARMONIC OPERATOR BY A "QUASI-DIRECT" METHOD AND VARIOUS ITERATIVE METHODS

R. Glowinski

Université de Paris VI, Analyse Numérique, L.A.
189, Tour 55.65, 5° étage, 4 place Jussieu, 75230
PARIS CEDEX 05, FRANCE.

O. Pironneau

IRIA-LABORIA, Domaine de Voluceau, B.P. n°5, 78150
LE CHENAY, FRANCE

1. INTRODUCTION. ORIENTATION.

Let Ω be a *bounded* domain of $\mathbb{R}^2$ with a smooth boundary Γ. We would like to discuss in this report several methods for solving numerically the *Dirichlet problem for the biharmonic operator*

$$(1.1) \qquad (P_0) \quad \begin{cases} \Delta^2\Psi = f \quad \text{in } \Omega \\ \Psi|_\Gamma = g_1 \quad , \\ \frac{\partial\Psi}{\partial n}|_\Gamma = g_2 \end{cases}$$

The results presented here (without proof) were announced in GLOWINSKI-PIRONNEAU [1], [2] ; complete proofs could be found in GLOWINSKI-PIRONNEAU [3]. Extensions to time dependent problems, Navier-Stokes equations in the $\{\Psi,\omega\}$ formulation (1), and numerical experimentations will be given in reports in preparation of GLOWINSKI-PIRONNEAU and BOURGAT-GLOWINSKI-PIRONNEAU. If Ω is n-connected ($n \geq 1$) the $\{\Psi,\omega\}$ formulation of Stokes and Navier-Stokes equations is more complicated but the methods of this report could be extended to these cases (Cf. GLOWINSKI [2] and the above reports in preparation).

---

(1) Ψ : stream function , ω : vorticity function.

Let us describe briefly the content of this report :

In Sec.2 we give some results related to the continuous problem. In Sec. 3,4 it is shown that using a convenient *mixed finite element approximation of* (P_0), introduced in CIARLET-RAVIART [1], solving (P_0) is equivalent to solving a *finite number of approximate Dirichlet problems for* $-\Delta$, *plus a "small" linear system* the matrice of which is *symmetric* and *positive definite*. This last system arises from the variational discretization of an *"integral equation"* on Γ. In Sec.5 several iterative methods for solving (P_0) will be discussed, among them the *conjugate-gradient* method.

The point of view given here seems to give a natural and general framework for solving (P_0), on rather general domains Ω, as a system of *coupled* Dirichlet problems for $-\Delta$. In particular it containts several methods[2] for solving (P_0) on *rectangles*, using *finite differences*, since the usual *13 points-formula* approximation of (P_0) is a special case (actually the simplest one) of the mixed finite element approximation used here (Cf. Sec.4.7, Rem. 4.5 and also GLOWINSKI [1], GLOWINSKI-LIONS-TREMOLIERES [1, Ch.4], GLOWINSKI-PIRONNEAU [3], etc...).

(2) References are given in Sec.6

2. THE CONTINUOUS PROBLEM.

2.1. Functional context and notations.

The following functional spaces are essential in the study of (P_o) :

$$H^2(\Omega) = \{v \mid v \in L^2(\Omega)\ ,\ \frac{\partial v}{\partial x_i} \in L^2(\Omega)\ ,\ \frac{\partial^2 v}{\partial x_i \partial x_j} \in L^2(\Omega)\ ,\ 1 \leq i,j \leq 2\}$$

$$V = H^2(\Omega) \cap H^1_o(\Omega) = \{v \mid v \in H^2(\Omega)\ ,\ v = 0 \text{ on } \Gamma\}$$

$$H^2_o(\Omega) = \overline{\mathcal{D}(\Omega)}^{H^2(\Omega)} = \{v \mid v \in H^2(\Omega)\ ,\ v = \frac{\partial v}{\partial n} = 0 \text{ on } \Gamma\}\ .$$

The space $H^2(\Omega)$ is an Hilbert space for the following inner-product

$$(u,v)_{H^2(\Omega)} = (u,v)_{L^2(\Omega)} + \sum_{1\leq i,j\leq 2} (\frac{\partial u}{\partial x_i}, \frac{\partial v}{\partial x_i})_{L^2(\Omega)} + \sum_{1\leq i,j\leq 2} (\frac{\partial^2 u}{\partial x_i\, \partial x_j}, \frac{\partial^2 v}{\partial x_i\, \partial x_j})_{L^2(\Omega)}\ ,$$

moreover since Ω is bounded with a smooth boundary we have the

PROPOSITION 2.1. : *The mapping* $v \to \|\Delta v\|_{L^2(\Omega)}$ *defines on* V *a norm which is equivalent to the norm induced by* $H^2(\Omega)$. ∎

The following spaces are also essential when studying (P_o) :

$$H(\Omega;\Delta) = \{v \mid v \in L^2(\Omega)\ ,\ \Delta v \in L^2(\Omega)\}\ ,$$

$$\mathcal{H} = \{v \mid v \in H(\Omega;\Delta)\ ,\ \Delta v = 0\}\ .$$

The space $H(\Omega;\Delta)$ is an Hilbert space for the following inner-product

$$(u,v)_{H(\Omega;\Delta)} = (u,v)_{L^2(\Omega)} + (\Delta u, \Delta v)_{L^2(\Omega)}$$

and for the related norm

(2.1) $$\|v\|_{H(\Omega;\Delta)} = (\|v\|^2_{L^2(\Omega)} + \|\Delta v\|^2_{L^2(\Omega)})^{1/2}\ .$$

From (2.1) it follows

PROPOSITION 2.2. : *On* $\mathcal{H}$, *the topologies induced by* $H(\Omega;\Delta)$ *and* $L^2(\Omega)$ *are identical.*

2.2. Properties of the traces.

Let us denote by γ_o and γ_1 the two *trace mappings* defined by

$$\gamma_o v = v|_\Gamma\ ,\quad \gamma_1 v = \frac{\partial v}{\partial n}\Big|_\Gamma\ .$$

Then (Cf. LIONS-MAGENES [1])

PROPOSITION 2.3. : *The mapping* $\{\gamma_0,\gamma_1\}$ *is linear, continuous and surjective from* $H(\Omega;\Delta)$ *to* $H^{-\frac{1}{2}}(\Gamma)\times H^{-\frac{3}{2}}(\Gamma)$.

PROPOSITION 2.4. : *The mapping* $\{\gamma_0,\gamma_1\}$ *is linear, continuous and surjective from* $H^2(\Omega)$ *to* $\times H^{\frac{3}{2}}(\Gamma)\times H^{\frac{1}{2}}(\Gamma)$.

PROPOSITION 2.5.: *The mapping* γ_1 *is surjective from* V *to* $H^{\frac{1}{2}}(\Gamma)$.

PROPOSITION 2.6.: *The mapping* γ_0 *restricted to* $\mathcal{H}$ *is an isomorphism (algebraical and topological) from* $\mathcal{H}$ *to* $H^{-\frac{1}{2}}(\Gamma)$.

2.3. A Green formula.

Let us denote by $<.,.>$ (resp. $<<.,.>>$) *the bilinear form of the duality* beetween $H^{\frac{1}{2}}(\Gamma)$ and $H^{-\frac{1}{2}}(\Gamma)$ (resp. $H^{\frac{3}{2}}(\Gamma)$ and $H^{-\frac{3}{2}}(\Gamma)$) which extends $(.,.)_{L^2(\Gamma)}$, i.e. $<v,w> = \int_\Gamma vw\, d\Gamma \;\; \forall\, v\in H^{\frac{1}{2}}(\Gamma)$, $w\in L^2(\Gamma)$ (resp. $<<v,w>> = \int_\Gamma vw\, d\Gamma \;\; \forall\, v\in H^{\frac{3}{2}}(\Gamma)$, $w\in L^2(\Omega)$). Then (Cf. LIONS-MAGENES, loc. cit.) we have the following *Green formula*

$$(2.2)\qquad \begin{cases} \int_\Omega \Delta u\, v\, dx - \int_\Omega u\, \Delta v\, dx = <\gamma_1 u,\gamma_0 v> - <<\gamma_0 u,\gamma_1 v>> \\ \forall\, u\in H^2(\Omega)\ ,\ \forall\, v\in H(\Omega;\Delta). \end{cases}$$

2.4. Existence, uniqueness and decomposition results for (P_0)

If, in (1.1), we assume

$$(2.3)\qquad f\in L^2(\Omega)\ ,\ g_1\in H^{\frac{3}{2}}(\Gamma)\ ,\ g_2\in H^{\frac{1}{2}}(\Gamma)$$

then it follows from LIONS-MAGENES [1]

THEOREM 2.1. : *The problem* (P_0) *has a unique solution in* $H^2(\Omega)$. ∎

Then we can easily prove the following

PROPOSITION 2.7. : (P_0) *is equivalent to*

$$(2.4)\qquad \begin{cases} -\Delta\omega = f\ , \\ -\Delta\Psi = \omega\ , \\ \gamma_0\Psi = g_1\ ,\ \gamma_1\Psi = g_2\ . \end{cases}$$ ∎

REMARK 2.1. The decomposition (2.4) is well-known in *Fluid Mechanics* where

ω is the *vorticity* function ,
Ψ is the *stream* function. ∎

In the sequel the *trace of* ω on Γ will play a fundamental role, theoretically *and numerically*.

The existence of the trace $\gamma_o\omega$ follows, via the results of sec. 2.2., from

PROPOSITION 2.8. : *If* f, g_1, g_2 *obey* (2.3), *then* ω *has on* Γ *a trace* $\gamma_o\omega \in H^{-\frac{1}{2}}(\Gamma)$. ∎

2.5. On the relation beetween $\gamma_o\omega$ and $\gamma_1\Psi$.

Various iterative methods for solving (P_o) (Cf. CIARLET-GLOWINSKI [1], BOURGAT [1], BOSSAVIT [1], GLOWINSKI [1], GLOWINSKI-PIRONNEAU [3], Sec.5 of this report, etc...) and also the "almost direct" method described in Sec. 4 are actually based on the results of this Sec. 2.5.

The following lemma is essential and follows from the results of Sec. 2.1., 2.2., 2.3., 2.4. :

LEMMA 2.1. : *Let* $\lambda \in H^{-\frac{1}{2}}(\Gamma)$. *Then*

(i) *Problem*

(2.5) $$\begin{cases} \Delta^2\Psi = 0 , \\ \Psi|_\Gamma = 0 \\ -\Delta\Psi|_\Gamma = \lambda \end{cases}$$

has a unique solution in $V = H^2(\Omega) \cap H^1_o(\Omega)$.

(ii) *Let* Ψ *be the solution of* (2.5) *in* V ; *then the operator* A *defined by*

(2.6) $$A\lambda = -\gamma_1\Psi$$

is an isomorphism from $H^{-\frac{1}{2}}(\Gamma)$ *to* $H^{\frac{1}{2}}(\Gamma)$.

(iii) *The bilinear form* $a(.,.) : H^{-\frac{1}{2}}(\Gamma)\times H^{-\frac{1}{2}}(\Gamma) \to \mathbb{R}$ *defined by*

(2.7) $$a(\lambda,\mu) = < A\lambda,\mu >$$

is continuous, symmetric and $H^{-\frac{1}{2}}(\Gamma)$- *elliptic*. ∎

Let go back to (P_o) with $f \in L^2(\Omega)$, $g_1 \in H^{\frac{3}{2}}(\Gamma)$, $g_2 \in H^{\frac{1}{2}}(\Gamma)$; we have seen (Cf. Theorem 2.1) that (P_o) has a unique solution in $H^2(\Omega)$ and that $\omega = -\Delta\Psi$ has a *trace* $\lambda = \gamma_o\omega$ belonging to $H^{-\frac{1}{2}}(\Gamma)$. It will be shown in a moment that λ is solution of a *linear variational equation* in $H^{-\frac{1}{2}}(\Gamma)$.

Let $\overline{\Psi}$ be the unique solution in V of

$$(2.8) \qquad \begin{cases} \Delta^2\overline{\Psi} = 0 \,, \\ \overline{\Psi}_{|\Gamma} = 0 \,, \\ -\Delta\overline{\Psi}_{|\Gamma} = \lambda \,; \end{cases}$$

problem (2.8) is equivalent to

$$(2.9) \qquad \begin{cases} -\Delta\overline{\omega} = 0 \\ \overline{\omega}_{|\Gamma} = \lambda \\ -\Delta\overline{\Psi} = \overline{\omega} \,, \\ \overline{\Psi}_{|\Gamma} = 0 \,. \end{cases}$$

Let Ψ_o be the unique solution in $H^2(\Omega)$ of

$$(2.10) \qquad \begin{cases} \Delta^2\Psi_o = f \,, \\ \Psi_{o|\Gamma} = g_1 \\ -\Delta\Psi_{o|\Gamma} = 0 \,; \end{cases}$$

problem (2.10) is equivalent

$$(2.11) \qquad \begin{cases} -\Delta\omega_o = f \,, \\ \omega_{o|\Gamma} = 0 \,, \\ -\Delta\Psi_o = \omega_o \\ \Psi_{o|\Gamma} = g_1 \,. \end{cases}$$

We obviously have $\Psi = \Psi_o + \overline{\Psi}$, $\omega = \omega_o + \overline{\omega}$ and we notice that *the computation of* Ψ_o *requires the solution of two Dirichlet problems for* $-\Delta$. *We have a similar result for* $\overline{\Psi}$, *once* λ *is known.*

The *key result* of this section is stated in the following.

THEOREM 2.2. : *Let* Ψ *be the solution of* (P_o) ; *the trace* λ *of* $-\Delta\Psi$ *on* Γ *is the unique solution of the linear variational equation*

$$(2.12)\quad \begin{cases} < A\lambda,\mu > = < \dfrac{\partial\Psi_o}{\partial n} - g_2,\mu > \quad \forall\, \mu \in H^{-\frac{1}{2}}(\Gamma) \;, \\ \lambda \in H^{-\frac{1}{2}}(\Gamma) \;. \end{cases} \quad ■$$

REMARK 2.2. Since $a(.,.)$ is symmetric, (2.12) is equivalent to the *minimization* problem

$$(2.13)\quad \begin{cases} J(\lambda) \leq J(\mu) \quad \forall\, \mu\in H^{-\frac{1}{2}}(\Gamma) \;, \\ \lambda \in H^{-\frac{1}{2}}(\Gamma) \;, \end{cases}$$

where

$$J(\mu) = \frac{1}{2} < A\mu,\mu > - < \frac{\partial\Psi_o}{\partial n} - g_2,\mu > . \quad ■$$

REMARK 2.3. Considering the boundary condition $\frac{\partial\Psi}{\partial n}\Big|_\Gamma = g_2$ as a *linear constraint*, we can associate to (P_o) the *Lagrangian functional* : $\mathcal{L}$: $H^2(\Omega)\times H^{-\frac{1}{2}}(\Gamma) \to \mathbb{R}$, defined by

$$\mathcal{L}(v,\mu) = \frac{1}{2}\int_\Omega |\Delta v|^2 dx - \int_\Omega fv \; dx + < \frac{\partial v}{\partial n} - g_2,\mu > .$$

If $\hat{V} = \{v \in H^2(\Omega),\ v = g_1 \text{ on } \Gamma\}$, it can be proved that $\{\Psi,-\gamma_o\Delta\Psi\}$ is the unique saddle-point of $\mathcal{L}$ on $\hat{V}\times H^{-\frac{1}{2}}(\Gamma)$ and that

$$J(\mu) = - \underset{v\in\hat{V}}{\mathrm{Min}}\, \mathcal{L}(v,\mu).$$

It follows from these results that (2.13) is the *dual problem of* (P_o) *related to the Lagrangian* $\mathcal{L}$. We refer to CIARLET-GLOWINSKI [1], GLOWINSKI [1], GLOWINSKI-LIONS-TREMOLIERES [1, Ch.4] for a more detailed analysis of (P_o) by duality methods related to Lagrangians of the same kind than $\mathcal{L}$. ■

REMARK 2.4. The data f and g_1 occur in (2.12) through Ψ_o (Cf. 2.10) ■

REMARK 2.5. Let $\tilde{\mu}$ be an extension of μ in Ω. Then it follows from the *Green formula*

$$(2.14)\quad \begin{cases} a(\lambda,\mu) = < A\lambda,\mu > = - < \dfrac{\partial\Psi}{\partial n} \,,\, \mu > = \\ = - \displaystyle\int_\Omega \Delta\Psi\,\tilde{\mu}\; dx - \int_\Omega \nabla\Psi\cdot\nabla\tilde{\mu}\; dx = \int_\Omega \omega\,\tilde{\mu}\; dx - \int_\Omega \nabla\Psi\cdot\nabla\tilde{\mu}\; dx \end{cases}$$

where Ψ is the solution of (2.5) and $\omega = -\Delta\Psi$.

Similarly we have

$$(2.15) \qquad < \frac{\partial \Psi_o}{\partial n}, \mu > = \int_\Omega \nabla\Psi_o \cdot \nabla\tilde{\mu}\, dx + \int_\Omega \Delta\Psi_o\, \tilde{\mu}\, dx = \int_\Omega \nabla\Psi_o \cdot \nabla\tilde{\mu}\, dx - \int_\Omega \omega_o \tilde{\mu}\, dx$$

where $\{\omega_o, \Psi_o\}$ is the solution of (2.11).

We have to notice that (2.14),(2.15) are not true in general, since $\tilde{\mu}$ is not enough regular. However if μ is smooth enough ($\mu \in H^{\frac{1}{2}}(\Gamma)$ for instance) then we can take $\mu \in H^1(\Omega)$ and (2.14),(2.15) are *true*.

The interest of (2.14),(2.15) is that they give mathematical expressions of the two sides of (2.12) *in which* $\frac{\partial \Psi}{\partial n}$ *and* $\frac{\partial \Psi_o}{\partial n}$ *don't occur explicitely*. We shall make use of this remark in the Sec.3 and 4 when (P_o) and (2.12) will be approximated using a *mixed finite element method*. ■

2.6. Summary on the resolution of (P_o)

Let Ψ be the solution of

$$(P_o) \qquad \begin{cases} \Delta^2 \Psi = f \quad \text{in } \Omega , \\ \Psi_{|\Gamma} = g_1 , \\ \frac{\partial \Psi}{\partial n}\Big|_\Gamma = g_2 \end{cases}$$

and $\omega = -\Delta\Psi, \lambda = \omega_{|\Gamma}$.

It has been shown in Sec. 2.5 that solving (P_o) is equivalent to solving sequentially the following problems

$$(2.16) \qquad \begin{cases} -\Delta\omega_o = f \quad \text{in } \Omega \\ \omega_o|_\Gamma = 0 \end{cases}$$

$$(2.17) \qquad \begin{cases} -\Delta\Psi_o = \omega_o \quad \text{in } \Omega , \\ \Psi_o|_\Gamma = g_1 , \end{cases}$$

$$(2.18) \qquad A\lambda = \frac{\partial \Psi_o}{\partial n} - g_2$$

$$(2.19) \qquad \begin{cases} -\Delta\omega = f \quad \text{in } \Omega , \\ \omega_{|\Gamma} = \lambda , \end{cases}$$

$$(2.20)\qquad \begin{cases} -\Delta\Psi = \omega \quad \text{in } \Omega , \\ \Psi_{|\Gamma} = g_1 . \end{cases}$$

Hence we have to solve *four Dirichlet problems for* $-\Delta$ *plus an "integral equation"* on Γ the variational formulation of which has been given in (2.12). In the following sections the approximation of (2.18) will be of particular interest.

REMARK 2.6. Actually A is a *pseudo-differential operator which is not explicitely known in general.* ■

2.7. Explicit form of A in the case of a circular domain.

The results of this section are not essential in the sequel. We assume in this section that

$$(2.21)\qquad \Omega = \{x \mid x \in \mathbb{R}^2 , \ x_1^2 + x_2^2 < R^2\}$$

then

THEOREM 2.3. : *Let* A *be the isomorphism from* $H^{-1/2}(\Gamma)$ *to* $H^{1/2}(\Gamma)$ *defined in Sec. 2.5. Then if* Ω *is given by* (2.21), *and if* λ *is a function defined on* Γ *and smooth enough* ($\lambda \in L^2(\Gamma)$), *for instance)* , *we have*

$$(A\lambda)(x) = \int_\Gamma A(x,y)\lambda(y)\,d\Gamma(y) \ , \ \forall\ x \in \Gamma ,$$

the kernel $A(x,y)$ *being given by*

$$(2.22)\qquad \begin{cases} A(x,y) = \dfrac{1}{2\pi} \left[\left(1 - \dfrac{|y-x|^2}{2R^2}\right) \mathrm{Log}\, \dfrac{R}{|y-x|} + \dfrac{|y-x|}{R} \sqrt{1 - \dfrac{|y-x|^2}{4R^2}} \ \times \right. \\ \left. \times \ \mathrm{Arc\,cos}\, \dfrac{|y-x|}{2R} - \dfrac{1}{2} \right] , \end{cases}$$

with $|y-x|$ = *distance* (x,y) . ■

3. APPROXIMATION OF (P_o) BY A MIXED FINITE ELEMENT METHOD.

We shall assume in this section that Ω is a *polygonal domain* but the following results could be easily extended to the case of a *curved boundary* Γ , using *isoparametric finite elements* (cf. CIARLET-RAVIART [2]) .

3.1. Triangulation of Ω . Fundamental spaces.

Let $\mathcal{T}_h$ be a triangulation of Ω obeying

$$(3.1)\qquad \mathcal{T}_h \text{ is finite, } T \subset \bar{\Omega}\ \forall\ T \in \mathcal{T}_h , \quad \bigcup_{T \in \mathcal{T}_h} T = \bar{\Omega} ,$$

$$(3.2)\qquad \begin{cases} T, T' \in \mathcal{T}_h, T \neq T' \Rightarrow \mathring{T} \cap \mathring{T}' = \emptyset \text{ and } T \cap T' = \emptyset \text{ or } T \text{ and } T' \\ \text{have either only one common vertex or a whole common edge.} \end{cases}$$

As usual h will be lenght of the largest edge of the $T \in \mathcal{T}_h$.

Let P_k be the space of the polynomials of degree $\leq k$; then we define the following *finite dimensional* spaces

(3.3) $$V_h = \{v_h \in C^0(\bar{\Omega}) \ , \ v_h|_T \in P_k \quad \forall\ T \in \mathcal{T}_h\}$$

(3.4) $$V_{oh} = V_h \cap H_0^1(\Omega) = \{v_h \in V_h \ , \ v_h = 0 \ \text{ on } \ \Gamma\} \ .$$

(3.5) $$M_h \subset V_h \ \text{ such that } \ V_h = V_{oh} \oplus M_h \ ,$$

(3.6) $$\begin{cases} W_{gh} = \{(v_h,q_h) \in V_h \times V_h \ , \ v_h|_\Gamma = g_{1h}, \int_\Omega \nabla v_h \cdot \nabla \mu_h \, dx = \\ = \int_\Omega q_h \mu_h \, dx + \int_\Gamma g_{2h} \mu_h \, d\Gamma \quad \forall\ \mu_h \in V_h\}. \end{cases}$$

In (3.5) M_h is *not uniquely* defined but a natural choice for M_h will be given later. In (3.6), g_{1h} is an approximation of g_1 belonging to $\gamma_o V_h$ and g_{2h} is an approximation of g_2 such that $\int_\Gamma g_{2h}\mu_h \, d\Gamma$ is "easy" to compute.

3.2. Approximation of (P_o).

We approximate (P_o) by

$$(P_h) \begin{cases} j_h(\Psi_h,\omega_h) \leq j_h(v_h,q_h) \quad \forall\ (v_h,q_h) \in W_{gh} \ , \\ (\Psi_h,\omega_h) \in W_{gh} \ , \end{cases}$$

where

(3.7) $$j_h(v_h,q_h) = \frac{1}{2}\int_\Omega |q_h|^2 \, dx - \int_\Omega f_h v_h \, dx \ .$$

In (3.7), f_h is an approximation of f such that $\int_\Omega f_h v_h \, dx$ is easy to compute. Such a finite element approximation is said to be mixed (cf. CIARLET-RAVIART [1], CIARLET-GLOWINSKI [1]). We have

PROPOSITION 3.1. : (P_h) *has a unique solution.*

3.3. Convergence results $(k \geq 2)$

We assume that the angles of $\mathcal{T}_h$ are *bounded below, uniformly in* h , by $\theta_o > 0$ and that $\mathcal{T}_h$ satisfies

(3.8) $$\frac{\displaystyle\max_{T \in \mathcal{T}_h} h(T)}{\displaystyle\min_{T \in \mathcal{T}_h} h(T)} \leq \tau \ , \quad \forall\ \mathcal{T}_h \ , \ \tau \ \textit{independent} \text{ of } h \ ,$$

where in (3.8), $h(T)$ = diameter of T . Then if $k \geq 2$ it follows from CIARLET-RAVIART [1] that under the above hypotheses we have

$$(3.9) \quad \|\Psi_h - \Psi\|_{H^1(\Omega)} + \|\omega_h - (-\Delta\Psi)\|_{L^2(\Omega)} \leq C \|\Psi\|_{H^{k+2}(\Omega)} h^{k-1}$$

with C independent of h and Ψ [3]. For a discussion of the case $k = 1$ we refer to GLOWINSKI [1], GLOWINSKI-LIONS-TREMOLIERES [1,ch.4]. We refer also to SCHOLZ [1] where under the above assumptions on $\mathcal{T}_h$, it is proved, for $k \geq 3$, that

$$\|\Psi_h - \Psi\|_{L^2(\Omega)} + h^2 \|\omega_h - (-\Delta\Psi)\|_{L^2(\Omega)} \leq C h^{k+1} \|\Psi\|_{H^{k+1}(\Omega)}$$

with C independent of h and ψ .

REMARK 3.1. The above results still hold if instead of *triangular* finite elements we use *quadrilateral* finite elements. ∎

3.4. Decomposition of (P_h) .

We have $V_h = V_{oh} \oplus M_h$. Let $\{\Psi_h, \omega_h\}$ be the solution of (P_h) and λ_h the component of ω_h in M_h , then

$$(3.10) \quad \begin{cases} \omega_h = (\omega_h - \lambda_h) + \lambda_h \quad , \text{ with} \\ \omega_h - \lambda_h \in V_{oh} \ , \ \lambda_h \in M_h \ . \end{cases}$$

In CIARLET-GLOWINSKI [1], GLOWINSKI-PIRONNEAU [3] is proved the following

THEOREM 3.1. : *Let* $\{\Psi_h, \omega_h\}$ *be the solution of* (P_h) , λ_h *be the component of* ω_h *in* M_h ; *Then* $\{\Psi_h, \omega_h, \lambda_h\}$ *could be characterized as being the unique element of* $V_h \times V_h \times M_h$ *such that*

$$(3.11) \quad \begin{cases} \int_\Omega \nabla\omega_h \cdot \nabla v_h dx = \int_\Omega f_h v_h dx \quad \forall v_h \in V_{oh} \ , \\ \omega_h - \lambda_h \in V_{oh} \end{cases}$$

$$(3.12) \quad \begin{cases} \int_\Omega \nabla\Psi_h \cdot \nabla v_h dx = \int_\Omega \omega_h v_h dx \quad \forall v_h \in V_{oh} \ , \\ \Psi_h \in V_h \ , \ \Psi_h|_\Gamma = g_{1h} \ , \end{cases}$$

$$(3.13) \quad \int_\Omega \nabla\Psi_h \cdot \nabla\mu_h dx = \int_\Omega \omega_h \mu_h dx + \int_\Gamma g_{2h} \mu_h d\Gamma \quad \forall \mu_h \in M_h . \ ∎$$

---

(3) The estimate (3.9) supposes that f, g_1, g_2 have been correctly approximated.

REMARK 3.2. Relations (3.11)-(3.13) are the discretized analogous of (2.19), (2.20), (3.14) where

$$(3.14)\quad \begin{cases} \int_\Omega \nabla\Psi\cdot\nabla\mu dx = \int_\Omega \omega\mu dx + \int_\Gamma g_2\mu d\Gamma \quad \forall\, \mu \in M \text{ , with} \\ M \subset H^1(\Omega) \text{ , } H^1(\Omega) = H^1_o(\Omega) \oplus M \text{ .} \end{cases}$$ ∎

3.5. Discrete analogous lemma 2.1.

Let us consider $\lambda_h \in M_h$. Then we define ω_h , ψ_h as the solutions of the two following approximate Dirichlet problems

$$(3.15)\quad \begin{cases} \int_\Omega \nabla\omega_h\cdot\nabla v_h dx = 0 \quad \forall\, v_h \in V_{oh} \text{ ,} \\ \omega_h \in V_h \text{ , } \omega_h - \lambda_h \in V_{oh} \text{ ,} \end{cases}$$

$$(3.16)\quad \begin{cases} \int_\Omega \nabla\Psi_h\cdot\nabla v_h dx = \int_\Omega \omega_h v_h dx \quad \forall\, v_h \in V_{oh} \text{ ,} \\ \Psi_h \in V_{oh} \text{ .} \end{cases}$$

Then let us define the bilinear form $a_h : M_h \times M_h \to \mathbb{R}$ by

$$(3.17)\quad a_h(\lambda_h,\mu_h) = \int_\Omega \omega_h\mu_h dx - \int_\Omega \nabla\Psi_h\cdot\nabla\mu_h dx \quad \forall\, \mu_h \in M_h \text{ .}$$

We notice that the definition of $a_h(.,.)$ follows the view point of Remark 2.5. We can easily prove the following

LEMMA 3.1. : *The bilinear form* $a_h(.,.)$ *is symmetric positive definite ; moreover*

$$(3.18)\quad a_h(\lambda_{1h},\lambda_{2h}) = \int_\Omega \omega_{1h}\omega_{2h} dx \quad \forall\, \lambda_{1h}, \lambda_{2h} \in M_h$$

where ω_{jh} (j = 1,2) *is the solution of* (3.15) *corresponding to* λ_{jh} .

3.6. Using Lemma 3.1 to solve (P_h).

Let $\{\Psi_h,\omega_h\}$ be the solution of (P_h) and λ_h be the component of ω_h in M_h .

Let $\overline{\omega}_h$ and $\overline{\Psi}_h$ be the respective solutions of

$$(3.19)\quad \begin{cases} \int_\Omega \nabla\overline{\omega}_h\cdot\nabla v_h dx = 0 \quad \forall\, v_h \in V_{oh} \text{ ,} \\ \overline{\omega}_h \in V_h \text{ , } \overline{\omega}_h - \lambda_h \in V_{oh} \text{ ,} \end{cases}$$

$$(3.20)\quad \begin{cases} \int_\Omega \nabla\overline{\Psi}_h \cdot \nabla v_h \, dx = \int_\Omega \overline{\omega}_h v_h \, dx & \forall\, v_h \in V_{oh} , \\ \overline{\Psi}_h \in V_{oh} . \end{cases}$$

Let ω_{oh} and Ψ_{oh} be the respective solutions of

$$(3.21)\quad \begin{cases} \int_\Omega \nabla\omega_{oh} \cdot \nabla v_h \, dx = \int_\Omega f_h v_h \, dx & \forall\, v_h \in V_{oh} , \\ \omega_{oh} \in V_{oh} , \end{cases}$$

$$(3.22)\quad \begin{cases} \int_\Omega \nabla\Psi_{oh} \cdot \nabla v_h \, dx = \int_\Omega \omega_{oh} v_h \, dx & \forall\, v_h \in V_{oh} , \\ \Psi_{oh} \in V_h , \; \Psi_{oh} = g_{1h} \text{ on } \Gamma . \end{cases}$$

We have $\Psi_h = \overline{\Psi}_h + \Psi_{oh}$, $\omega_h = \overline{\omega}_h + \omega_{oh}$ and relations (3.19) - (3.22) are the discrete analogous of (2.8) - (2.11).

The finite dimensional analogous of Theorem 2.2 is

<u>THEOREM 3.2.</u> : *Let $\{\Psi_h, \omega_h\}$ be the solution of (P_h) and λ_h the component in M_h of ω_h then λ_h is the unique solution of the linear variational problem*

$$(3.23)\quad \begin{cases} a_h(\lambda_h, \mu_h) = \int_\Omega \nabla\Psi_{oh} \cdot \nabla\mu_h \, dx - \int_\Omega \omega_{oh}\mu_h \, dx - \int_\Gamma g_{2h}\mu_h \, d\Gamma & \forall\, \mu_h \in M_h , \\ \lambda_h \in M_h \end{cases}$$

which is equivalent to a linear system, the matrix of which is symmetric and positive definite.

<u>REMARK 3.3.</u> To compute the right hand side of (3.23) we have to solve the *two approximate Dirichlet problems* (3.21),(3.22). In the same way, once λ_h is known we have to solve the *two approximate Dirichlet problems* (3.11), (3.12) to know ω_h and Ψ_h .

3.7. On the conditioning of $a_h(\cdot,\cdot)$.

In practical applications the linear system equivalent to (3.23) will have to be solved by *direct* or *iterative* methods. For these reasons it is important to know the *conditioning* of the matrix of this system (this matrix will be explicited in Sec. 4). The following Theorem 3.3 will provide an estimate of the condition number of this matrix (cf. Sec. 4.6).

THEOREM 3.3. : *We assume that* Ω *is convex ; then under the assumptions on* $\mathcal{C}_h$ *of Sec.3.3 , we have for* $k \geq 1$ *and* h *sufficiently small*

(3.24) $$\alpha h \,\|\gamma_o \lambda_h\|^2_{L^2(\Gamma)} \leq a_h(\lambda_h,\lambda_h) \leq \beta \,\|\gamma_o \lambda_h\|^2_{L^2(\Gamma)} \quad \forall\, \lambda_h \in M_h\,,$$

with $\alpha,\beta > 0$ *and independent of* h *and* λ_h . ∎

REMARK 3.4. Using Theorem 3.3 we can prove

(3.25) $$\lim_{h\to 0} \sup_{\lambda_h \in M_h - \{0\}} \frac{a_h(\lambda_h,\lambda_h)}{\|\gamma_o \lambda_h\|^2_{L^2(\Gamma)}} = \sup_{\lambda \in L^2(\Gamma) - \{0\}} \frac{a(\lambda,\lambda)}{\|\lambda\|^2_{L^2(\Gamma)}} = |A|$$

where $|A|$ is the *largest eigenvalue* of A . ∎

3.8. Summary on the approximate problem.

Let $\{\Psi_h,\omega_h\}$ be the solution of (P_h) and λ_h the component in M_h of ω_h . This vector λ_h is solution of a *linear system* the matrix of which is *symmetric* and *positive definite*. An equivalent variational formulation is given by (3.23) but the bilinear form $a_h(.,.)$ *is not explicitly known*, and unfortunatly the situation is the same for the matrix of the above linear system. The construction of this matrix and techniques for solving the linear system will be discussed in the next section. Iterative methods for solving this system and avoiding the construction of the matrix will be discussed in Sec.5.

4. CONSTRUCTION AND SOLUTION OF THE LINEAR SYSTEM EQUIVALENT TO (3.23).

4.1. Generalities.

Let $N_h = \dim(M_h)$ and let $\mathcal{B}_h = \{w_i\}_{1\leq i\leq N_h}$ be a *basis* of M_h .

If $\lambda_h \in M_h$ we have

(4.1) $$\lambda_h = \sum_{j=1}^{N_h} \lambda_j w_j \quad .$$

Then

PROPOSITION 4.1. : *Problem (3.23) is equivalent to the linear system in* $\{\lambda_1\, \lambda_2 \,,\ldots,\, \lambda_{N_h}\}$

$$(E_h)\quad \begin{cases} \displaystyle\sum_{j=1}^{N_h} a_h(w_j,w_i)\lambda_j = \int_\Omega \nabla\Psi_{oh}\cdot\nabla w_i\,dx - \int_\Omega \omega_{oh} w_i\,dx - \int_\Gamma g_{2h} w_i\,d\Gamma\,, \\ 1 \leq i \leq N_h\,. \end{cases}$$ ∎

If we denote $a_{ij} = a_h(w_j,w_i)$ and $A_h = (a_{ij})_{1\leq i,j\leq N_h}$ we can easily prove

PROPOSITION 4.2. : *The matrix* A_h *is a* $N_h \times N_h$, *symmetric and positive definite matrix.* ∎

In the next subsections we shall discuss the construction of A_h and of the right hand side of (E_h) , once a convenient choice has been done for M_h and $\mathcal{B}_h$.

4.2. Choice of M_h

A space M_h will be convenient if the computation of the a_{ij} and right hand sides of (E_h) is of little cost. It will be the case if the functions $w_i \in \mathcal{B}_h$ have a "*small*" *support*. It follows from CIARLET-GLOWINSKI [1],GLOWINSKI-LIONS-TREMOLIERES [1, ch.4] that an optimal choice for M_h seems to be the following :

$$(4.2)\qquad \begin{cases} M_h \subset V_h \quad , \quad M_h \oplus V_{oh} = V_h , \\ v_h \in M_h \Rightarrow v_{h|T} = 0 \quad \forall\, T \in \mathcal{C}_h \ , \ T \cap \Gamma = \emptyset \ . \end{cases}$$

In particular, if V_h is defined by *Lagrange finite elements* (cf. Figure 4.1 for $k = 2$) , M_h consists of those functions of V_h *vanishing* at the nodes of $\mathcal{C}_h$ *don't belonging to* Γ . We obviously have

$$N_h = \dim(M_h) = \mathrm{Card}(\Sigma_h)$$

where

$$\Sigma_h = \{P \mid P \in \Gamma \ , \ P \text{ is a node of } \mathcal{C}_h\} ;$$

then the *canonical* choice for $\mathcal{B}_h$ is

$$\mathcal{B}_h = \{w_i\}_{1 \le i \le N_h}$$

where w_i is defined by

$$(4.3)\qquad \begin{cases} w_i \in V_h , \\ w_i(P_i) = 1,\ P_i \in \Sigma_h, w_i(Q) = 0 \quad \forall Q \text{ node of } \mathcal{C}_h \ , \ Q \ne P_i \ . \end{cases}$$

In (4.3) we suppose that Σ_h has been *arranged* from 1 to N_h . With this choice for M_h and $\mathcal{B}_h$, the coefficients λ_j in the expension (4.1) of λ_h are precisely the values taken by λ_h at the boundary nodes P_j , $1 \le j \le N_h$, hence

$$(4.4)\qquad \lambda_j = \lambda_h(P_j) \quad \forall\, P_j \in \Sigma_h \ , \ 1 \le j \le N_h \ .$$

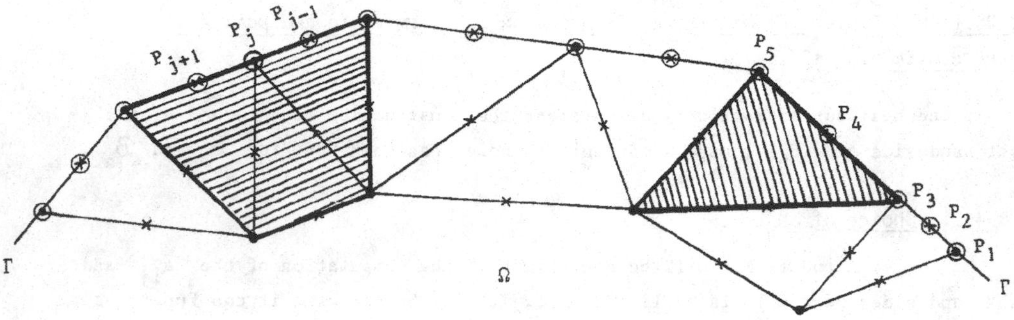

Figure 4.1.

(k = 2 ; boundary nodes are circled and supports of w_4 , w_j are hachured).

4.3. Computation of the right hand sides of (E_h) .

Let $b_h = \{b_1, b_2, \ldots, b_{N_h}\}$ be the right hand side of (E_h).
Therefore we have

(4.5) $$b_i = \int_\Omega \nabla\Psi_{oh}\cdot\nabla w_i dx - \int_\Omega \omega_{oh} w_i dx - \int_\Gamma g_{2h} w_i dx \ , \ 1 \leq i \leq N_h \ .$$

To compute b_h we have to compute ω_{oh} and Ψ_{oh} by solving (3.21), (3.22).

REMARK 4.1. *Smaller* is the support of w_i , *faster* is the computation of b_i . Moreover if M_h obeys (4.2), in the computation of b_h it suffices to know ω_{oh} and Ψ_{oh} on the union of the $T \in \mathcal{T}_h$, such that $T \cap \Gamma \neq \emptyset$. We can take advantage from this fact to reduce the storage requirements on the computer we are using.

4.4. Computation of the matrix A_h .

Let $w_j \in \mathcal{B}_h$; omitting h we denote by ω_j and Ψ_j the solutions of (3.15),(3.16) corresponding to w_j , therefore

(4.6) $$\begin{cases} \int_\Omega \nabla\omega_j\cdot\nabla\mu_h dx = 0 \quad \forall \ \mu_h \in V_{oh} \ , \\ \omega_j \in V_h \ , \ \omega_j - w_j \in V_{oh} \ , \end{cases}$$

$$(4.7)\quad \begin{cases} \int_\Omega \nabla\Psi_j\cdot\nabla\mu_h dx = \int_\Omega \omega_j\mu_h dx \quad \forall\ \mu_h \in V_{oh}\ , \\ \Psi_j \in V_{oh}\ . \end{cases}$$

It follows from (3.17) that

$$(4.8)\quad \begin{cases} a_{ij} = a_h(w_j,w_i) = \int_\Omega \omega_j w_i dx - \int_\Omega \nabla\Psi_j\cdot\nabla w_i dx\ , \\ 1 \leqslant i,j \leqslant N_h\ . \end{cases}$$

It follows from (4.8) that to compute the jth column of A_h we have to solve the two approximate Dirichlet problems (4.6),(4.7), then, w_i describing $\mathcal{B}_h$, compute the integrals occuring in (4.8). Remark 4.1 still holds for the computation of the a_{ij}. We have also to notice that, taking account of the symmetry of A_h, we just have to consider the a_{ij} such that $1 \leqslant j \leqslant i$.

In Sec. 4.5.2 we shall take advantage of these properties of A_h when solving (E_h) by the *Cholesky Method.*

REMARK 4.2. It follows from (3.18) that

$$(4.9)\quad a_{ij} = \int_\Omega \omega_i\omega_j dx \quad \forall\ 1 \leqslant i,j \leqslant N_h$$

It seems, from (4.9) that A_h can be computed by solving "only" N_h approximate Dirichlet problems, instead of $2\ N_h$ if we use (4.8). Actually this simplification is not real. Indeed the use of (4.9) will need the storage of $\omega_1,\omega_2 \ldots \omega_{N_h}$; this is always possible using tapes on disks but it will increase the computational time of A_h rather considerably. Moreover the integrals occuring in (4.9) have to be computed on the whole domain Ω and not only in the neighbourhood of Γ, as it is the case when using (4.8) (cf. Remark 4.1).

4.5. Solution of (E_h).

4.5.1. : Generalities.

Let $\check{\lambda}_h \in \mathbb{R}^{N_h}$ be the vector $\{\lambda_1,\lambda_2,\ldots,\lambda_{N_h}\}$; then (E_h) could be written

$$(4.10)\quad A_h\check{\lambda}_h = b_h\ .$$

Since the matrix A_h is *symmetric* and *positive definite* we can use a *direct method* like *Cholesky* to solve (4.10). We can also use an *iterative method* like S.O.R , S.S.O.R. (cf.VARGA [1], D.M. YOUNG [1] or *gradient* and *conjugate gradient* (cf. J.W. DANIEL [1], CEA [1], E. POLAK [1], CONCUS-GOLUB [1]). We shall discuss the Cholesky method in Sec. 4.5.2. For *gradient* and *conjugate gradient methods* it will be seen in Sec. 5 that they can be used *without knowing* A_h .

4.5.2. Solution of (E_h) by a Cholesky method.

Since A_h is *symmetric* and *positive definite*, it is well known that it exists a *regular, lower triangular matrix* $L_h = (\ell_{ij})_{1\leq i,j\leq N_h}$,and only one such that

$$(4.11)\quad \begin{cases} A_h = L_h L_h^t \ , \\ \ell_{ii} > 0 \ , \ 1 \leq i \leq N_h \ . \end{cases}$$

Since L_h is upper triangular we have

$$(4.12)\qquad \ell_{ij} = 0 \quad \forall\ 1 \leq i < j \leq N_h \ .$$

Let us recall the basic formulae of Cholesky method :

If $j = 1$, we have :

$$(4.13)\quad \begin{cases} \ell_{11} = \sqrt{a_{11}} \\ \ell_{i1} = \dfrac{a_{i1}}{\ell_{11}} \qquad \forall\ 2 \leq i \leq N_h \ . \end{cases}$$

If $2 \leq j \leq N_h$, we have :

$$(4.14)\quad \begin{cases} \ell_{jj} = \sqrt{a_{jj} - \sum\limits_{k=1}^{j-1} \ell_{jk}^2} \ , \\ \ell_{ij} = \dfrac{a_{ij} - \sum\limits_{k=1}^{j-1} \ell_{ik}\,\ell_{jk}}{\ell_{jj}} \qquad \forall\ j+1 \leq i \leq N_h \ . \end{cases}$$

From (4.13),(4.14) it appears that, when computing L_h , it is not necessary to store A_h (or at least the a_{ij} such that $1 \leq j \leq i \leq N_h$).Actually if we assume that the $(j-1)$ first columns, of L_h are known, then to compute the jth column of L_h we solve first (4.6),(4.7) in order to know $\{\omega_j,\Psi_j\}$. Then, thank to ω_j,Ψ_j , we can compute a_{jj} using (4.8) , ℓ_{jj} using (4.14). Similarly we can compute, for $j+1 \leq i \leq N_h$, a_{ij} by using (4.8) , ℓ_{ij} by using (4.14). The same method could be used for the first column of A_h. Then it is clear that for computing the jth column of L_h it is not necessary to store the a_{ij} such that $j \leq i \leq N_h$, since the knowledge of the $(j-1)$ first columns of L_h give us the possibility of computing ℓ_{ij} once a_{ij} is known (this supposes that in the jth

column we compute first a_{jj} , ℓ_{jj}). ■

Once L_h is known, it is well-known that computing $\check{\lambda}_h$ is equivalent to solve

(4.15) $$\begin{cases} L_h y_h = b_h \ , \\ L_h^t \check{\lambda}_h = y_h \ . \end{cases}$$

Obtaining λ_h from $\check{\lambda}_h$ is trivial, then we compute ω_h and Ψ_h by solving the *two approximate Dirichlet problems* (3.11), (3.12). ■

REMARK 4.2. Once L_h is known we can easily solve problems (P_h) corresponding to various functions f , g_1, g_2 , the other parameters remaining the same.

4.5.3. Summary on the use of Cholesky method.

For solving (P_h) , via (E_h) , by Cholesky method we have to solve

- The *two* approximate Dirichlet problems (3.21), (3.22) to obtain b_h ,
- The $2\ N_h$ problems (4.6), (4.7) , $1 \leqslant j \leqslant N_h$, to compute L_h ,
- The *two* linear systems (4.15), to compute λ_h ,
- The *two* approximate Dirichlet problems (3.11), (3.12) to compute ω_h and Ψ_h .

We have in particular to solve $2\ N_h + 4$ *approximate Dirichlet problems.*

4.6. On the conditioning of A_h .

The *condition number* $\nu(M)$ of $N \times N$ square matrix M is given by

(4.16) $$\nu(M) = \|M\| \, \|M^{-1}\| \ ,$$

where $\|M\|$ denote the matricial norm of M related to the standard euclidian norm of $\mathbb{R}^N$. If M is *symmetric* and *positive definite* then

(4.17) $$\nu(M) = \frac{\mu_{max}}{\mu_{min}}$$

where μ_{max} (resp. μ_{min}) is the *largest* (resp. *smallest*) *eigenvalue* of M . The smaller $\nu(A_h)$ will be, the easier to solve (directly or iteratively) will be (E_h). If the hypotheses on $\mathcal{C}_h$ are those of Sec. 3.3 it follows from Theorem 3.3 and (4.17)

THEOREM 4.1. : *In the case of Lagrange finite elements, and with the same hypotheses than in Theorem 3.3 we have for* $k \geq 1$

(4.18) $$\nu(A_h) = O(\frac{1}{h}) .$$

REMARK 4.3. The matrices occuring in the classical approximations of Δ (resp.Δ^2) have a condition number in $O(h^{-2})$(resp.$O(h^{-4})$). ∎

4.7. Some Remarks.

REMARK 4.4. The $2N_h+4$ approximate Dirichlet problems we have seen in Sec.4.5 can be formulated by

(4.19) $$(-\Delta)_h u_h = c_h$$

where $(-\Delta)_h$ is a $N'_h \times N'_h$ symmetric and positive definite matrix (approximating $-\Delta$), with $N'_h = \dim (V_{oh})$.

Since the $2\,N_h+4$ systems (4.19) are only different by their right hand sides, a *Cholesky factorization* taking account of the *sparsity* of $(-\Delta)_h$ can be done. Doing so, we shall obtain

(4.20) $$(-\Delta)_h = \Lambda_h \Lambda_h^t .$$

The matrix Λ_h is a *regular, lower triangular matrix*. Once Λ_h has been computed and stored, solving the $2\,N_h+4$ systems (4.19) is not very costly. ∎

REMARK 4.5. All the results obtained in Sec. 4.5 still hold if we use *numerical integration* to define W_{gh} and (P_h). Particularly if $k = 1$ and if we use *special triangulations*, approximating $\int_\Omega q_h \mu_h dx$ by

(4.20) $$\begin{cases} \frac{1}{3} \sum_{T \in \mathcal{C}_h} \text{Measure } (T) \sum_{i=1}^{3} q_h(M_{iT})\mu_h(M_{iT}) , \\ \\ M_{iT} , 1 \leq i \leq 3 , \text{ vertices of } T , \end{cases}$$

gives the *classical 13-points finite difference approximation* of (P_o) (cf. GLOWINSKI [2] , GLOWINSKI-LIONS-TREMOLIERES [1,ch.4]).

5. ON THE USE OF ITERATIVE METHODS. THE CONJUGATE GRADIENT METHOD.

5.1. Généralités.

The use of *gradient* or *conjugate gradient* methodswill give the possibility of solving (E_h) without knowing explicitly A_h. Actually we shall have "only" to solve *two* approximate Dirichlet problems at each iteration. Being given the limited number of pages at our disposal we shall give only a brief description of the methods, refering to GLOWINSKI-PIRONNEAU [3], BOURGAT-GLOWINSKI-PIRONNEAU [1], BOURGAT [1] for a much more detailled analysis. In order to describe these algorithms it is convenient to introduce $r_h : M_h \to \mathbb{R}^{N_h}$ defined as follows :

Let $\mathcal{B}_h = \{w_i\}_{i=1}^{N_h}$ be the basis of M_h of Sec. 4.1. If $\mu_h \in M_h$ we have

(5.1) $$\mu_h = \sum_{i=1}^{N_h} \mu_i w_i \ ;$$

then r_h is the *isomorphism* defined by

(5.2) $$r_h\mu_h = \{\mu_1, \mu_2, \ldots, \mu_{N_h}\} \qquad \forall\, \mu_h \in M_h \ .$$

If we denote by $(\cdot,\cdot)_h$ the canonical euclidian inner-product of $\mathbb{R}^{N_h}$ and $\|\cdot\|_h$ the corresponding norm, we have

(5.3) $$a_h(\lambda_h,\mu_h) = (A_h r_h\lambda_h, r_h\mu_h)_h \qquad \forall\, \lambda_h,\mu_h \in M_h \ ,$$

(5.4) $$\int_\Omega \nabla\Psi_{oh}\cdot\nabla\mu_h dx - \int_\Omega \omega_{oh}\mu_h dx - \int_\Gamma g_{2h}\mu_h d\Gamma = (b_h, r_h\mu_h)_h \quad \forall\, \mu_h \in M_h \ ,$$

where A_h and b_h have been defined in Sec.4.1 and 4.3 respectively.

5.2. Gradient methods with constant step .

5.2.1. Description of the algorithm

Let $s_h : M_h \times M_h \to \mathbb{R}$ be a symmetric and positive definite bilinear form and let consider $\rho > 0$. Then the gradient method with a constant step is given by

(5.5) $$\lambda_h^o \in M_h \ \text{arbitrary given,}$$

λ_h^n known, we compute λ_h^{n+1} by

$$(5.6)\quad \begin{cases} \int_\Omega \nabla\omega_h^n\cdot\nabla v_h\, dx = \int_\Omega f_h v_h\, dx \quad \forall\, v_h \in V_{oh}\ , \\ \omega_h^n - \lambda_h^n \in V_{oh}\ , \end{cases}$$

$$(5.7)\quad \begin{cases} \int_\Omega \nabla\psi_h^n\cdot\nabla v_h dx = \int_\Omega \omega_h^n v_h\, dx \quad \forall\, v_h \in V_{oh}\ , \\ \psi_h^n \in V_h\ ,\ \psi_h^n = g_{1h} \text{ on } \Gamma\ , \end{cases}$$

$$(5.8)\quad \begin{cases} s_h(\lambda_h^{n+1},\mu_h) = s_h(\lambda_h^n,\mu_h) + \rho\left(\int_\Omega \nabla\psi_h^n\cdot\nabla\mu_h dx - \int_\Omega \omega_h^n\mu_h dx - \int_\Gamma g_{2h}\mu_h\, d\Gamma\right) \\ \forall\, \mu_h \in M_h\ ,\ \lambda_h^{n+1} \in M_h\ . \end{cases}$$

Since $s_h(\cdot,\cdot)$ is symmetric and positive definite, there exists a unique matrix S_h symmetric and positive definite such that

$$(5.9)\qquad s_h(\lambda_h,\mu_h) = (S_h r_h\lambda_h, r_h\mu_h)_h\ .$$

Then (5.8) could be written

$$(5.10)\qquad r_h\lambda_h^{n+1} = r_h\lambda_h^n - \rho\, S_h^{-1}(A_h r_h\lambda_h^n - b_h).$$

It follows from (5.10) that the computation of λ_h^{n+1} will require the solution of a linear system the matrix of which is S_h .

5.2.2. Choice of s_h .

The choice of $s_h(\cdot,\cdot)$ is a very important matter and we refer to GLOWINSKI-PIRONNEAU [3], BOURGAT-GLOWINSKI-PIRONNEAU [1] for a detailled discussion of this problem.

If we assume that M_h is defined by (4.2) and that we use *Lagrange finite elements* , then for $s_h(\cdot,\cdot)$ we can use (cf. CIARLET-GLOWINSKI [1], GLOWINSKI-LIONS-TREMOLIERES [1, ch. 4])

$$(5.11)\qquad s_h(\lambda_h,\mu_h) = \int_\Gamma \lambda_h\mu_h\, d\Gamma\ ,$$

$$(5.12)\qquad s_h(\lambda_h,\mu_h) = \int_\Omega \lambda_h\mu_h\, dx$$

$$(5.13)\qquad s_h(\lambda_h\mu_h) = \int_\Omega \nabla\lambda_h\cdot\nabla\mu_h\, dx$$

The matrices related to (5.11), (5.12) , (5.13) are sparse but non diagonal.

Using numerical integration it is easy to approximate forms (5.11), (5.12) by bilinear forms for which S_h is diagonal (cf. GLOWINSKI-PIRONNEAU [3], LIONS-GLOWINSKI-TREMOLIERES, loc. cit.) .

5.2.3. Convergence of the gradient algorithm (5.5)-(5.8)

We can prove

THEOREM 5.1. : *Let* $(\lambda_h^n)_n$ *be the sequence defined by* (5.5)-(5.8) *and* λ_h *be the solution of* (E_h). *Then* $\forall \lambda_h^o \in M_h$

$$\lim_{n \to +\infty} \lambda_h^n = \lambda_h$$

if and only if

$$0 < \rho < \frac{2}{\Lambda_{N_h}} ,$$

where Λ_{N_h} *is the largest eigenvalue of* $S_h^{-1} A_h$. ∎

5.3. Gradient methods with variable step.

When using algorithm (5.5)-(5.8) the optimal choice of ρ is an important and non easy problem. In order to avoid this kind of difficulties we can use at the price of some extra *computations* the two following variants of (5.5)-(5.8) :

• Method of steepest descent.

(5.14) $$\lambda_h^o \in M_h$$

(5.15) $$g_h^o = A_h r_h \lambda_h^o - b_h ,$$

then for $n \geq 0$

(5.16) $$\rho_n = \frac{(S_h^{-1} g_h^n, g_h^n)_h}{(A_h S_h^{-1} g_h^n, S_h^{-1} g_h^n)_h}$$

(5.17) $$r_h \lambda_h^{n+1} = r_h \lambda_h^n - \rho_n S_h^{-1} g_h^n ,$$

(5.18) $$g_h^{n+1} = g_h^n - \rho_n A_h S_h^{-1} g_h^n . \quad \blacksquare$$

• Method of minimal residual.

It is like the above algorithm except that we replace (5.16) by

(5.19) $$\rho_n = \frac{(A_h S_h^{-1} g_h^n , S_h^{-1} g_h^n)_h}{(S_h^{-1} A_h S_h^{-1} g_h^n, A_h S_h^{-1} g_h^n)_h}$$

In these two algorithms the step (5.18) will require if A_h is not known the solution of two approximate Dirichlet problems.

REMARK 5.1. Using for instance the methods of MARCHOUK-KUZNETSOV [1], it will be interesting to estimate the rate of convergence of the two above algorithms.

5.4. Solution of (E_h) by a conjugate gradient method.

Since A_h is symmetric and positive definite it is rather natural to solve (E_h) by a conjugate gradient method. We recall that this method is quadratically convergent and theorically convergent in a finite number of iteration if there are no round-off errors.

In GLOWINSKI-PIRONNEAU [3] it is proved that when solving (E_h) , the conjugate gradient method takes the following form :

(5.20) $$\lambda_h^o \in M_h \ ,$$

(5.21) $$g_h^o = A_h r_h \lambda_h^o - b_h \ ,$$

(5.22) $$z_h^o = g_h^o \ ,$$

and for $n \geqslant 0$,

(5.23) $$\rho_n = \frac{(z_h^n, g_h^n)_h}{(A_h z_h^n, z_h^n)_h}$$

(5.24) $$r_h \lambda_h^{n+1} = r_h \lambda_h^n - \rho_n z_h^n \ ,$$

(5.25) $$g_h^{n+1} = g_h^n - \rho_n A_h z_h^n$$

(5.26) $$\gamma_n = \frac{(g_h^{n+1}, g_h^{n+1})_h}{(g_h^n, g_h^n)_h}$$

(5.27) $$z_h^{n+1} = g_h^{n+1} + \gamma_n z_h^n \ . \ \blacksquare$$

We notice that for $n \geqslant 0$ and if we assume that λ_h^n , g_h^n , z_h^n are known , then we have to solve *two approximate Dirichlet problems* to compute $A_h z_h^n$. Once this vector is known we compute ρ_n , λ_h^{n+1} , g_h^{n+1} , then γ_n and z_h^{n+1} . ■

6 COMMENTS.

From our numerical experimentations it appears that the two most efficient methods are :

(i) The *conjugate gradient method* described in Sec. 5 if (P_h) has to be solved only once.

(ii) The "*quasi direct*" *method* described in Sec.4 if we need a *biharmonic solver* to be used many times. This situation could occur when solving iteratively *Navier Stokes equations* in the $\{\Psi,\omega\}$ formulation or a time dependent Stokes (or Navier-Stokes) problem since in that last case the various methods discussed in Sec. 4.5 could be easily extended (cf. BOURGAT-GLOWINSKI-PIRONNEAU [3], GLOWINSKI-PIRONNEAU [4]). ■

In conclusion we would like to mention that there exists in the litterature a collection of paper related to the solution of the Dirichlet problem for Δ^2 through a sequence of pairs of Dirichlet problems for $-\Delta$. Let us mention among others SMITH [1],[2],[3], BOSSAVIT [1], EHRLICH [1], [2], [3], Mc LAURIN [1], EHRLICH-GUPTA [1].These papers are related to *finite differences approximations on rectangles* and are not using the fact that actually the discrete problem is equivalent to solving a linear system related to the *discrete vorticity trace* with a *symmetric* and *positive definite matrix* . It follows from this, that to our knowledge the "quasi direct" method of Sec.4 , the gradient methods with variable step and the conjugate gradient method of Sec. 5 seems to be new. Obviously the good tool to derive these algorithms, was the *mixed finite element approximation* of Sec. 3.

Applications of the gradient method with constant steps (cf. Sec. 5.2) are given in BOURGAT [1].

BIBLIOGRAPHY.

BOSSAVIT, A. [1] Une méthode de décomposition de l'opérateur biharmonique. Note HI 585/2, Electricité de France, (1971).

BOURGAT, J.F. [1] Numerical study of a dual iterative method for solving a finite element approximation of the biharmonic equation, LABORIA Report 156, and to appear in Comp. Meth. Applied Mech. Eng.

BOURGAT, J.F., GLOWINSKI, R., PIRONNEAU, O. [1] Numerical methods for the Dirichlet problem for the biharmonic equation and applications (to appear)

CEA, J. [1] Optimisation. Théorie et Algorithmes, Dunod, 1971.

CIARLET, P.G., GLOWINSKI, R. [1] Dual iterative techniques for solving a finite element approximation of the biharmonic equation, Comp. Meth. Applied Mech. Eng. 5, (1975), pp. 277-295.

CIARLET, P.G., RAVIART, P.A. [1] A mixed finite element method for the biharmonic equation, in Mathematical aspects of finite elements in partial differential equations. C. de Boor, Ed. Acad. Press, (1974), pp. 125-145.

[2] Interpolation theory over curved element with application to finite element methods. Comp. Meth. Applied Mech. Eng. 1, (1972), pp. 217-249.

CONCUS, P., GOLUB G.H. [1] Monography on conjugate gradient (to appear).

DANIEL, J.W. [1] The approximate minimization of functionals. Prentice Hall (1970).

EHRLICH, L.W. [1] Solving the biharmonic equation as coupled difference equations Siam J. Num. Anal. 8 (1971), pp. 278-287

[2] Coupled harmonic equations, SOR and Chebyshef acceleration, Math. Comp. 26, (1972), pp. 335-343.

[3] Solving the biharmonic equation in a square. Comm. ACM, 16 (1973), pp. 711-714.

EHRLICH, L.W., GUPTA, M.M. [1] Some difference schemes for the biharmonic equation Siam J. Num. Anal. 12, (1975), pp. 773-790.

GLOWINSKI, R. [1] Approximations externes par éléments finis d'ordre un et deux du problème de Dirichlet pour Δ^2. In Topics in Numerical Analysis, J.J.H. Miller Ed, Academic Press, (1973), pp. 123-171.

[2] Sur l'équation biharmonique dans un domaine multi-connexe C.R.A.S. Paris. (to appear).

GLOWINSKI, R., LIONS, J.L., TREMOLIERES, R. [1] Analyse Numérique des Inéquations variationnelles (Tome 2), Dunod-Bordas, (1976).

GLOWINSKI, R., PIRONNEAU O. [1] Sur la résolution numérique du problème de Dirichlet pour l'opérateur biharmonique par une méthode "quasi-directe". C.R.A.S. Paris, t. 282 A, pp. 223-226, (1976).

[2] Sur la résolution numérique du problème de Dirichlet pour Δ^2 par la méthode du gradient conjugué. Applications. C.R.A.S. Paris, t. 282 A, p. 1315-1318 (1976)

[3] Sur la résolution par une méthode "quasi directe", et par diverses méthodes itératives, d'une approximation par éléments finis mixtes du problème de Dirichlet pour Δ^2. Report 76010, Laboratoire d'Analyse Numérique, Université Paris6, (1976).

[4] Stanford University report (to appear).

LIONS, J.L., MAGENES, E. [1] Problèmes aux limites non homogènes, (T.1), Dunod, 1968.

MARCHOUK, G.I., KUZNETSOV, J.A. [1] Méthodes Itératives et Fonctionnelles Ouadratiques dans : Sur les Méthodes Numériques en Sciences Physiques et Economiques, LIONS J.L., MARCHOUK G.I., Ed. Dunod, 1974, pp. 1-132.

Mc LAURIN, J.W. [1] A genral coupled equation approach for solving the biharmonic boundary value problem, Siam J. Num. Anal. 11, (1974) pp. 14-33.

POLAK, E. [1] Computational Methods in Optimization, Acad. Press, 1971.

SHOLZ, R. [1] Approximation Von sattelpunkten mit finiten elementen (to appear).

VARGA,R.S. [1] Matrix iterative Analysis, Prentice-Hall, 1962.

YOUNG, D.M. [1] Iterative solution of large linear systems, Acad. Press, 1971

SUR L'APPROXIMATION DE PROBLEMES A FRONTIERE LIBRE DANS LES MATERIAUX INHOMOGENES

J.L. LIONS
Collège de France et Laboria

INTRODUCTION.

On va décrire ci-après quelques résultats très partiels (et quelques problèmes ouverts) entrant dans le thème général suivant:

comment approcher numériquement la solution de problèmes à frontière libre (problèmes avec obstacle, problèmes unilatéraux, etc.) dans des milieux à structure du type "matériaux composites"?

On va appliquer dans ce qui suit les techniques d'homogénéisation et d'Inéquations Variationnelles (I.V); appliquant ensuite les méthodes numériques d'approximation des solutions des I.V (cf. Glowinski-Lions-Trémolières [1]), on obtient aussi des méthodes donnant la convergence des solutions; la convergence de l'"approximation" des frontières libres n'est pas établie, mais semble vérifiée: elle l'est en tous cas sur les exemples traités numériquement (cf. Bourgat [1]).

L'homogénéisation est rappelée au N°. 1. Elle est due à plusieurs auteurs (cf. Spagnolo [1], de Giorgi-Spagnolo [1], Sbordone [1], Babuska [1], Bahbalov [1], Sanchez-Palencia [1] et la bibliographie de ces travaux) dans le cas d'opérateurs symetriques; le cas opérateurs non symétriques et des opérateurs d'ordre quelconque peut être abordé par les méthodes d'échelles multiples (cf. Bensoussan-Lions-Papanicolaou [1] ...[4]) et par des méthodes d'énergie très élégantes dues à Tartar [1].

Nous utilisons aux N°. 2 et 3 l'homogénéisation et les I.V pour la résolution de deux problèmes à frontière libre.

Les résultats numériques présentés ici ont été obtenus par Bourgat [1] auquel nous renvoyons pour de nombreux compléments.

Le plan est le suivant:

1. EXEMPLE D'HOMOGENEISATION

1.1. Position du problème

On considère des fonctions a_{ij} ayant les propriétés suivants:

$$(1.1)\quad \begin{cases} y\to a_{ij}(y),(i,j=1,\dots,n) \text{ est dans } L^\infty(Y), \\ Y=\prod_{i=1}^{n} \left]0,y_i^o\right[, \text{ à valeurs réelles,} \end{cases}$$

$$(1.2)\quad \sum_{i,j=1}^{n} a_{ij}(y)\xi_i\xi_j \geq \alpha \sum_{i=1}^{n} \xi_i^2,\ \alpha>0, \text{ p.p. dans } Y.$$

On supposera que $y\to a_{ij}(y)$ est prolongée à R^n par Y-périodicité (périodo de y_i^o en y_i).

On considère également une fonction a_o avec

$$(1.3)\quad a_o \in L^\infty(Y), \text{ à valeurs réelles } \geq \alpha_o > 0,$$

et l'on supposera que a_o est prolongée à R^n par Y-périodicité.

On considère l'opérateur A^ε défini, pour $\varepsilon>0$ "petit", par

$$(1.4)\quad A^\varepsilon\phi = -\sum_{i,j=1}^{n} \frac{\partial}{\partial x_i}\left(a_{ij}\left(\frac{x}{\varepsilon}\right)\frac{\partial\phi}{\partial x_j}\right) + a_o\left(\frac{x}{\varepsilon}\right)\phi.$$

Il s'agit donc d'un opérateur elliptique à coefficients très rapidement oscillants, tels qu'il s'en rencontre dans la modélisation de problèmes physiques à structure périodique "fine".

On considère maintenant des problèmes aux limites attachés à A^ε sur un ouvert Ω borné de R^n.

On introduit dans ce but les espaces de Sobolev usuels: soit $H^1(\Omega) = \{v|v, \frac{\partial v}{\partial x_i} \in L^2(\Omega)\}$, muni de sa structure hilbertienne usuelle:

$$||v||^2_{H^1(\Omega)} = \int_\Omega \left(v^2 + \sum_{i=1}^{n}\left(\frac{\partial v}{\partial x_i}\right)^2\right)dx;$$

soit $H_o^1(\Omega)$ l'adhérence dans $H^1(\Omega)$ du sous espace $\mathcal{D}(\Omega)$ des fonctions C^∞ à support compact; soit ensuite V avec

(1.5) $H^1_o(\Omega) \subset V \subset H^1(\Omega)$, V fermé dans $H^1(\Omega)$,

les inclusions dans (1.5) étant strictes ou non.

Pour $u,v \in H^1(\Omega)$ on pose

(1.6) $$a^\varepsilon(u,v) = \sum_{i,j} \int_\Omega a_{ij}(\frac{x}{\varepsilon}) \frac{\partial u}{\partial x_j} \frac{\partial v}{\partial x_i} dx + \int_\Omega a_o(\frac{x}{\varepsilon})\, u\, v\, dx.$$

Grâce à (1.2)(1.3), il existe $u_\varepsilon \in V$ unique tel que

(1.7) $a^\varepsilon(u_\varepsilon,v) = (f,v) \quad \forall\, v \in V,$

(où $(f,v) = \int_\Omega f\, v\, dx$ et où, par example, $f \in L^2(\Omega)$).

REMARQUE 1.1.

La forme $a^\varepsilon(u,v)$ n'est pas supposée symétrique.

Notre object est maintenant l'étude de u_ε lorsque $\varepsilon \to 0$.

1.2 Opérateurshomogénéisés.

Pour $\phi,\psi \in H^1(Y)$, on pose

(1.8) $$a_Y(\phi,\psi) = \sum \int_Y a_{ij}(y) \frac{\partial \phi}{\partial y_j} \frac{\partial \psi}{\partial y_i} dy.$$

On introduit le sous espace W des fonctions $\psi \in H^1(Y)$ et qui sont périodiques au sens: les traces de ψ sur des faces opposées de Y sont égales.

Désignant par y_j la fonction $y \to y_j$, on introduit $\chi^j \in W$ comme la solution (définie à l'addition d'une constante près) de

(1.9) $a_Y(\chi^j,\psi) = a_Y(y_j,\psi) \quad \forall\, \psi \in W.$

On définit alors (de manière unique)

(1.10) $$q_{ij} = \frac{1}{|Y|} a_Y\, (\chi^j - y_j, \chi^i - y_i)$$

où $|Y|$ = volume de Y.

REMARQUE 1.2.

On prendra garde aux ordres des indices i et j dans (1.10). Si l'on veut le même ordre des indices dans les deux termes de l'égalité, on introduit l'adjoint $a^*_Y(\phi,\psi) = a_Y(\psi,\phi)$; alors

(1.10 bis) $$q_{ij} = \frac{1}{|Y|} a^*_Y(\chi^i - y_i,\ \chi^j - y_j).$$

On appelle alors opérateur homogénéisé $\mathcal{A}$ de A^ε l'operateur à coefficients constants

(1.11) $$\mathcal{A} = - \sum_{i,j=1}^{n} q_{ij} \frac{\partial^2}{\partial x_i \partial x_j} + \overline{a_o},$$

où $\overline{a_o}$ est la valeur moyenne de a_o sur Y:

(1.12) $\overline{a_o} = \frac{1}{|Y|} \int_Y a_o(y)dy.$

On vérifie facilement que $\mathcal{A}$ est elliptique.

1.3 Problèmes aux limites homogénéisés.

Pour $u,v \in H^1(\Omega)$ on pose

(1.13) $\mathcal{A}(u,v) = \Sigma \int_\Omega q_{ij} \frac{\partial u}{\partial x_j} \frac{\partial v}{\partial x_i} dx + \int_\Omega \overline{a_o}\ u\ v\ dx.$

Il existe alors un élément $u \in V$ et un seul tel que

(1.14) $\mathcal{A}(u,v) = (f,v).$

On démontre alors (cf. L. Tartar [1], A. Bensoussan , J.L. Lions et G. Papanicolaou [4]) que

(1.15) la solution u_ε de (1.7) converge, lorsque $\varepsilon \to 0$, dans V faible, vers la solution u du problème aux limites homogénéisé (1.14).

Nous allons maintenant voir comment ce résultat s'étend à des problèmes à frontière libre.

2. PROBLEMES AVEC OBSTACLES

2.1. Position du problème

Les notations sont celles du N°. 1. On cherche maintenant u_ε solution de

(2.1) $A^\varepsilon u_\varepsilon - f \geqslant 0,\ u_\varepsilon \geqslant 0,\ (A^\varepsilon u_\varepsilon - f)u_\varepsilon = 0$ dans Ω,

(2.2) $u_\varepsilon = 0$ sur Γ = frontière de Ω.

Plus généralement, on appelle "problème aux limites avec obstacle" le problème:

(2.3) $A^\varepsilon u_\varepsilon - f \geqslant 0,\ u_\varepsilon - \psi \geqslant 0,\ (A^\varepsilon u_\varepsilon - f)(u - \psi) = 0$

avec (2.2), ψ représentant l'obstacle. On s'intéresse ici au cas "$\psi = 0$" pour simplifier.

D'après (2.1) il y a dans Ω deux régions:

dans une région C_ε (ensemble de coincidence) on a:

$u_\varepsilon = 0$,

et dans $\Omega - C_\varepsilon$ on a, $Au_\varepsilon = f$.

L' interface S_ε entre les deux régions est une surface libre.

REMARQUE 2.1.

On renvoie à l'exposé de C. Baiocchi [1] à ce colloque, et à la bibliographie de ce travail, pour des problèmes de ce genre relatifs à un opérateur elliptique.

Tout ce qui suit vaut avec des conditions aux limites différentes de (2.2) et notamment avec des dérivées obliques (ce qui correspond aux conditions aux limites rencotrées par C. Baiocchi [1],

Notre object est l'étude de u_ε et de S_ε lorsque $\varepsilon\to 0$.

2.2 Formulation en I.V.

On peut formuler (2.1)(2.2) sous forme d'une I.V.. On Définit

(2.4) $K = \{v \mid v \in H_0^1(\Omega),\ v \geq 0 \text{ p.p. dans } \Omega\}$;

alors (2.1)(2.2) (avec les notations du N° 1) équivaut à l'I.V. suivante: trouver u_ε telle que

(2.5) $\begin{cases} u_\varepsilon \in K, \\ a^\varepsilon(u_\varepsilon, v-u_\varepsilon) \geq (f, v-u_\varepsilon) \quad \forall v \in K. \end{cases}$

On sait (Lions-Stampacchia [1]) que ce problème admet une solution unique.

2.3 Homogénéisation.

La forme $\mathcal{A}(u,v)$ étant la même que celle introduite au N° 1, on appel le "I.V. homogénéisée" l'I.V. suivante:

(2.6) $\begin{cases} u \in K \\ \mathcal{A}(u, v-u) \geq (f, v-u) \quad \forall v \in K. \end{cases}$

On démontre alors (cf. Bensoussan-Lions-Papanicolau [1]) que:

(2.7) lorsque $\varepsilon\to 0$, la solution u_ε de l'I.V. (2.5) converge dans V faible vers la solution u de l'I.V. homogénéisée (2.6).

REMARQUE 2.2.

Le résultat (2.7) a été démontré, avec des variantes diverses, par plusieurs auteurs: Boccardo-Marcellini [1], Boccardo-Capuzzo-Dolcetta [1], Attouch [1], Konishi [1].

REMARQUE 2.3.

Il n'est pas vrai qu'un résultat du type (2.7) soit vrai pour toute inéquation attachée à A^ε, i.e. quel que soit l'ensemble convexe fermé (non vide) K dans (2.5). Voici un contre exemple très simple, dû à L. Tartar: supposons que K vérifie

(2.8) K est un ensemble convexe compact (non vide) de $H^1(\Omega)$;

soit u_ε la solution du problème (2.5) correspondant; introduisons alors

(2.9) $\bar{a}(u,v) = \Sigma \int_\Omega \bar{a}_{ij} \frac{\partial u}{\partial x_j} \frac{\partial v}{\partial x_j} dx + \int_\Omega \bar{a}_0\, u\, v\, dx,$

où

(2.10) $\overline{a_{ij}} = \frac{1}{|Y|} \int_Y a_{ij}(y)dy;$

soit u la solution de

$$(2.11)\quad \begin{cases} \bar{a}(u,v-u) \geq (f,v-u) \quad \forall v \in K, \\ u \in K; \end{cases}$$

on a alors

$$(2.12)\quad \begin{cases} u_\varepsilon \to u \text{ dans } V, \text{ où } u \text{ est la solution de } (2.11) \\ \underline{(\text{et non pas de } (2.6) \text{ avec } K \text{ donné par } (2.8))}. \end{cases}$$

REMARQUE 2.4.

On peut d'ailleurs construire des exemples (Carbone [1] , Tartar [2]) où le problème homogénéisé n'est donné ni par $\mathcal{A}(u,v)$ ni par $\bar{a}(u,v)$.

REMARQUE 2.5. (Problème ouvert)

Le problème (2.6) correspond à

$$(2.13)\quad \mathcal{A}u-f \geq 0,\ u \geq 0,\ (\mathcal{A}u-f)u = 0 \ \text{dans } \Omega$$

$$(2.14)\quad u = 0 \text{ sur } \Gamma;$$

On a encore $\Omega = C_o \cup \{\Omega - C_o\}$, $C_o = \{x | u(x) = 0\}$, avec une interface S_o entre C_o et $\Omega - C_o$, qui est la surface libre du problème homogénéisé.

Le problème suivant est ouvert:

$$(2.15)\quad \begin{cases} \underline{S_o \text{ réalise-t-elle une bonne approximation, lorsque } \varepsilon \to 0,} \\ \underline{\text{des surfaces libres } S_\varepsilon} \ ? \end{cases}$$

On va indiquer maintenant brièvement les résultats d'expériences numériques.

2.4. Exemples numériques

On prend $\Omega =]0,1[^2$ dans R^2.

Le parallélotope Y dans R^2_y est $]0,1[^2$: les coefficients $a_{ij}(y)$ sont des fonctions discontinues dans Y données comme suit:

$$(2.16)\quad \begin{cases} Y = Y_\alpha \cup Y_\beta \\ Y_\alpha = \left]\frac{1}{3}, \frac{2}{3}\right[^2 \quad , Y_\beta = Y - Y_\alpha; \end{cases}$$

les a_{ij} sont constantes par morceaux, avec

$$(2.17)\quad \begin{cases} a_{ij} = \alpha_{ij} \quad \text{dans } Y_\alpha,\ \beta_{ij} \text{ dans } Y_\beta \\ \alpha_{11} = -5,\ \alpha_{22} = 10,\ \alpha_{12} = \alpha_{21} = 2, \\ \beta_{11} = 2,\ \beta_{22} = 1;\ \beta_{12} = \beta_{21} = 1 \end{cases}$$

L'opérateur A^ε est donné par (1.14) avec

$$a_o(y) = 1.$$

Du point de vue numérique, on va calculer 1°) u_ε pour diverses valeurs de ε; 2°) les coefficients de l'opérateur homogénéisé et 3°) la solution u de l'I.V. homogénéisé.

La fonction f est choisie par

(2.18) $f(x) = 360\ x_1\ x_2(1-x_1)(1-x_2)-5x_1-10.$

Calcul de u_ε

On utilise pour la discrétisation des éléments finis linéaires avec une triangulation qui s'appuie sur les discontinués des coefficients $a_{ij}(\frac{x}{\varepsilon})$ et pour la résolution de l'I.V. une méthode d'itération avec projection (comme dans Glowinsky, Lions, Trémolières [1]).

Le calcul est fait pour

$\varepsilon = \frac{1}{2}, \frac{1}{4}, \frac{1}{8}.$

Afin de pouvoir tenir compte des discontinuités des a_{ij}, le nombre de triangles à introduire est déjà de l'ordre de 1.152 pour $\varepsilon = \frac{1}{8}$.

Les résultats numériques montrent la convergence (prévue par la théorie) de u_ε (en fait de $u_{\varepsilon h}$ = solution approchée de l'I.V. relative à ε) lorsque $\varepsilon \to 0$ et une convergence très régulière de la surface libre S_ε (en fait de la surface libre $S_{\varepsilon h}$ approchée). (La notation $S_{\varepsilon h}$ est symbolique, car pour pouvoir effectuer un calcul significatif, h doit décroitre avec ε).

Calcul de $\mathcal{A}$.

Les coefficients q_{ij} sont colculés sur Y par les formules (1.9)(1.10); on choisit $\chi^j = 0$ aux sommets de Y.

On utilise encore des éléments finis du 1er ordre avec une triangulation s'appuyant sur les interfaces Y_α et Y_α. On prend 400 triangles. On obtient:

$$(2.19)\qquad \begin{cases} q_{11} = 2,205 \quad , \quad q_{22} = 1,25 \\ q_{12} = q_{21} = 1,035. \end{cases}$$

Calcul de la solution u de l'I.V. homogénéisée.

On calcule u par la même méthode que les u_ε, avec une triangulation fixe de 400 triangles.

On aboutit ainsi aux conclusions suivantes:

CONCLUSIONS

1) l'approximation de u_ε par u (en fait de $u_{\varepsilon h(\varepsilon)}$ par u_h) est très bonne, dès $\varepsilon = \frac{1}{2}$, excellente pour $\varepsilon = \frac{1}{8}$;

2) dès que ε est très petit (par example $\varepsilon < \frac{1}{20}$) la méthode proposée semble être la seule raisonablement possible;

3) pour $\varepsilon = \frac{1}{8}$ le gain de temps est de l'ordre d'un facteur 4 (temps de calcul des q_{ij} non compris; mais ce calcul est effectué une

fois pour toutes; en outre le gain de temps est encore de l'ordre d'un facteur 2 même en tenant compte du temps de calcul des q_{ij}).

L'approxiamtion de la surface libre.

Comme déjà dit, nous n'avons pas de résultat de convergence établi pour l'approximation de S_ε par S_o (cf. (2.15)); les résultats numériques semblent suggérer que

(2.20) $\quad$ distance $(S_\varepsilon, S_o) = O(\varepsilon)$

mais les expériences numériques faites ne sont pas encore assez nombreuses pour pouvoir énoncer (2.20) comme une conjecture.

Pour une étude plus complète de tous ces aspects, nous renvoyons à J.F. Bourgat [1].

3. PROBLEMES UNILATERAUX

3.1. Position du problème

Les notations étant toujours celles du N°. 1, on cherche maintenant u_ε solution de

(3.1) $\quad A^\varepsilon u_\varepsilon = f$ dans Ω,

avec les conditions aux limites unilatérales

(3.2) $\quad u_\varepsilon \geqslant 0, \quad \dfrac{\partial u_\varepsilon}{\partial \nu_{A^\varepsilon}} \geqslant 0, \quad u_\varepsilon \dfrac{\partial u_\varepsilon}{\partial \nu_{A^\varepsilon}} = 0$ sur Γ

où $\dfrac{\partial}{\partial \nu_{A^\varepsilon}}$ désigne de la dérivée conormale associée à A^ε.

Si l'on introduit

(3.3) $\quad K = \{v \mid v \in H^1(\Omega),\ v \geqslant 0 \text{ sur } \Gamma\}$

alors (3.1)(3.2) équivaut à l'I.V.

(3.4) $\quad \begin{cases} u_\varepsilon \in K, \\ a^\varepsilon(u_\varepsilon, v-u_\varepsilon) \geqslant (f, v-u_\varepsilon) \quad \forall v \in K. \end{cases}$

D'après (3.2) la frontière Γ de Ω est divisée en deux régions: la région $\Gamma_{o\varepsilon}$ (ensemble de contact) où $u_\varepsilon = 0$, et la région $\Gamma - \Gamma_{o\varepsilon}$ où $\dfrac{\partial u_\varepsilon}{\partial \nu_{A^\varepsilon}} = 0$. L'interface σ_ε entre ces deux régions est une surface libre (sur Γ).

On veut étudier u_ε et σ_ε lorsque $\varepsilon \to 0$.

3.2 Homogénéisation

Les notations étant celles du N°. 1, on introduit "l'I.V. homogénéisée".

$$(3.5)\quad \begin{cases} u_\varepsilon \in K, \\ \mathcal{A}(u,v-u) \geq (f,v-u) \quad \forall v \in K. \end{cases}$$

On a alors (cf. Bensoussan-Lions-Papanicolau [4])

(3.6) $u_\varepsilon \to u$ dans $H^1(\Omega)$ faible.

REMARQUE 3.1. (Problème ouvert).

Le problème (3.5) équivant à

(3.7) $\quad u = f$ dans Ω .

(3.8) $\quad u \geq 0$, $\frac{\partial u}{\partial \nu} \geq 0$, $\; u \frac{\partial u}{\partial \nu} = 0$ sur Γ;

on a encore $\Gamma = \Gamma_o \bigcup \{\Gamma - \Gamma_o\}$, Γ_o = ensemble de Γ où $u = 0$, avec une interface σ_o.

Nous ignorons si σ_o réalise une bonne approximation de σ_ε lorsque ε est "petit".

3.3. Résultats numériques

Les résultats numériques (cf. J.F. Bougart [1]) ont été obtenus dans des conditions analogues à celles du N° 2.4.

Les conclusions sont tout à fait comparables à celles du N° 2.4. Là encore, il samble y avoir une "très bonne" convergence des ensembles $\Gamma_{o\varepsilon}$ vers Γ_o.

BOBLIOGRAPHIE

ATTOUCH [1] Thèse, Paris, 1976.

I. BABUSKA [1] Reports, Univ. of Maryland, 1974.

C. BAIOCCHI [1] Ces proceedings.

N.V. BAKBALOV [1] Averaged characteristics of bodies with periodic structure . Sov. Phys. Doklady, 19, N° 10, 1975, p. 650-651.

A. BENSOUSSAN, J.L. LIONS, G. PAPANICOLAU [1] Sur quelques phénomènes asymptotiques stationnaires. C.R.A.S. 281 (1975), p. 89-94.

[2] Sur quelques phénomènes asymptotiques d'évolution C.R.A.S. 281(1975), p. 317-322.

[3] Sur de nouveaux phénomènes asymptotiques. C.R.A.S. 281 (1975).

[4] Livre en préparation.

L. BOCCARDO, I. CAPUZZO DOLCETTA [1] G-convergenze e problema di Dirichlet unilaterale. A paraitre.

L. BOCCARDO, P. MARCELLINI [1] Sulla convergenza delle soluzioni di disequazioni variazionali. Istituto Mat. U. Dini, Firenze, Rapport Avril 1975.

BOURGAT [1] Rapport Laboria 1975.

L. CARBONE CRISTIANO [1] Sur la convergence des intégrales du type de l'énergie sur des fonctions à gradient borné. J. de M. P.A. 1976.

E. Di GIORGI, S. SPAGNOLO [1] Sulla convergenza degli integrali della energia per operatori ellittici del 2° ordine. Boll. U.M.I. (4) 8 (1973), 391-411.

R. GLOWINSKI, J.L. LIONS, R. TREMOLIERES |1| Analyse Numérique des Inéquations Variationnelles. Vol. 1 et 2, Paris, Dunod-Bordas, 1976.

Y. KONISHI [1] Une remarque sur la convergence des résolvantes non linéaires. L.A.N. 189, Université de Paris VI, 1976.

J.L. LIONS, G. STAMPACCHIA [1] Variational Inequalities. Comm. Pure Applied Math. 20 (1967), 493-519.

P. MARCELLINI [1] Un teorema di passaggio al limite per la somma di funzioni convesse. Boll. U.M.I. 4 (11) (1975).

E. SANCHEZ-PALENCIA [1] Comportaments local et macroscopique d'un type de milieux physiques hétérogènes. Int. J. Engeng. Sci. 12 (1974), 331-351.

C. SBORDONE [1] Sulla G-convergenza di equazioni ellittiche e paraboliche . Ricerche di Mat.

S. SPAGNOLO [1] Sulla convergonza di soluzioni di equazioni parabolicge ed ellittiche. Ann. S. Normale Sup. Pisa,XXII(1968),571-597.

L. TARTAT [1] C.R.A.S. Paris, 1975.

[2] C.R.A.S. Paris, 1975.

SUR LES PROBLEMES VARIATIONNELS NON COERCIFS ET L'EQUATION DU TRANSPORT

José Luis MENALDI y Edmundo ROFMAN
Instituto de Matemática "Beppo Levi"
Universidad Nacional de Rosario
ARGENTINA

(Ce travail est inclus dans le Programme de Cooperation avec l'IRIA-Rocquencourt-FRANCIA)

§1. PRESENTATION.

Dans cet exposé on désire présenter une application de la technique d'approximation interne (dans notre cas, la méthode d'éléments finis) avec régularisation,pour une classe de *problèmes variationnels elliptiques* pas nécessairement coercives. A titre d'exemple on résoud *l'équation du transport de neutrones*, tel qu'il était présenté dans [1].

Soit Ω un ouvert convexe borné du plan (x,y) , avec frontière suffisamment régulière, et par n_x et n_y les composants du vecteur unitaire de la normale extérieure à Γ , et enfin soit Q le disque unité du plan (μ,ν). Le problème est:

$$\text{(A)} \quad \begin{cases} \textit{Trouver une fonction } U = U(x,y,\mu,\nu) \textit{ telle que} \\ \mu\dfrac{\partial u}{\partial x} + \nu\dfrac{\partial u}{\partial y} + \sigma u = f \qquad \textit{dans} \quad \Omega \times Q \quad , \\ u(x,y,\mu,\nu) = 0 \quad \textit{si} \quad (x,y) \in \Gamma \quad \textit{avec} \quad \left(\mu n_x + \nu n_y\right)(x,y) < 0 \quad . \end{cases}$$

Le problème sera étudié comme un cas particulier du problème suivante:

Soit $V \subset H$ deux espaces de Hilbert réels; V dense dans H avec injection continue.

Soit $a(.,.)$ une forme bilinéaire continue dans $V \times H$ et elliptique, c'est-à dire:

$$\text{(1.1)} \quad \begin{cases} \text{il existe } M \in \mathbb{R}^+ \text{ tel que } |a(v,\tilde{v})| \leq M\|v\|_V\|\tilde{v}\|_H \quad , \ \forall\, v \in V \ , \ \tilde{v} \in H \quad ; \\ a(v,v) > 0 \qquad\qquad v \in V \quad , \quad v \neq 0 \quad . \end{cases}$$

On prend un élément l du dual de H, $l \in H'$; le problème est

(B) *Trouver* $u \in V$ *tel que* $a(u,v) = \langle\, l,v \,\rangle \quad \forall\, v \in V$.

La résolution approximative du problème (B) se fera avec l'hypothèse d'existence de la solution.

§2. TRANSFORMATION DU PROBLEME (A) A UN PROBLEME DU TYPE (B).

On fixe le paramètre (μ,ν) , et on appelle

(2.1) $$\Gamma_- = \left\{(x,y) \in \Gamma \;/\; (\mu n_x + \nu n_y)(x,y) < 0\right\}$$

puis le problème (A) s'écrit

(C) $$\begin{cases} \textit{Trouver } U = U(x,y) \textit{ tel que} \\ \mu\dfrac{\partial u}{\partial x} + \nu\dfrac{\partial u}{\partial y} + \sigma u = f & \textit{dans } \Omega \;, \\ u = 0 & \textit{sur } \Gamma_- \;. \end{cases}$$

Mais le problème (C) peut aussi s'écrire sous la forme (B) pour $H = L^2(\Omega)$, $V = \left\{v \in H \;/\; \mu\frac{\partial v}{\partial x} + \nu\frac{\partial v}{\partial y} \in H \;,\; v/_{\Gamma_-} = 0\right\}$ avec le produit interne naturel et si on suppose:

(2.2) $f \in L^2(\Omega)$;

(2.3) il existent $\alpha,\beta \in \mathbb{R}^+$ tels que $\alpha \leq \sigma(x,y) \leq \beta$ (p.p) pour $(x,y) \in \Omega$,

et on prend

(2.4) $$a(v,\tilde{v}) = \int_\Omega \left(\mu\frac{\partial v}{\partial x} + \nu\frac{\partial v}{\partial y} + \sigma v\right)\tilde{v}\,dx\,dy \;,\quad v \in V \,,\; \tilde{v} \in H \;,$$

(2.5) $$\langle 1,\tilde{v}\rangle = \int_\Omega f \,.\, \tilde{v}\,dx\,dy \qquad \text{pour} \qquad \tilde{v} \in H \;.$$

Remarque: l'hypothèse (2.3) peut être remplacée par

(2.3)' $$\sigma \in L^\infty(\Omega)$$

avec un changement convenable de la fonction inconnue.

§3. RESOLUTION APPROXIMATIVE DU PROBLEME (B).

Soit $\left\{V_h \,,\, r_h\right\}_{h>0}$ une approximation interne de V convergente, c'est-à-dire

(3.1) $$V_h \subset V \;,\quad r_h : V \to V_h \text{ tel que } \|r_h v - v\|_V \to 0 \quad \text{si } h \downarrow 0 \;\; \forall\, v \in V$$

et on définit, pour $\varepsilon > 0$, la forme bilinéaire

(3.2) $$a_\varepsilon(v,\tilde{v}) = a(v,\tilde{v}) + \varepsilon(v,\tilde{v})_V \qquad \forall\, v,\tilde{v} \in V \;;$$

le problème approché est

($B_{\varepsilon h}$) *Trouver* $u_{\varepsilon h} \in V_h$ *tel que* $a_\varepsilon(u_{\varepsilon h},v_h) = \langle 1,v_h\rangle \quad \forall\, v_h \in V_h$.

On sait [2] que:

$$u_{\varepsilon h} \to u_\varepsilon \quad \text{fort dans } V \text{ si } h \downarrow 0$$
$$u_\varepsilon \to u \quad \text{fort dans } V \text{ si } \varepsilon \downarrow 0$$

u étant la solution de (B), et u_ε la solution de

(B_ε) *Trouver* $u_\varepsilon \in V$ *tel que* $a_\varepsilon(u_\varepsilon,v) = \langle 1,v\rangle \quad \forall\, v \in V$.

Pour le problème ($B_{\varepsilon h}$) on a le résultat suivant:

THEOREME.

On suppose que la solution du problème (B) existe; alors la solution $u_{\varepsilon h}$ *du problème* $(B_{\varepsilon h})$ *converge dans la norme de l'espace* V *vers la solution* u *du problème* (B) *quand* ε *et* h *tend vers zéro avec* $\varepsilon^{-1}\|r_h u - u\|_H$ *tendant aussi vers zéro, c'est-à-dire*

(3.3) $\quad \|u - u_{\varepsilon h}\|_V \longrightarrow 0 \quad$ *si* $\quad \varepsilon \downarrow 0 \,,\ h \downarrow 0$ *avec* $\varepsilon^{-1}\|r_h u - u\|_H \longrightarrow 0$.

Il faut remarquer que la condition $\varepsilon^{-1}\|r_h u - u\|_H \longrightarrow 0$ *dépend de la solution* u *que l'on ne connait pas.*

Mais si on a

(3.4) $\qquad V \subset H \qquad$ *avec injection compacte*

il en résulte

(3.5) $\qquad \|I - r_h\|_{V \to H} = \alpha(h) \longrightarrow 0 \qquad$ *si* $\qquad h \to 0$

et (3.3) *peut s'écrire*

(3.3)' $\quad \|u - u_{\varepsilon h}\|_V \longrightarrow 0 \quad$ *si* $\quad \varepsilon \downarrow 0 \,,\ h \downarrow 0 \quad$ *avec* $\quad \varepsilon^{-1}\alpha(h) \longrightarrow 0$.

DEMONSTRATION DU THEOREME.

On la fera en trois parties

1°) $\qquad \|u_{\varepsilon h}\|_V \leqq c \quad , \quad \forall\, \varepsilon, h \quad ; \quad$ c ne dépend pas de ε ni de h .

Dans (B) on prend $v = u_{\varepsilon h} - u$ et, dans $(B_{\varepsilon h})$, $v_h = r_h u - u_{\varepsilon h}$:

$$a(u, u_{\varepsilon h} - u) = \langle 1, u_{\varepsilon h} - u \rangle \qquad ,$$

$$a(u_{\varepsilon h}, r_h u - u_{\varepsilon h}) + \varepsilon(u_{\varepsilon h}, r_h u - u_{\varepsilon h})_V = \langle 1, r_h u - u_{\varepsilon h} \rangle \qquad ;$$

on additionne, utilisant l'ellipticité de $a(.,.)$,

$$a(u, u_{\varepsilon h} - u) + a(u_{\varepsilon h}, r_h u - u_{\varepsilon h}) \leqq a(u_{\varepsilon h}, r_h u - u) \qquad :$$

on a

$$a(u_{\varepsilon h}, r_h u - u) + \varepsilon(u_{\varepsilon h}, r_h u - u_{\varepsilon h})_V \geqq \langle 1, r_h u - u \rangle \qquad ;$$

ensuite

(3.6) $$\left(\|1\|_{H'} + M\|u_{\varepsilon h}\|_V\right)\varepsilon^{-1}\|r_h u - u\|_H \geqq (u_{\varepsilon h}, u_{\varepsilon h} - r_h u)_V$$

(3.7) $$\|u_{\varepsilon h}\|_V \|r_h u\|_V + \left(\|1\|_{H'} + M\|u_{\varepsilon h}\|_V\right)\varepsilon^{-1}\|r_h u - u\|_H \geqq \|u_{\varepsilon h}\|_V^2$$

alors de (3.7) résulte le 1°) et d'après (3.6) on a

(3.8) $$\overline{\lim_{\varepsilon,h}}\,(u_{\varepsilon h}, u_{\varepsilon h} - r_h u) \leqq 0 \quad . \ \blacksquare$$

2°) $\qquad u_{\varepsilon h} \longrightarrow u \qquad$ faible dans V .

D'après 1°), il existe $u_o \in V$ et une sous suite $\{u_{\varepsilon' h'}\}$ telle que

$$u_{\varepsilon' h'} \longrightarrow u_o \quad \text{faible dans } V \quad ;$$

alors comme on a pour tout v dans V les égalités suivantes:

(3.9) $$\lim_{\varepsilon' h'} a(u_{\varepsilon' h'}, r_{h'} v) = a(u_o, v)$$

(3.10) $$\lim_{h'} \langle 1, r_{h'} v \rangle = \langle 1, v \rangle$$

on utilise $(B_{\varepsilon h})$ et il résulte des (3.9) et (3.10)

$$a(u_o, v) = \langle 1, v \rangle \qquad \forall v \in V \quad .$$

Puis par unicité de la solution, on aura $u_o = u$ et, ensuite, 2°). ∎

3°) $u_{\varepsilon h} \longrightarrow u$ fort dans V .

On peut ecrire

$$\|u_{\varepsilon h} - u\|_V^2 = (u_{\varepsilon h}, u_{\varepsilon h} - r_h u)_V + (u_{\varepsilon h}, r_h u - u)_V - (u, u_{\varepsilon h} - u)_V \quad ;$$

si on utilise (3.8), (3.9) et 2°), on obtien 3°). ∎∎

On peut remarquer qu'en réalité on n'utilise pas (3.1) mais par contre on a besoin du

(3.9)' $$\begin{cases} \|r_h v - v\|_V \longrightarrow \text{ si } h \downarrow 0 \quad \forall v \text{ dans un dense de } V \quad \text{et} \\ \|r_h u - u\|_V \longrightarrow 0 \quad \text{si} \quad h \downarrow 0 \quad . \end{cases}$$

§4. RESOLUTION DU PROBLEME (A) AVEC LA TECHNIQUE ENONCEE.

On suppose Ω domaine avec sa fonction polygonal, on définit une triangulation τ_h (h étant un paramètre qui sera précisé après) comme étant une famille finie de triangle ayant les propriétés suivantes:

(4.1) $$\bigcup_{T \in \tau_h} T = \Omega$$

(4.2) $$\left\{ T, T' \in \tau_h \Longrightarrow T \cap T' = \phi \quad \begin{array}{l} \text{ou } T \text{ et } T' \text{ ont un sommet commun} \\ \text{ou } T \text{ et } T' \text{ ont un coté commun .} \end{array} \right.$$

Soient

(4.3) $$\Sigma_h = \left\{ \text{ensemble des sommets de } \tau_h \text{ qui ne sont pas dans } \Gamma_- \right\}$$

(4.4) $$h = \sup \left\{ \text{diam}(T) \;/\; T \in \tau_h \right\} \quad .$$

On suppose aussi que

(4.5) tous les angles de T dans τ_h sont supérieurs ou égaux a $\theta_o > 0$ pour tout h .

On définit

(4.6) $V_h = \left\{ v \in C^\circ(\Omega) \;/\; v \text{ est une fonction affine sur chaque triangle } T \in \tau_h \text{ et } v|_{\Gamma_-} = 0 \right\}$,

(4.7) $$V = \left\{ v \in H^1(\Omega) \;/\; v/_{\Gamma_-} = 0 \right\} \quad .$$

Alors avec

(4.8) $$w_{hM}(P) = \begin{cases} 1 & \text{si} \quad P = M \\ 0 & \text{si} \quad P \neq M \end{cases} \quad , \quad P \in \Sigma_h \quad \text{avec} \quad w_{hM} \in V_h$$

on a: $\{w_{hM}\}_{M\in\Sigma_h}$ une base de V_h et

(4.9) $$v_h = \sum_{M\in\Sigma_h} v_h(M)w_{hM} \qquad \forall\, v_h \in V_h \quad .$$

On définit l'opérateur

(4.10) $$r_h : V \longrightarrow V_h \quad \text{par} \quad r_h v = \sum_{M\in\Sigma_h} v(M) \qquad \forall\, v \in V \cap C^\circ(\overline{\Omega})$$

puis on a

(4.11) $$\|r_h v - v\|_H \leq c h^2 \|v\|_{H^2(\Omega)} \qquad \forall\, v \in H^2(\Omega) \quad , \quad h > 0 \quad .$$

Alors le problème $(B_{\varepsilon h})$ peut s'écrire

(C_h) $$\sum_{M\in\Sigma_h} \alpha_{MP}\, \xi_M = \beta_P \qquad \forall\, P \in \Sigma_h$$

équation linéaire sur R^N , $N = \dim(V_h) = \text{card}(\Sigma_h)$

avec $\alpha_{MP} = a(w_{hM}, w_{hP}) + h^\gamma (w_{hM}, w_{hP})_V \qquad 0 < \gamma < 2$

et $\beta_P = \langle 1, w_{hP} \rangle$; donc on trouve u_h , la solution de $(\beta_{\varepsilon h})$ pour $\varepsilon = h^\gamma$, de (4.9), avec $u_h(M) = \xi_M$.

Puis le théorème s'énonce par:

THEOREME.

Si on suppose que la solution u *du problème* (C) *appartient à* $H^2(\Omega)$ *il en résulte que la solution* u_h *du* (C_h) *converge dans* V *vers* u *lorsque* h *tend vers zéro, c'est-à-dire:*

(4.12) $$\|u_h - u\|_V \longrightarrow 0 \qquad \text{si} \qquad h \downarrow 0 \quad .$$

On doit remarquer que si on prend un autre opérateur Γ_h , on peut remplacer l'hypothèse $u \in H^2(\Omega)$ par $u \in H^1(\Omega)$ et on a (4.12) seulement por $0 < \gamma < 1$.

On remarque aussi que si on suppose que la solution de (C) appartient à $H^k(\Omega)$, k est un entier plus grand que 1 , on peut utiliser pour une régularisation dans (3.2) le produit interne dans $H^k(\Omega)$ et pour définir V_h , les éléments finis d'ordre k . On obtient ainsi une convergence similaire à (4.12) mais en norme de l'espace $H^k(\Omega)$.

REMARQUES.

- L'approximation présenté est aussi utile quand la forme bilinéaire est coercive, puisqu'on fait un changement d'espaces du travail et on obtient une meilleure convergence.

- On peut aussi utiliser l'approximation externe, mais il faut transformer le problème en un autre non-linéaire, ou on a une plus forte convergence. Cette technique sera le sujet d'une prochaine publication.

REFERENCES ET BIBLIOGRAPHIE

(Prochaine page).

REFERENCES ET BIBLIOGRAPHIE

[1] LESAINT, P. — RAVIART, P.A.. "*On a finite element method for solving the neutron transport equation*". Paper presented at the Symposium on Math. Aspects of Finite Elements in Partial Differential Equations, Madison Ap.1-3 1974.

[2] LIONS, J.L. — STAMPACCHIA, G.. "*Variational Inequalities*". Comm. on Pure and Ap. Math. Vol. XX. pp.493-519. (1967).

[3] CLEMENT, P.. "*Approximation by finite element functions using local regularization*". R.A.I.R.O. (9e. année, Août 1975, pp.77-84).

[4] LESAINT, P.. "*On Introduction to finite element methods*". Lecture notes of the Automn Course on Math. and Numerical Methods in Fluid Dynamics. SMR 13 A/35 , I. C.T.P. - (Trieste) - (Italy). 1973.

[5] MERCIER, B.. "*On the Boundary Conditions in the finite elements*". Lecture notes of the Automn Course on Math. and Numerical Methods in Fluid Dynamics. SMR 13 A/ 42. I.C.T.P. - (Trieste) - (Italy). 1973.

[6] TEMAM, R.. "*Analyse Numérique*". Presses Universitaires de France. 1970.

Los originales de este trabajo fueron preparados en el *Instituto de Matemática "Beppo Levi"* por la Sra. H. I. Warecki de MUTY.

APPLICATION OF A MIXED FINITE ELEMENT METHOD TO A NONLINEAR PROBLEM OF ELASTICITY

T. MIYOSHI
Department of Mathematics
Kumamoto University, Kumamoto (Japan)

Preface

Many new variational methods for solving boundary value problems were recently proposed. Some of these are very useful in practical applications. In fact, mathematical justifications of the methods too have been obtained for some ones. The mixed finite element method is one of such methods. In this paper we analyze a mixed finite element scheme for solving a boundary value problem which occurs in the analysis of nonlinear bending of elastic plates. This scheme can be regarded also as a generalized finite difference scheme, and in practical applications, it will be more convenient than the one proposed in the author's previous paper [5].

In this paper we consider the von Kármán equations. These equations may be inconvenient for solving the actual elastic problems, since the treatment of the boundary conditions is not so easy. However, the essential difficulty in solving nonlinear plate bending problems will consist in that the equations are fourth order and semi-linear. Therefore we believe that the method which is useful for these equations will be so for equations in other formulations.

1. Approximate scheme

Consider a thin elastic plate of arbitrary shape subjected to a lateral loard g. Let Ω be a bounded region of (x_1,x_2) - plane which represents the shape of the plate in its undeformed state. Then the system of equations

$$(1.1) \qquad \begin{cases} \Delta^2 f = -[w,w] \\ \Delta^2 w = [f,w] + g \end{cases}$$

is a mathematical model of the nonlinear bending of this plate, where w and f correspond to the normal deflection and Airy's stress function , respectively. Here, [f,w] denotes the nonlinear term :

$$[f,w] = D_{11}fD_{22}w + D_{22}fD_{11}w - 2\, D_{12}fD_{12}w \; ,$$

where $D_{ij}u$ denotes the second order derivative.

Our problem is to solve (1.1) under the boundary condition $w = dw/dn = f = df/dn = 0$, being n the outward normal to the boudary $\partial\Omega$. We assume through the present paper that $\partial\Omega$ and g are sufficiently smooth, so that the equation (1.1) has a sufficiently smooth solution (we refer to [1] or [2], for example, about the proof of the existence and smoothness of the solution).

In our formulation two space $Ł_2$ and H are essential. Let $W^k_\alpha(\Omega)$ (k; positive integer, $\alpha>1$) be the usual sobolev space of functions. Let $\mathring{W}^1_2(\Omega)$ be the completion of the space of all C^∞ - functions with support in Ω in the norm

$$|u|_1^2 = \sum_{|\alpha|=1} \int |D^\alpha u|^2 dx_1 dx_2 .$$

$Ł_2$ is the product space $W^1_2 \times L_2(\Omega) \times L_2(\Omega) \times L_2(\Omega)$ with the norm

$$\|W\|^2_{Ł_2} = |w|_1^2 + \sum_{i\leq j} \|W_{ij}\|^2_{L_2},$$

and H is the space $\mathring{W}^1_2 \times W^1_2 \times W^1_2 \times W^1_2$ with the norm obtained by changing the sufix L_2 in the right-side of the above expression to W^1_2, where $W = (w, W_{11}, W_{12}, W_{22})$. Let us define the following bi-linear form

$$L(W,\Phi) = \sum_{i\leq j} \{(D_j w, D_i \Phi_{ij})_{L_2} + (W_{ij}, \Phi_{ij})_{L_2}\} + \sum_{i,j} (D_i W_{ij}, D_j \phi)_{L_2}$$

for W, $\Phi \varepsilon H$, where $W_{12} = W_{21}$. By Sobolev's imbedding theorem we can define a weak solution of our problem as follows.

<u>DEFINITION.</u> Let $[F,W] = F_{11}W_{22} + F_{22}W_{11} - 2F_{12}W_{12}$. A pair $(F,W)\varepsilon H\times H$ is called a weak solution of the equation (1.1), if

$$(1.2)\quad \begin{cases} L(F,\Phi) = ([W,W],\phi)_{L_2} & \text{for all } \Phi \in H, \\ L(W,\Phi) + ([F,W],\phi)_{L_2} + (g,\phi)_{L_2} = 0 & \text{for all } \Phi \in H. \end{cases}$$

As pointed out in [5], equation (1.2) can be represented by a single operator equation. Let L, C and B be defined by

$$L(W,\Phi) = (LW,\Phi)_H \qquad \text{for all } \Phi\varepsilon H,$$

$$([F,W],\phi)_{L_2} = (C(F,W),\Phi)_H \qquad \text{for all } \Phi\varepsilon H,$$

$$(g,\phi)_{L_2} = (Bg,\Phi)_H \qquad \text{for all } \Phi \varepsilon H,$$

(these are well defined). Since L is invertible on

$$H_{1+\varepsilon} = (W^2_{1+\varepsilon} \cap \mathring{W}^1_2) \times \{0\} \times \{0\} \times \{0\} \quad (\varepsilon>0)$$

and $C(F,W)$ belongs to H_{1+1} for F,W in H, we have the following as abstract version of the von Karman equations.

$$(1.3)\qquad W + C(W) + L^{-1}Bg = 0,$$

where $C(W) = L^{-1}C(L^{-1}C(W,W),W)$. Take W_0, $W_1 \varepsilon H$ and set $Z = W_1 - W_0$. Then we can write

$$C(W_1) - C(W_0) = C'_{(W_0)}Z + D(W_0, Z).$$

where

$$C'_{(W_0)}Z = L^{-1}C(L^{-1}C(W_0,W_0),Z) + 2L^{-1}C(L^{-1}C(W_0,Z),W_0)$$

and $D(W_0,Z)$ is a nonlinear term of third order nonlinearity in Z.

The operator $C'_{(W_0)}$ is defined on H, but it can be extended to whole Ł as a compact operator [5]. In what follows we regard it as the extended operator and assume, for a fixed solution W_0 of (1.3), that the equation

$$(1.4)\qquad KZ = (I + C'_{(W_0)})Z = 0$$

has only a trivial solution, that is, we do not seek the solutions at which singular phenomena (like bifurcation etc.) occur.

Finite element subspaces : Let Ω_h $(h>0)$ be a triangulation of Ω. We assume Ω_h is a closed subregion of Ω satisfying the following four conditions.

(1) Any vertex of a triangle does not lie part way along the side of another.

(2) Adjecent nodes on $\partial\Omega_h$ do not lie together in Ω. If two adjecent nodes on $\partial\Omega_h$ are both on $\partial\Omega$, then the boudary $\partial\Omega$ must be nonconcave between these nodes. If p, q and r are serial nodes on $\partial\Omega_h$, being q in Ω, then the boundary $\partial\Omega$ must contain a concave part between p and r. The length of the perpendicular from q to the line segment connecting the boundary nodes on $\partial\Omega$ does not exceed $O(h^2)$ as h tends to 0, being h the largest side length of all triangles in Ω_h.

(3) The ratio of the smallest and the largest sides of triangles in Ω_h is bounded below by a positive constant as $h \to 0$.

(4) There is a closed subregion Ω_h' of Ω_h which is composed of square meshes of equal side length, and the number of grids in the set $\Omega_h - (\Omega_h')_{\text{interior}}$ is of order $O(h^{-1})$ as $h \to 0$. Each square in Ω_h' is triangulated by the diagonal of north-east direction.

Let $\{\hat{\phi}_p\}$ be the piecewise linear finite element basis belonging to $W_2^1(\Omega_h)$ and satisfying $\hat{\phi}_p = 1$ at the node p and $= 0$ at all other nodes. We extend $\hat{\phi}_p$ to the skin $\Omega - \Omega_h$ and regard it as a function in $W_2^1(\Omega)$. This is carried out by setting, for example, $d\hat{\phi}/d\nu = 0$, being ν the direction of the perpendicular to the line segment connecting the two boundary nodes on $\partial\Omega$. In what follows, $\{\hat{\phi}_p\}$ denotes the basis extended in this way. (Remark : This extention is only for rigorous theoretical treatment, and unnecessary for the actual computation.)

Corresponding to each $\hat{\phi}_p$ we define a piecewise constant function $\bar{\phi}_p$ as follows. Let $T_{p,k}$ ($k=1,2,\ldots, K_p$) be the set of all triangles in Ω_h with vertex p. Let $Q_{p,k}$ ($\subset T_{p,k}$) be the quadrilateral obtained by connecting the vertex p, middle points of the two sides containing p and the center of gravity of $T_{p,k}$. Then $\bar{\phi}_p$ is the characteristic function of the region

$$U_p = \sum_{k=1}^{K_p} Q_{p,k}$$

Subspaces used in the following discussion are

$\hat{S}_0$: subspace of $\overset{\circ}{W}{}_2^1(\Omega)$ spanned by $\{\hat{\phi}_p;\ p\varepsilon\Omega_h - \partial\Omega_h\}$,
$\hat{S}_1$: subspace of $W_2^1(\Omega)$ spanned by $\{\hat{\phi}_p;\ p\varepsilon\Omega_h\}$,
$\bar{S}$: linear space spanned by $\{\bar{\phi}_p;\ p\varepsilon\Omega_h\}$,
$\hat{H}$: $= S_0\times S_1\times S_1\times S_1$ (subspace of H).

<u>Finite element scheme</u> : The approximate scheme introduced and analyzed in [5] is

$$\text{(C)}\qquad \begin{cases} L(\hat{F},\hat{\Phi}) = ([\hat{W},\hat{W}],\hat{\phi})_{L_2} & \text{for all } \hat{\Phi}\ \varepsilon\ \hat{H}, \\ L(\hat{W},\hat{\Phi}) + ([\hat{F},\hat{W}],\hat{\phi})_{L_2} + (g,\hat{\phi})_{L_2} = 0 & \text{for all } \hat{\Phi}\ \varepsilon\hat{H}, \end{cases}$$

where $(\hat{F},\hat{W})\varepsilon\ \hat{H}\times\hat{H}$. This approximation is, in a certain sense, of consistent mass type, since we have to invert Gram matrices.
To describe more convenient approximation, let us introduce the following bilinear form on $\hat{H}\times\hat{H}$.

$$\bar{L}(\hat{W},\hat{\Phi}) \equiv \sum_{i\leq j}\{(D_j\hat{w},D_i\hat{\Phi}_{ij})_{L_2} + (\bar{W}_{ij},\bar{\Phi}_{ij})_{L_2}\} + \sum_{i,j}(D_i\hat{W}_{ij},D_j\hat{\phi})_{L_2},$$

where $\bar{W}_{ij}$ denotes the function which belongs to $\bar{S}$ and coincides with $\hat{W}_{ij}$ at all nodes in Ω_h (we define $\bar{W}_{ij} = 0$ outside of Ω_h).

The scheme proposed here is

$$\text{(L)}\qquad \begin{cases} \bar{L}(\hat{F},\hat{\Phi}) = ([\bar{W},\bar{W}],\bar{\phi})_{L_2} & \text{for all } \hat{\Phi}\ \varepsilon\hat{H}, \\ \bar{L}(\hat{W},\hat{\Phi}) + ([\bar{F},\bar{W}],\bar{\phi})_{L_2} + (g,\hat{\phi})_{L_2} = 0 & \text{for all } \hat{\Phi}\ \varepsilon\hat{H}, \end{cases}$$

where $[\bar{F},\bar{W}] = \bar{F}_{11}\bar{W}_{22} + \bar{F}_{22}\bar{W}_{11} - 2\bar{F}_{12}\bar{W}_{12}$. Note that this scheme is exactly a 13-points finite difference scheme in the interior of Ω_h'. Our problem is to study whether this can give reasonable approximate solution. Since this equation can not be treated under the same frame work as for (C) - because $\bar{L}(\ ,\ )$ etc. are not well defined in H in this form -, we have to change the point of view in this case.

In [5] the system (C) is represented in the form

$$\hat{L}\hat{W} + PC(\hat{L}^{-1}PC(\hat{W},\hat{W}),W) + PBg = 0,$$

where $\hat{L} = PLP$, being P the projection $H\to\hat{H}$. Now, Let us regard the equation (C) as an <u>original equation</u> defined only on $\hat{H}$, and write this in the form

$$(1.5) \qquad \hat{L}\hat{W} + \hat{C}(\hat{L}^{-1}\hat{C}(\hat{W},\hat{W}),\hat{W}) + \hat{B}g = 0.$$

In this expression, the operators $\hat{L}$, $\hat{C}$ and $\hat{B}$ work, of course, only in the space $\hat{H}$. The bilinear form $\bar{L}(\ ,\ )$ is now well defined on $\hat{H}\times\hat{H}$, and by the same reason as for $\hat{L}$, can be represented by a bounded operator on $\hat{H}$, say by $\bar{L}$:

$$\bar{L}(\hat{F},\hat{\Phi}) = (\bar{L}\hat{F},\hat{\Phi})_{\hat{H}} \qquad \text{for all } \hat{\Phi} \in \hat{H}.$$

On the other hand, the representation of the nonlinear terms is not selfevident. We have to prepare the following inequalities which can be proved by an elementally calculation.

<u>LEMMA 1.</u> For any $\hat{u}$ in $\hat{S}_1$ and any $p \geq 2$, holds

$$(1.6) \qquad \|\hat{u}\|_{L_p(\Omega)} \leq c_1\|\bar{u}\|_{L_p(\Omega_h)} \leq c_2\|\hat{u}\|_{L_p(\Omega)},$$

where c_1 and c_2 are constants independent of $\hat{u}$, h and p.

We can now estimate the nonlinear term as follows.

$$\left|([\bar{F},\bar{W}],\bar{\Phi})_{L_2}\right| \leq c\|\hat{F}\|_{\mathcal{L}_2}\|\hat{W}\|_{\hat{H}}|\hat{\Phi}|_1.$$

Therefore, for fixed $\hat{F}$ and $\hat{W}$ there is a unique $\bar{C}(\hat{F},\hat{W})\in \hat{H}$ such that

$$([\bar{F},\bar{W}],\bar{\Phi})_{L_2} = (\bar{C}(\hat{F},\hat{W}),\hat{\Phi})_{\hat{H}} \qquad \text{for all } \hat{\Phi} \in \hat{H}.$$

Taking into account the invertibility of $\bar{L}$, our equation (L) too can be represented by the similar from as for (C) :

$$(1.7) \qquad \bar{L}\hat{W} + \bar{C}(\bar{L}^{-1}\bar{C}(\hat{W},\hat{W}),\hat{W}) + \hat{B}g = 0.$$

Once the discrete system is expressed in this form, the analysis of the scheme can be performed by the same method used for (C). In what follows we shall describe the process briefly.

2. Some results from linear problems

Let us first consider the equation

$$\bar{L}(\hat{W},\hat{\Phi}) = (g, \bar{\phi})_{L_2} \qquad \text{for all } \hat{\Phi} \in \hat{H}. \tag{2.1}$$

This is , of course, an approximate scheme for the Dirichlet problem of the biharmonic equation $\Delta^2 w = g$. Careful reading of the proof given in [4] yields

THEOREM 1. Let w be the exact solution of the Dirichlet problem of the biharmonic equation for loard term $g \varepsilon L_2(\Omega)$ and W the solution of (2.1). Then, hold the following error estimates.

$$|w - \hat{w}|_1 , \quad \| D_{ij}w - \hat{W}_{ij}\|_{L_2} \leqslant ch^{1/2} \|g\|_{L_2} .$$

By operating $\bar{L}^{-1}$ to the both sides of (1.7),we rewrite the equation as follows.

$$\hat{W} + \bar{C}(\hat{W}) + \bar{L}^{-1}\hat{B}g = 0. \tag{2.2}$$

Let be defined $\bar{C}'_{(\hat{W})}$ by

$$\bar{C}'_{(\hat{W})}\hat{Z} = \bar{L}^{-1}\bar{C}(\bar{L}^{-1}\bar{C}(\hat{W},\hat{W}),\hat{Z}) + 2\bar{L}^{-1}\bar{C}(\bar{L}^{-1}\bar{C}(\hat{W},\hat{Z}),\hat{W}).$$

Then we can write

$$\bar{C}(\hat{W}_1) - \bar{C}(\hat{W}_0) = \bar{C}'_{(\hat{W}_0)}(\hat{W}_1 - \hat{W}_0) + \bar{D}(\hat{W}_0,\hat{W}_1 - \hat{W}_0), \tag{2.3}$$

where $\bar{D}$ is a nonlinear operator of third order nonlinearity defined by

$$\bar{D}(\hat{W}_0,\hat{Z}) = 2\bar{L}^{-1}\bar{C}(\bar{L}^{-1}\bar{C}(\hat{W}_0,\hat{Z}),\hat{Z}) + \bar{L}^{-1}\bar{C}(\bar{L}^{-1}\bar{C}(\hat{Z},\hat{Z}),\hat{W}_0 + \hat{Z}). \tag{2.4}$$

Let $\hat{C}'_{(\hat{W})}$ be the linear operator which is derived from equation (1.5) - the <u>original equation</u> for (2.2) - and has the same form as for $\bar{C}'_{(\hat{W})}$. Let W_0 be the solution of (1.3) - exact one for our problem - and $\hat{W}_0 \varepsilon \hat{H}$ be its interpolate. In [5] it is proved that if h is sufficiently small and (1.4) has no nontrivial solution, then the operator $I + \hat{C}'_{(\hat{W}_0)}$ is invertible on $\hat{H}$ and holds

(2.5) $$\sup_{\hat{W}\in\hat{H}} \|(I + \hat{C}'_{(\hat{W}_0)})^{-1}\hat{W}\|_{Ł_2}/\|\hat{W}\|_{Ł_2} \leq c < \infty \quad \text{as } h \to 0.$$

Our problem in this section is to prove that this is true for the equation

(2.6) $$\bar{K}\hat{Z} \equiv (I + \bar{C}'_{(\hat{W}_0)})\hat{Z} = \hat{G} \qquad \hat{G} \in \hat{H}$$

under the same assumption. The basic tools in treating this problem are the following estimates.

LEMMA 2.

(1) Let w be a function in W_2^1. Then holds

$$\left|(\hat{u} - \bar{u}, w)_{L_2}\right| \leq ch\|\hat{u}\|_{L_2} \|w\|_{W_2^1} .$$

(2) Let $\hat{U} = \bar{L}^{-1}\bar{C}(\hat{V},\hat{W})$ and $0<\varepsilon<1$ ($\hat{V},\hat{W} \in\hat{H}$). Then hold

a. $$\|\hat{U}\|_{Ł_2} \leq c_\varepsilon \left\|[\bar{V},\bar{W}]\right\|_{L_{1+\varepsilon}} ,$$

b. $$\|\hat{U}\|_{Ł_2} \leq c_\varepsilon \|\hat{V}\|_{\hat{H}} \|\hat{W}\|_{Ł_2} ,$$

c. $$\|\hat{U}\|_{Ł_2} \leq c\left|\hat{V}\right|_{max} \|\hat{W}\|_{Ł_2} ,$$

d. $$\left\|[\bar{V},Z]\right\|_{L_{1+\varepsilon}} \leq ch^{-2\varepsilon/(1+\varepsilon)} \|\hat{V}\|_{Ł_2} \|Z\|_{Ł_2} \qquad (Z \in Ł_2),$$

where $|\hat{V}|_{max} = \max(\max|\hat{v}|, \max|\hat{V}_{ij}|)$ and c_ε is a constant dependent possiblly on ε but not on h.

Proof of this lemma is not difficult. (1) is a consequence of the approximation theory in W_2^1 (prove first that the inequality holds for $\hat{w}$ in $\hat{S}_1$). The first 3 inequalities in (2) can be proved by the same way. We shall show (a) as an example.

The equation $\bar{L}^{-1}\hat{U} = \bar{C}(\hat{V},\hat{W})$ is equivalent to

(2.7) $$\sum_{i\le j}\{(D_j\hat{u},D_i\hat{\Phi}_{ij})_{L_2}+(\bar{U}_{ij},\bar{\Phi}_{ij})_{L_2}\}+\sum_{i,j}(D_i\hat{U}_{ij},D_j\hat{\phi})_{L_2}$$

$$=([\bar{V},\bar{W}],\bar{\phi})_{L_2}\qquad\text{for all } \hat{\Phi}\in\hat{H}.$$

Therefore, by Lemma 1 and Sobolev's imbedding theorem we have

$$\sum_{i,j}(\hat{U}_{ij},\hat{U}_{ij})_{L_2}\le c\,|([\bar{V},\bar{W}],\bar{u})_{L_2}|\le c_\varepsilon\|[\bar{V},\bar{W}]\|_{L_{1+\varepsilon}}|\hat{u}|_1.$$

At the same time (2.7) implies $|\hat{u}|_1\le c\,\|\hat{U}\|_{Ł_2}$. (a) is thus proved. To prove (d) we remind the inverse relation. Then it is easy to prove

$$\|\bar{u}\|_{L_{2+\varepsilon}}\le ch^{-\varepsilon'/(2+\varepsilon')}\|\bar{u}\|_{L_2},$$

where $2+\varepsilon' = 2(1+\varepsilon)/(1-\varepsilon)$. Therefore by Lemma 1 we have

$$\|[\bar{V},Z]\|_{L_{1+\varepsilon}}\le c\ \Sigma\|\bar{V}_{ij}\|_{L_{2+\varepsilon'}}\|Z_{kl}\|_{L_2}\le ch^{-\varepsilon'/(2+\varepsilon')}\|\hat{V}\|_{Ł_2}\|Z\|_{Ł_2},$$

which is the desired one.

The following theorem is the conclusion of this section.

<u>THEOREM 2.</u> Assume that the equation (1.4) has no nontrivial solution. Then, if h is sufficiently small, there is a function $\hat{H}^*$ for any $\hat{G}\in\hat{H}$ such that $\|\hat{H}^*\|_{Ł_2}\le c\|\hat{G}\|_{Ł_2}$ and holds

(2.8) $$\|\bar{K}\hat{H}^*-\hat{G}\|_{Ł_2}\le q\|\hat{G}\|_{Ł_2}\qquad(0<q<1),$$

where q is independent of h and $\hat{G}$. Therefore $\bar{K}$ is invertible for such small h and the norm of its inverse is uniformly bounded as h tends to 0.

<u>PROOF.</u> The inequality is satisfied by the function

$$\hat{H}^*=(I+\hat{C}'_{(\hat{W}_0)})^{-1}\hat{G}.$$

First we note that this is well defined for small h and the inverse operator used is uniformly bounded. In fact this was one of the main theorem of the previous paper [5]. If the existence of q in the theorem is shown, then the existence and uniqueness of the solution to (2.6), and the uniform boundedness of $\bar{K}^{-1}$ are a direct consequence of a theorem in approximate method [3].

Let us rewrite the left side of (2.8) as

$$\bar{K}\hat{H}^* - \hat{G} = \bar{C}'_{(\hat{W}_0)}\hat{H}^* - \hat{C}'_{(\hat{W}_0)}\hat{H}^* = S_1 + R_1 + S_2 + R_2,$$

where

$$\begin{aligned} S_1 &= \bar{L}^{-1}\bar{C}(\bar{L}^{-1}\bar{C}(\hat{W}_0,\hat{W}_0),\hat{H}^*) - L^{-1}C(L^{-1}C(W_0,W_0),\hat{H}^*), \\ R_1 &= L^{-1}C(L^{-1}C(W_0,W_0),\hat{H}^*) - \hat{L}^{-1}\hat{C}(\hat{L}^{-1}\hat{C}(\hat{W}_0,\hat{W}_0),\hat{H}^*), \\ S_2 &= 2\bar{L}^{-1}\bar{C}(\bar{L}^{-1}\bar{C}(\hat{W}_0,\hat{H}^*),\hat{W}_0) - 2L^{-1}C(L^{-1}C(W_0,\hat{H}^*),W_0), \\ R_2 &= 2L^{-1}C(L^{-1}C(W_0,\hat{H}^*),W_0) - 2\hat{L}^{-1}\hat{C}(\hat{L}^{-1}\hat{C}(\hat{W}_0,\hat{H}^*),\hat{W}_0). \end{aligned}$$

The quantities R_1 and R_2 are exactly those estimated in [5], and they satisfy

$$\|R_1\|_{Ł_2} \leq c_\varepsilon h^{1/2-2\varepsilon/(1+\varepsilon)} \|\hat{H}^*\|_{Ł_2},$$

$$\|R_2\|_{Ł_2} \leq ch^{1/2} \|\hat{H}^*\|_{Ł_2},$$

where ε (<1) is an arbitraly positive constant and c_ε is a constant depending possiblly on ε but not on h and $\hat{H}^*$. To estimate S_2, Let us introduce

$$V = L^{-1}C(W_0,\hat{H}^*), \qquad \hat{V} = \bar{L}^{-1}\bar{C}(\hat{W}_0,\hat{H}^*).$$

Then by Lemma 2 and the error estimate in Theorem 1 we have

$$\begin{aligned} \|S_2\|_{Ł_2} &\leq 2\|\bar{L}^{-1}\bar{C}(\hat{V},\hat{W}_0) - L^{-1}C(\bar{V},\bar{W}_0)\|_{Ł_2} \\ &\quad + 2\|L^{-1}C(\bar{V},\bar{W}_0) - L^{-1}C(V,W_0)\|_{Ł_2} \\ &\leq ch^{1/2} \|\hat{H}^*\|_{Ł_2}. \end{aligned}$$

S_1 is estimated by the same way too. In this case, however

$$\|S_1\|_{Ł_2} \leq c_\varepsilon h^{1/2-2\varepsilon/(1+\varepsilon)} \|\hat{H}^*\|_{Ł_2}.$$

Summarizing all estimates we finally get

$$\|\bar{K}\hat{H}^* - \hat{G}\|_{Ł_2} \leq c_\varepsilon h^{1/2-2\varepsilon/(1+\varepsilon)} \|\hat{G}\|_{Ł_2},$$

for sufficiently small h. Taking ε sufficiently small, and then making h small if necessary, we have

$$c_\varepsilon h^{1/2-2\varepsilon/(1+\varepsilon)} < 1,$$

which is the desired inequality.

3. Existence and convergence of the approximate solutions

The operator $\bar{C}'_{(\hat{W})}$ is the Frechet derivative of $\bar{C}(\hat{W})$. Theorem 2 thus suggests the applicability of the Newton's iteration which starts from $\hat{W}_0$. Let us rewrite the equation (2.2) as

$$\begin{aligned}\hat{W} &= R\hat{W} \\ &= \hat{W}_0 - (I + \bar{C}'_{(\hat{W}_0)})^{-1}[\hat{E} + \bar{D}(\hat{W}_0,\hat{Z})],\end{aligned} \tag{3.1}$$

where $\hat{E}$ denotes the residual given by substituting $\hat{W}_0$ into (2.2), $\bar{D}$ is the one defined by (2.4) and $\hat{Z} = \hat{W} - \hat{W}_0$. We first estimate the residual $\hat{E}$. The next lemma can be proved by introducing the function

$$L^{-1}C(L^{-1}C(W_0,W_0),\hat{W}_0),$$

Lemma 3. Let $W_0 \varepsilon H$ be a solution of the original (Karman's) problem and $\hat{W}_0 \varepsilon \hat{H}$ be its interpolate. Then holds, for any $1>\varepsilon>0$

$$\|\hat{E}\|_{Ł_2} = \|\hat{W}_0 + \bar{C}(\hat{W}_0) + \bar{L}^{-1}\hat{B}p\|_{Ł_2} \leq c_\varepsilon h^{1/2-2\varepsilon/(1+\varepsilon)}. \tag{3.2}$$

We can now prove our main theorem which states the existence and convergence of the approximate solutions.

Theorem 3. Let W_0 be a solution of (1.3). Assume that the linear, homogenious equation (1.4) has no nontrivial solution at W_0. Consider the iteration

$$\hat{W}_n = R\hat{W}_{n-1} \qquad (n=1,2,\dots). \tag{3.3}$$

Then, if h is sufficiently small, there is a closed ball

$$S_\delta = \{ \hat{W}\varepsilon \hat{H};\ \|\hat{W} - \hat{W}_0\|_{Ł_2} \leq \delta\} \qquad \delta = h^{1/2-3/(n+1)} \quad (n\geq 11,\text{integer})$$

such that

(A) $\|R\hat{W} - \hat{W}_0\|_{Ł_2} \leq \delta$ for all $\hat{W}\varepsilon S_\delta$,

(B) $\|R\hat{W}_1 - R\hat{W}_2\|_{Ł_2} \leq q\ \|\hat{W}_1 - \hat{W}_2\|_{Ł_2}$ $(0<q<1)$

for all $\hat{W}_1,\hat{W}_2\varepsilon S_\delta$. Therefore the iteration (3.3) defines a function which is a solution of the discrete equation (2.2). This solution is unique in the δ-neighbourhood of $\hat{W}_0$.

<u>PROOF.</u> (A): Assume that $\hat{W}$ belongs to S_δ for some constant δ. The key is the estimate of the nonlinear term $\overline{D}$ appearing in the identity (3.1). Let us devide $\overline{D}$ as

$$\overline{D}(\hat{W}_0,\hat{Z}) = D_1 + D_2 + D_3,$$

where

$$D_1 = 2\overline{L}^{-1}\overline{C}(\overline{L}^{-1}\overline{C}(\hat{W}_0,\hat{Z}),\hat{Z}),$$
$$D_2 = \overline{L}^{-1}\overline{C}(\overline{L}^{-1}\overline{C}(\hat{Z},\hat{Z}),\hat{W}_0),$$
$$D_3 = \overline{L}^{-1}\overline{C}(\overline{L}^{-1}\overline{C}(\hat{Z},\hat{Z}),\hat{Z}).$$

By employing Lemma 2 we can estimate these temrs as follows.

$$\|D_1\|_{Ł_2} \leq c_\varepsilon h^{-2\varepsilon/(1+\varepsilon)}\ \delta^2,$$
$$\|D_2\|_{Ł_2} \leq c_\varepsilon h^{-2\varepsilon/(1+\varepsilon)}\ \delta^2,$$
$$\|D_3\|_{Ł_2} \leq c_\varepsilon h^{-4\varepsilon/(1+\varepsilon)}\ \delta^3,$$

where ε is arbitrary constant satisfying $0<\varepsilon<1$. Substituting this into (3.1) we have

$$\|\hat{W}_0 - R\hat{W}\|_{Ł_2} \leq c_\varepsilon(h^{1/2-2\varepsilon/(1+\varepsilon)} + h^{-2\varepsilon/(1+\varepsilon)}\ \delta^2 + h^{-4\varepsilon/(1+\varepsilon)}\ \delta^3), \tag{3.4}$$

provided that $\hat{W}$ belongs to S_δ. If we choose $\varepsilon=1/n$ for positive integer n (≥ 11) and set $\delta= h^{1/2-3/(n+1)}$, then (3.4) implies

$$\| \hat{W}_0 - R\hat{W} \|_{Ł_2} \leq c_n h^{1/(n+1)} \delta .$$

Therefore the condition (A) is satisfied by this δ, provided that h is sufficiently small.

(B) : The way of proof is exactly the same for (A). In this case, however, we devide as follows.

$$\bar{D}(\hat{W}_0,\hat{Z}_1) - \bar{D}(\hat{W}_0,\hat{Z}_2) = \sum_{i=1}^{3} (D_i^{(1)} - D_i^{(2)}),$$

where

$$D_1^{(k)} = 2\bar{L}^{-1}\bar{C}(\bar{L}^{-1}\bar{C}(\hat{W}_0,\hat{Z}_k),\hat{Z}_k),$$
$$D_2^{(k)} = \bar{L}^{-1}\bar{C}(\bar{L}^{-1}\bar{C}(\hat{Z}_k,\hat{Z}_k),\hat{W}_0),$$
$$D_3^{(k)} = \bar{L}^{-1}\bar{C}(\bar{L}^{-1}\bar{C}(\hat{Z}_k,\hat{Z}_k),\hat{Z}_k),$$

and $\hat{Z}_i = \hat{W}_i - W_0$, being $\hat{W}_i$ in S_δ for some constant δ. The result is

$$\| R\hat{W}_1 - R\hat{W}_2 \|_{Ł_2} \leq c_\varepsilon(\alpha + \alpha^2) \| \hat{W}_1 - \hat{W}_2 \|_{Ł_2},$$

where $\alpha = h^{-2\varepsilon/(1+\varepsilon)}\delta$. In this case too we can set $\varepsilon = 1/n$ ($n \geq 11$, integer) and $\delta = h^{1/2-3/(n+1)}$, and (B) holds for sufficiently small h.

The last assertion will be evident, since the iteration (3.3) is contractive under these conditions.

<u>REMARK.</u> We may assert, pure-theoretically, that the order of convergence is almost the same as in the biharmonic case. However the numerical constants appearing in the proofs depend on the <u>n</u> (this n reflects the ε in Lemma 2), and if we want large n then we have to make h small inevitablly. This means that the actual accuracy of the approximate solution obtained will not be better than that of biharmonic case. Our numerical experience shows this is the case(for numerical examples and more detailed proof, see [6]). The scheme treated here is one of the simples mixed method. Therefore, If we use more complicated elements, then the convergence rate will increase ,of course, corresponding to its degree.

REFERENCES

[1]. Berger,M.S. : On von Kármán's equations and the buckling of a thin elastic plate I, The clamped plate. Comm. Pure and Appl. Math. 20,687-720(1967).

[2]. Knightly,G.H. : An existence theorem for the von Kármán equations. Arch. Rational Mech. Anal. 27,233-242(1967).

[3]. Kantrovich,L.V. and G.P.Akilov : Functional Analysis in Normed Spaces : Pergamon press 1964.

[4]. Miyoshi,T. : A finite element method for the solutions of fourth order partial differential equations. Kumamoto J. Sci. (Math.) 9,87-116(1972).

[5]. Miyoshi,T. : A mixed finite element method for the solutions of the von Kármán equations. Numer Math. (to appear).

[6]. Miyoshi,T. : Lumped mass approximation to the nonlinear bending of elastic plates. (to appear)

ERROR ESTIMATES FOR SOME VARIATIONAL INEQUALITIES

U. Mosco
Università di Roma

1. We suppose that Ω is a convex bounded open subset of $\mathbb{R}^n$, whose boundary we denote by Γ, and that $a(u,v)$ is the bilinear form

(1) $$a(u,v) = \sum_{i,j=1}^{n} \int_{\Omega} (a_{ij} u_{x_i} v_{x_j} + b_j u_{x_j} + cuv)\,dx$$

whose coefficients are real-valued functions and verify

(2) $$a_{i_j} \in C^1(\bar{\Omega}) \;,\quad b_j \;, c \in L^\infty(\Omega) \;,\qquad i,j = 1,\dots,n$$

(3) $$c(x) \geq c_0 > 0 \quad \text{a.e.}\, x \in \Omega$$

The form $a(u,v)$ is continuous on the Sobolev space $H^1(\Omega)$ and we denote by M a positive constant such that

(4) $$|a(u,v)| \leq M \|u\|_{1,\Omega} \|v\|_{1,\Omega} \qquad \forall u, v \in H^1(\Omega) .$$

By

$$\|\cdot\|_{s,\Omega}$$

we denote the usual norm of the Sobolev space $H^s(\Omega)$, $s = 0,1,2$, $H^0(\Omega) \equiv L^2(\Omega)$.

We also suppose that the form $a(u,v)$ is coercive on $H^1(\Omega)$, that is, there exists a constant $\gamma > 0$ such that

(5) $$a(u,u) \geq \gamma \|u\|_{1,\Omega} \quad \text{for all } u \in H^1(\Omega).$$

We suppose that two real-valued functions f and ψ are given on Ω, satisfying the conditions

(6) $$f \in L^2(\Omega)$$

(7) $$\psi \in H^2(\Omega).$$

We consider first the following *variational inequality*

(8) $$\begin{cases} u \in K \\ a(u,u-v) \leq (f,u-v) \qquad \forall v \in K \end{cases}$$

where K is the convex cone

(9) $$K = \left\{ v \in H^1_0(\Omega) \;:\; v \geq \psi \quad \text{a.e. in } \Omega \right\}$$

and ψ is supposed to be non-positive a.e. on Γ.

We use the notation

$$(f, w) = \int_{\Omega} f\, w\, dx \quad \text{for every } w \in L^2(\Omega).$$

Problem (8) is one of the most fundamental examples of variational inequalities, introduced by Lions and Stampacchia [9], and corresponds to the weak formulation of the so-called *unilateral Dirichlet problem* (or, *obstacle problem*), whose strong formulation is given by the following system of inequalities

$$\begin{cases} u - \psi \geq 0 \\ Lu - f \geq 0 \\ (u - \psi)\cdot(Lu - f) = 0 \end{cases}$$

to be satisfied in Ω, together with the boundary condition

$$u = 0 \qquad \text{on } \Gamma.$$

The operator L above is the 2nd order partial differential operator in divergence form associated with the form (1), i.e.,

$$Lu = \sum_{i,j=1}^{n} \left[-\frac{\partial}{\partial x_j}\left(a_{ij} \frac{\partial u}{\partial x_i}\right)\right] + \sum_j b_j \frac{\partial u}{\partial x_j} + cu \ .$$

For arbitrary $u \in H^1(\Omega)$, the identity

(10) $$\langle Lu, w \rangle = a(u,w) \quad , \qquad w \in H^1_0(\Omega)$$

defines Lu as an element of the dual space $H^{-1}(\Omega)$ of $H^1_0(\Omega)$.

We shall also consider the problem

(11) $$\begin{cases} u \in K_1 \\ a(u, u - v) \leq (f, u - v) \qquad \forall v \in K_1 \end{cases}$$

where

(12) $$K_1 = \left\{ v \in H^1(\Omega) \ : \ v \geq \psi \text{ on } \Gamma \right\} .$$

Problem (11) is the weak formulation of the problem

(13) $$Lu = f \quad \text{in } \Omega$$

with *Signorini boundary conditions*

$$u \geq \psi \qquad \frac{\partial u}{\partial \nu_L} \geq 0 \qquad (u - \psi)\frac{\partial u}{\partial \nu_L} = 0 \quad \text{on } \Gamma$$

where

$$\frac{\partial u}{\partial \nu_L} = \sum_{i,j=1}^{n} a_{ij} \frac{\partial}{\partial x_i} \cos(\vec{n}, x_j)$$

is the conormal derivative of u relative to L ($\vec{n}$ being the exterior normal to Γ).

2. As it has been shown by Lions and Stampacchia, [9], both problems (8) and (11) admit a unique solution u. What we want to discuss here is the *approximation* of u by continuous piece-wise affine approximate solutions u_h, the discretization parameter $h > 0$ being associated with the maximum mesh size of a suitable "triangularization" of the region Ω.

We shall deal mainly with problem (8) and we shall first describe some results due to R. Falk and to G.Strang and the author concerning the estimate of the discretization error

$$\|u - u_h\| \tag{14}$$

in the energy norm.

The approximate solution u_h, for each given h, is obtained by solving the problem

$$\left[\begin{array}{l} u_h \in K_h \\ a(u_h, u_h - v_h) \leq (f, u_h - v_h) \quad \forall v_h \in K_h \end{array}\right. \tag{15}$$

where K_h is a *finite-dimensional* approximation of K that we shall describe later. In any case, K_h will be taken to be a subset of the space

$$V = H_0^1(\Omega) \tag{16}$$

where the given K also lies in. In this sense our approximation scheme is an *internal* one, what enables us to consider the error (14) in the norm of the space V itself (the energy norm) or in that of the space

$$H = L^2(\Omega) . \tag{17}$$

Indeed, the basic question we want to answer is whether the standard estimates

$$\|u - u_h\|_{1,\Omega} \leq ch\|u\|_{2,\Omega} \tag{18}$$

and

$$\|u - u_h\|_{0,\Omega} \leq ch^2 \|u\|_{2,\Omega} \tag{19}$$

that we know to be true for the solution of the usual Dirichlet problem for the operator L and a continuous piece-wise affine approximate solution of it, or a variant of then, hold true for the unilateral Dirichlet problem we are dealing with.

As we shall see below, the answer is affirmative as far as (18) is concerned. As to (19), only partial results have been obtained so far, which leave the question essentially open as soon as $n > 1$.

3. As we know from the approximation theory of variational boundary value problems with no unilateral constraints (as the Dirichlet problem mentioned above), three are the main tools we must expect to be relevant to the estimates we are looking for: *regularity properties* of the solution u, *a priori estimates* of the error (14) in the chosen norm, *approximation results* by trial functions belonging to K_h.

As we shall see below, all these aspects of the problem present peculiar features in the unilateral case at hand. Two basic differences with respect to the Dirichlet problem can be point out since now. First, the set K and K_h are not linear subspaces of V, but only convex *cones*. Second, the approximation scheme we shall consider is not an internal one with respect to the constraint cone K itself, since we do not assume K_h to be a subset of K. Indeed, u_h will not be required to satisfy the constrained $u_h \geq \psi$ exactly, but only an approximate constraint $u_h \geq \psi_h$.

4. REGULARITY

It is well known, and it can be checked by trivial one-dimensional examples, that no matter how regular the coefficients of L, the boundary of Ω and the data f and ψ are, the solution u of problem (8) may well not be a function of class C^2.

However, under the assumptions of section 1, the following estimate for the solution u of problem (8) holds true

(20) $$\|u\|_{2,\Omega} \leq c\,(\|f\|_{0,\Omega} + \|\psi\|_{2,\Omega})\,,$$

which is the basic one needed in order to achieve the error estimates mentioned above.

Estimate (20) is due to H.Brezis and G.Stampacchia [3] when Ω has a smooth boundary. For domain Ω as in section 1, it can be derived from the corresponding regularity results for the Dirichlet problem due to P.Grisvard [6], by using the following dual estimate

(21) $$\|Lu\|_{0,\Omega} \leq \|\inf\{f,0\}\|_{0,\Omega} + \|\mathrm{Sup}\{f,L\psi,0\}\|_{0,\Omega}$$

for which we refer to [10],[12].

For problem (11), the estimate

(22) $$\|u\|_{2,\Omega} \leq c(\|f\|_{0,\Omega} + \|\psi\|_{\frac{3}{2},\Gamma})$$

has been obtained by H.Brezis [1], again for smooth Γ, and, recently, by P.Grisvard [7] for arbitrary convex Ω and $\psi = 0$.

5. A PRIORI ESTIMATES OF $\|u - u_h\|$.

When no unilateral constraints are involved and, say, $K = V$, the role of K_h is taken by some finite-dimensional subspace V_h of V and the following estimate

(23) $$\|u - u_h\|_{1,\Omega} \leq c\,\|u - v_h\|_{1,\Omega} \qquad \forall v_h \in V_h$$

is shown to hold. That is, the discretization error is of the same order, in the energy norm, than the *best approximation* error of u,

(24) $$\|u - \tilde{u}_h\|_{1,\Omega} = \inf_{v_h \in V_h} \|u - v_h\|_{1,\Omega} \quad ,$$

in terms of all trial functions $v_h \in V_h$.

This is no more true in general when unilateral constraints are involved and K, K_h are not linear subspaces of V. A trivial counter-example , due to G.Strang, is illustrated in the following self-explai ning picture

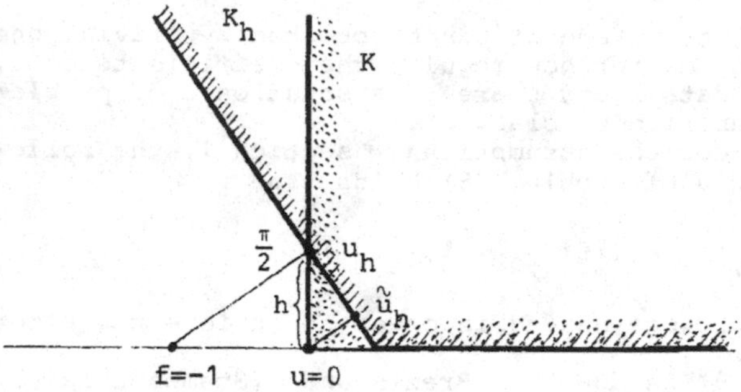

where $|u - u_h| \sim h$, whereas $|u - \tilde{u}_h| \sim h^2$.

Two kinds of a priori estimates replacing (23) have been considered so far, the first one, given in Lemma 1 below, by G.Strang and the author [11], the second one, given in Lemma 2, by R.Falk [5].

In Lemma 1 below we use the fact that

(24)' $$K = \psi + C$$

with $\psi \in V$ and C a convex cone in V, and

(24)" $$K_h = \psi_h + C_h$$

with $\psi_h \in V$ and

(25) $$C_h \subset C \; .$$

Lemma 1. *Let* $u = \psi + U$, $U \in C$, *be the solution of* (8) *and* $u_h = \psi + U_h$, $U_h \in C_h$, *the solution of* (15). *Then,*

$$(26) \qquad \|u - u_h\|_{1,\Omega} \leq \frac{M}{\gamma}\left[\|\psi - \psi_h\|_{1,\Omega} + \|U - V_h\|_{1,\Omega}\right]$$

for all $V_h \in C_h$ *such that* $2U - V_h \in C$.

When K is given by (9) and $\psi = 0$ a.e. on Γ, we have $V = H^1_0(\Omega)$,

$$(27) \qquad C = \left\{V \in H^1_0(\Omega) \ : \ V \geq 0 \quad \text{a.e. in } \Omega\right\}$$

and we obtain from (26) the estimate

$$(28) \qquad \|u - u_h\|_{1,\Omega} \leq \frac{M}{\gamma}\left[\|\psi - \psi_h\|_{1,\Omega} + \inf \|U - V_h\|_{1,\Omega}\right]$$

where the infimum is extended over all $V_h \in C_h$ such that $0 \leq V_h \leq U$ in Ω.

When $K = K_1$ is given by (12), we have $V = H^1(\Omega)$,

$$(29) \qquad C = \left\{V \in H^1(\Omega) \ : \ V \geq 0 \quad \text{a.e. on } \Gamma\right\} \quad ,$$

and we again obtain from (26) the estimate (28) where the infimum is now extended over all $V_h \in C_h$ such that $0 \leq V_h \leq U$ on Γ.

The basic remark yielding to (26) is that equality holds in (8) whenever $v \in K$ is such that $2u - v$ also belongs to K. Now, for any V_h as in Lemma 1 we have indeed both $v = \psi + V_h \in K$ and $2u - v = \psi + 2U - V_h \in K$, thus

$$a(u, U - V_h) = (f, U - V_h) .$$

By subtracting each member of this equation from the corresponding member of the inequality

$$a(u, U - U_h) \leq (f, U - U_h)$$

obtained by replacing $v = \psi + U_h$ into (8), we get

$$(30) \qquad a(u, V_h - U_h) \leq (f, V_h - U_h) .$$

On the other hand, by replacing $v_h = \psi_h + U_h$ into (15), we find

$$(31) \qquad a(u_h, U_h - V_h) \leq (f, U_h - V_h) .$$

Therefore, from (30) and (31) we get

$$a(u - u_h, V_h - U_h) \leq 0$$

hence

$$a(u - u_h, u - u_h) \leq a(u - u_h, \psi - \psi_h) + a(u - u_h, U - V_h)$$

from which (28) follows, by taking (4) and (5) into account.

The estimate (28) does not require any regularity information about the solution u. If such an information is available, such as (20), then an a priori estimate of $\|u-u_h\|_{1,\Omega}$ can be given in which no further restriction is imposed to the trial functions $v_h \in K_h$.

Lemma 2 [5]. *If* u *is the solution of problem* (8) *and* u_h *the solution of* (15), *where* $K_h \subset H^1_0(\Omega)$ *then*

$$(32)\qquad \|u-u_h\|_{1,\Omega} \le \left\{\frac{M^2}{\gamma^2}\|u-v_h\|^2_{1,\Omega} + \frac{2}{\gamma}\|f-Lu\|_{0,\Omega}\left[\|u-v_h\|_{0,\Omega} + \|u_h - v\|_{0,\Omega}\right]\right\}^{1/2}$$

for every $v \in K$ *and all* $v_h \in K_h$.

The estimate above actually holds for arbitrary K and K_h in $H^1(\Omega)$, provided we interpret Lu as an element of $H^{-1}(\Omega)$, according to (10), and we assume that $f - Lu$ belongs to some Hilbert space W densely injacted into $H^{-1}(\Omega)$, we refer to [5] for more details.

When adapted to problem (11), as in F.Scarpini, M.A.Vivaldi [14], the estimate of Falk's type becomes

Lemma 3 [14]. *Let* u *be the solution of problem* (11) *and* u_h *the solution of* (15), *where* $K_h \subset H^1_0(\Omega)$ *Then,*

$$(33)\qquad \|u-u_h\|_{1,\Omega} \le \left\{\frac{M^2}{\gamma^2}\|u-v_h\|^2_{1,\Omega} + \int_\Gamma (v_h - u)\frac{\partial u}{\partial \nu}d\sigma + \int_\Gamma (v-u_h)\frac{\partial u}{\partial \nu_L}\,d\sigma\right\}^{1/2}$$

for every $v \in K$ *and all* $v_h \in K_h$.

In order to discuss the specific approximation results all these a priori estimates call for, we must describe how the approximate K_h is obtained.

6. ESTIMATES IN THE ENERGY NORM

We suppose that $n = 1,2$ and Ω_h is a polygon inscribed in Ω, whose sizes do not exceed $h > 0$. We also suppose that a *regular* triangularization of Ω_h has been made, such that all sides of the triangles that decompose Ω_h are $\le h$, the ratio of any two sides is never less than some fixed $\ell > 0$ and no angle is less than some fixed $\alpha > 0$, ℓ and α independent on h.

Then, when $V = H^1_0(\Omega)$ as in problem (8), we take V_h to be the subspace of V of all piece-wise affine continuous functions on Ω_h that vanish at boundary nodes and are extended (continuosly) to be identically zero on $\Omega - \Omega_h$.

We now assume that

$$(34)\qquad \psi \in H^1_0(\Omega) \cap H^2(\Omega)\ ,$$

in particular, that ψ is continuous on $\bar{\Omega}$, and we take $\psi_h = \psi_I$, ψ_I the affine interpolate of ψ, on Ω_h , $\psi_h \equiv 0$ in $\Omega - \Omega_h$. Note that $\psi_h \in V_h$. We then define

$$(35)\qquad K_h = \left\{v_h \in V_h \,:\, v_h \ge \psi_h \ \text{ in } \ \Omega\right\}.$$

Clearly K_h is finite-dimensional since it is defined by the finitely many conditions

$$v_h(x^h_q) \ge \psi_h(x^h_q) = \psi(x^h_q)\ ,$$

x^h_q being any vertex of the given triangularization of Ω_h.

The set K given by (9) and K_h above are clearly of the type (24)' and (24)", with C given by (27) and

$$C_h = \left\{ V_h \in U_h : V_h \geq 0 \text{ in } \Omega \right\} ;$$

moreover, (25) holds.

We can thus use the a priori estimate of Lemma 1, what reduces the estimate of $\|u - u_h\|_{1,\Omega}$ for problem (8) to that of the two terms appearing at the right member of (28).

The first term, due to our choice of $\psi_h = \psi_I$, is easily estimated by well known results as

(36) $$\|\psi - \psi_h\|_{1,\Omega} \leq c\, h\, \|\psi\|_{2,\Omega} .$$

The second term , however, cannot be estimated by the standard approximation results, due to the unilateral constraint inposed to the trial functions V_h. Wat is used here is the following *unilateral approximation* result

Lemma 4 [11],[15]. *Assume* $U \in H^2(\Omega)$, $U \geq 0$ *on* Ω. *Then, there exists* $\tilde{U}_h \in V_h$ *satisfying*

(37) $$0 \leq \tilde{U}_h \leq U$$

such that

(38) $$\|U - \tilde{U}_h\|_{1,\Omega} \leq c\, h\, \|U\|_{2,\Omega} .$$

Let us remark that while this result can be extended to n = 3, see [15], it is not known whether it is true if n = 4 and it is certainly false if n = 5. The best approximants $\tilde{U}_h$ above is defined to be the *maximal* non-negative $U_h \in V_h$ satisfying the constraint $U_h \leq U$: the set of such U_h is not empty for $U_h \equiv 0$ clearly belongs to it. When n = 5, however, there exists non-negative functions $U \in H^2(\Omega)$, of non-zero norm, such that $U_h \equiv 0$ is the *unique* non-negative continuous piece-wise affine function satisfying $U_h \leq U$, thus contradicting Lemma 4.

The estimate (18) we are looking for follows immediately from Lemma 1 and Lemma 4 above. This is indeed the way it was proved in [11].

An alternative way of obtaining (18), which is that of [5], relies on the a priori estimate of Lemma 2. The two terms $\|u - v_h\|_{1,\Omega}$ and $\|u - v_h\|_{0,\Omega}$ appearing at the right hand of (32) can be optimally estimated by choosing , in both of them, v_h to be the affine interpolate, u_I, of u on Ω_h and $\equiv 0$ on $\Omega - \Omega_h$ so that

(39) $$\|u - u_h\|_{1,\Omega} \leq c\, h\, \|u\|_{2,\Omega}$$

and

(39)' $$\|u - u_h\|_{0,\Omega} \leq c\, h\, \|u\|_{2,\Omega} .$$

As to the remaining term $\|u_h - v\|_{0,\Omega}$, it can be shown, see [5] that by choosing $v = \sup\{u_h, \psi\}$ we have

(40) $$\|u_h - v\|_{0,\Omega} \leq c\, h^2 .$$

By combining (32) with (39),(39)' and (40) above, again we find the estimate

(41) $$\|u-u_h\|_{1,\Omega} \leq c\,h \,.$$

As to problem (11), energy estimates of $\|u-u_h\|$ have been also recently obtained, with different methods by F.Scarpini,M.A.Vivaldi[14] and Hlavacek [8], always for $n = 2$.

In [14], by relying on the a priori estimate of Lemma 3 above and estimating the boundary terms by only exploiting the $H^2(\Omega)$ regularity of u, the authors prove the following estimate

(42) $$\|u-u_h\|_{1,\Omega} \leq c\,h^{3/4}(\|u\|_{2,\Omega} + \|\psi\|_{2\Omega}) \,.$$

The trial space U_h is in this case that of all continuous piece-wise affine functions on Ω_h, extended to be affine on the whole of each boundary element of type

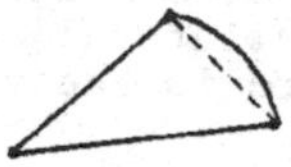

In [8], by relying on unilateral a priori estimate such as that of Lemma 1, the following estimate is obtained

(43) $$\|u-u_h\|_{1,\Omega} \leq c\,h\,(\|u\|_{2,\Omega} + \sum_m \|u\|_{2,\Gamma_m})$$

when Ω is a polygon with sides Γ_m, and $\psi = 0$.

However, while the solution u of problem (11), under the assumption of section 1,is known to belong to $H^2(\Omega)$, see [1][2],this solution cannot be expected in general to have *traces* on each side Γ_m belonging to $H^2(\Gamma_m)$, as required in the estimate above. Thus the question whether the estimate $\|u-u_h\|_{1,\Omega} \leq O(h)$ holds for arbitrary solutions $u \in H^2(\Omega)$ is still open.

Finally, we mention that F.Brezzi and G.Sacchi [4], dealing with an important case of unilateral Dirichlet problem arising in hydraulics, have shown that by using *quadratic elements*, in place of the affine ones considered above, a better estimate such as $\|u-u_h\|_{1,\Omega} \leq O(h^{3/2-\varepsilon})$, $\forall\,\varepsilon > 0$, can be achieved.

7. ESTIMATES IN THE L^2 NORM

There is a standard device that makes it possible, in absence of unilateral constaint, say for the usual Dirichlet problem, to obtain an order h^2 estimate of $u-u_h$ in the L^2 norm once the h order estimate in the energy norm is known. This device, known also as 'Nitsche's Trick", consists in relating the two errors $\|u-u_h\|_{0,\Omega}$ and $\|u-u_h\|_{1,\Omega}$ together, by the estimate

(44) $$\|u-u_h\|_{0,\Omega} \leq c\,h\,\|u-u_h\|_{1,\Omega} \;,$$

what is achieved by solving for

$$\varphi = u - u_h$$

the auxiliary problem

$$(45)\qquad \begin{cases} \Phi \in V \\ a(\Phi,w) = (\varphi,w) \qquad \forall w \in V . \end{cases}$$

The natural question can be arised, whether the estimate (44) holds true for the unilateral Dirichlet problem considered above.

Partial results have been recently given by Natterer [13], under the assumption - whose relevance is not evident in the general case - that the solution of an auxiliary problem taking the role of problem (45) above is smooth enough. The explicit example in which this assumption is satisfied is a one-dimensional version of problem (11).

As $n = 1$, estimate (44) is indeed true and we shall sketch a proof below that relies on the unilateral approximation result of Lemma 4. This proof seems to show some evidence that (44) may be true in higher dimensions, provided the *contact set* $\{u = \psi\}$ is smooth enough.

We thus suppose that Ω is an interval of the real line and for sake of simplicity we take $f = 0$ and $\psi_h = \psi \in H^1_0(\Omega) \cap H^2(\Omega)$ (in general we must have $0 \le \psi_h \le \psi$). Moreover, we assume that $a(u,v)$ is symmetric.

We shall prove *separately* the two estimates

$$(46)\qquad \|(u-u_h)^+\|_{0,\Omega} \le c\,h\,\|u-u_h\|_{1,\Omega}$$

$$(47)\qquad \|(u-u_h)^-\|_{0,\Omega} \le c\,h\,\|u-u_h\|_{1,\Omega} ,$$

where $(u-u_h)^+ = \sup\{u-u_h, 0\}$, $(u-u_h)^- = \inf\{u-u_h, 0\}$. Each one of the estimates above will be obtained *via* an auxiliary problem of unilateral kind taking the role of (45).

In order to prove (46), we consider for

$$(48)\qquad \varphi = (u-u_h)^+$$

the problem

$$(49)\qquad \begin{cases} \Phi \le 0 \quad \text{on } E , \quad \Phi \in H^1_0(\Omega) \\ a(\Phi, z-\Phi) \ge (\varphi, z-\Phi) \\ \forall z \le 0 \quad \text{on } E , \quad z \in H^1_0(\Omega) \end{cases}$$

where E is the closed set

$$(50)\qquad E = \{u = \psi\} .$$

Such a Φ exists, belongs to $H^2(\Omega - E)$, with

$$(51)\qquad \|\Phi\|_{2,\Omega-E} \le c\,\|\varphi\|_{0,\Omega}$$

and by the maximum principle we have $\Phi \ge 0$ on Ω, hence $\Phi \equiv 0$ on E. Thus, by Lemma 4, there exists $\Phi_h \in U_h$ satisfying $0 \le \Phi_h \le \Phi$ in Ω, with $\Phi_h \equiv 0$ on E, such that

$$(52)\qquad \|\Phi - \Phi_h\|_{1,\Omega-E} \le c\,h\,\|\Phi\|_{2,\Omega-E} .$$

Since $\Phi_h = 0$ on E, we have $a(u,\Phi_h) = 0$ and since $v_h = u_h + \Phi_h \in K_h$, $a(u_h, \Phi_h) \ge 0$. Therefore,

$$a(u-u_h, \tilde{\Phi}_h) \leq 0.$$

On the order hand, since $u-u_h \leq 0$ on E, $z = \Phi + u - u_h \leq 0$ on E, hence

$$(\varphi, u-u_h) \leq a(\Phi, u-u_h) .$$

Thus,

$$(\varphi, u-u_h) \leq a(\Phi - \tilde{\Phi}_h, u-u_h)$$

that implies

$$\|(u-u_h)^+\|^2_{0,\Omega} \leq M\|\Phi - \tilde{\Phi}_h\|_{1,\Omega} \|u-u_h\|_{1,\Omega} ,$$

hence, by (52) and (51),

$$\|(u-u_h)^+\|^2_{0,\Omega} \leq c\,h\,\|(u-u_h)^+\|_{0,\Omega} \|u-u_h\|_{1,\Omega}$$

and (46) is proved.

Note that the same proof applies when n = 2,3, provided the set $\Omega - E$ has a boundary smooth enough to ensure that Φ, which solves the Dirichlet problem

$$\begin{cases} L\Phi = \dot{\varphi} & \text{in } \Omega - E \\ \Phi = 0 & \text{on } \partial(\Omega - E) , \end{cases}$$

belongs to $H^2(\Omega - E)$.

In order to prove (47), we solve, with

$$\varphi = (u-u_h)^- ,$$

the problem

$$(53) \qquad \begin{cases} \Phi \geq 0 \text{ on } E_h , \quad \Phi \in H^1_0(\Omega) \\ a(\Phi, z-\Phi) \geq (\varphi, z-\varphi) \\ \forall z \geq 0 \text{ on } E_h , \quad z \in H^1_0(\Omega) \end{cases}$$

where

$$E_h = \{u_h = \psi\} .$$

Note that $\Omega - E_h$ is the union of a finite number of open intervals.

The solution Φ of (53), that depends on h, thus exists and belongs to $H^2(\Omega - E_h)$, with

$$(54) \qquad \|\Phi\|_{2,\Omega-E_h} \leq c\,\|\varphi\|_{0,\Omega}$$

and a constant c independent on h. Moreover, $\Phi \leq 0$ in Ω, hence $\Phi \equiv 0$ on E_h. By Lemma 4, there exists $\tilde{\Phi}_h \equiv 0$ on E_h, such that

$$(55) \qquad \|\Phi - \tilde{\Phi}_h\|_{1,\Omega-E_h} \leq c\,h\,\|\Phi\|_{2,\Omega-E_h} .$$

The constant c here does not depend on h, due to the special form of $\Omega - E_h$ in the case at hand (see the proof of Lemma 4 in [11], when $n = 1$).

Now, since $\phi_h = 0$ on E_h, we have $a(u_h, \phi_h) = 0$ and since $\phi_h \leq 0$, $v = u - \phi_h \in K$, hence $a(u, \phi_h) \leq 0$.
Therefore,

$$a(u - u_h, \phi_h) \leq 0 .$$

On the other hand, $u - u_h \geq 0$ on E_h, therefore $z = \Phi + u - u_h \geq 0$ on E_h and we have

$$(\varphi, u - u_h) \leq a(\Phi, u - u_h)$$

Thus

$$(\varphi, u - u_h) \leq a(\Phi - \phi_h, u - u_h)$$

that implies

$$\|(u - u_h)^-\|^2_{0,\Omega} \leq M\|\Phi - \phi_h\|_{1,\Omega} \|u - u_h\|_{1,\Omega}$$

hence, by (55) and (54)

$$\|(u - u_h)^-\|_{0,\Omega} \leq c\, h \|u - u_h\|_{1,\Omega} .$$

This proof does not extend, as such, to $n > 1$, since we cannot say then that $\Phi \in H^2(\Omega - E_h)$ nor that (54) and (55) hold with constants c independent on h.

Let us mention, finally, that L^∞ estimates for problem (8) have been recently given by C.Baiocchi and Nitsche and we refer to the lectures of these authors at this meeting.

REFERENCES

[1] H.Brezis : *Problèmes unilatéraux.* (Thèse) J.Math. Pures Appl.(1972).

[2] H.Brezis : *Seuil de régularité pour certains problèmes unilatéraux.* C.R.A.S. Paris t.273 pag. 35-37 (5 juillet 1971).

[3] H.Brezis, G.Stampacchia : *Sur la régularité de la solution d'inéquations elliptiques.* Bull. Soc. Math. France 96 (1968), 153-180.

[4] F.Brezzi, G.Sacchi : *A finite element approximation of a variational inequality related to hydraulics,* pre print.

[5] R.S.Falk : *Error estimates for the approximation of a class of Variational inequalities.* Math. of Comp. 28(1974), 963-971.

[6] P.Grisvard : *Alternative de Fredholm relative au problème de Dirichlet dans un polygone ou un polyedre.* Bolletino U.M.I. 5(1972), 132-164.

[7] P.Grisvard : *Régularité de la solution d'un problème aux limites unilateral dans un domaine convexe.* Seminaire Goulaouic-Schwartz, 1975-1976, Exp.XVI, 9 mars 1976.

[8] Ivan Hlaváček : *Dual finite element analysis for unilateral boundary value problems.* Matematický ústavčsav, Praha 1 Žitná 25 ČSSR.

[9] J.L.Lions, G.Stampacchia : *Variational inequalities.* Comm. Pure Appl. Math. 20(1967), 493-519.

[10] U.Mosco : *Implicit Variational problems and quasivariational inequalities.* Lectures at the Nato Advanced Study Institute on Nonlinear Operators and the Calculus of Variations. Bruxelles Sept. 1975, (to appear, Springer Verlag Lectures Notes).

[11] U.Mosco, G.Strang : *One sided approximation and variational inequalities.* Bull. A.M.S. 80(1974), 308-312.

[12] U.Mosco, G.M.Troianiello : *On the smootheness of solutions of unilateral Dirichlet problems.* Boll. UMI(8), (1973), 56-57.

[13] F.Natterer : *Optimale* L_2*-Konvergenz finiter Elemente bei variationsungleichungen,* pre print.

[14] F.Scarpini, M.A.Vivaldi : *Error estimates for the approximation of some unilateral problems,* to appear in RAIRO.

[15] G.Strang : *One-sided approximation and plate bending.* Proceedings of the Symposium on Computing Methods, IRIA France.

CERTAINS PROBLEMES NON LINEAIRES

DE LA PHYSIQUE DES PLASMAS

J. MOSSINO - R. TEMAM

INTRODUCTION : L'objet de cet article est d'étudier l'existence et l'approximation des solutions de certains problèmes non linéaires intervenant en physique des plasmas [1].

En bref, il s'agit de résoudre des problèmes du type :

$$-\Delta u = g \circ \bar{\beta}(u),$$

où $\bar{\beta}$ est un opérateur qui n'est ni monotone ni local :

$$\bar{\beta}(u)(x) = \text{mes}\,\{y \mid u(y) \leq u(x)\},$$

et l'on est amené à formuler un problème P plus général :

$$-\Delta u \in g \circ \beta(u),$$

où β est maintenant un opérateur multivoque.

---

[1] Problème posé par C. MERCIER, Service des Plasmas, Commissariat à l'Énergie Atomique, Fontenay-aux-Roses, communication personnelle, cf aussi C. MERCIER [6] et H. GRAD [3].

On utilise ici une méthode de régularisation :

l'opérateur multivoque β est approché par un opérateur univoque β_ε ; la démonstration de l'existence de solutions pour le problème correspondant P_ε repose sur le théorème de Leray-Schauder, et le passage à la limite lorsque ε tend vers zéro permet de conclure à l'existence de solutions pour P.

Cette technique de régularisation est bien adaptée à l'approximation numérique : on définit un problème discret régularisé $P_{\varepsilon h}$ ayant au moins une solution $u_{\varepsilon h}$ et on montre la convergence de $u_{\varepsilon h}$ vers une solution du problème "continu".

Le plan de l'exposé est le suivant :

I - Problèmes dans un domaine fixé.

II - Approximation interne par éléments finis.

III - Problèmes à frontière libre.

I - PROBLEMES DANS UN DOMAINE FIXE

Soit Ω un ouvert borné "régulier" de $\mathbb{R}^N$ dont on notera $|\Omega|$ la mesure, et soit g $(\Omega \times [0, |\Omega|] \longrightarrow \mathbb{R})$ une fonction de Carathéodory [2] qui vérifie :

- (i) pour presque tout x de Ω, $g(x,.)$ est monotone (croissante ou décroissante).

Soit $B=\{\varphi \in L^\infty(\Omega) \mid 0 \leqslant \varphi \leqslant |\Omega|$ p.p. dans $\Omega\}$. Pour tout u de B, on définit $g \circ u$ par :

$(g \circ u)(x) = g(x, u(x))$, p.p. $x \in \Omega$, et on fait l'hypothèse :

- (ii) l'application : $u \longrightarrow g \circ u$ est continue et bornée de B dans $L^p(\Omega)$ pour un exposant $p > \frac{N}{2}$.

Maintenant pour u dans $C^0(\overline{\Omega})$, on note $\underline{\beta}(u)$ et $\overline{\beta}(u)$ les fonctions de Ω dans $\mathbb{R}$ définies partout par:

---

[2] C'est à dire :

- pour presque tout x de Ω, $g(x, .)$ est continue sur $[0, |\Omega|]$,
- pour tout s dans $[0, |\Omega|]$, $g(., s)$ est mesurable sur Ω.

$$\underline{\beta}\,(u)\,(x) = \text{mes}\,\{y \mid u(y) < u(x)\},$$

$$\bar{\beta}\,(u)\,(x) = \text{mes}\,\{y \mid u(y) \leq u(x)\},$$

et l'on note $\beta(u)$ la fonction multivoque de Ω dans 2^R donnée par :

$$\beta(u)\,(x) = [\,\underline{\beta}\,(u)\,(x)\,,\,\bar{\beta}\,(u)\,(x)\,].$$

Il est alors clair que $\underline{\beta}\,(u)\,(x)$, $\bar{\beta}\,(u)\,(x)$ et $\beta\,(u)\,(x)$ sont dans $[0,\,|\Omega|]$ pour tout x de Ω.

Enfin, par convention, on notera $\varphi \in g \circ \beta\,(u)$ une fonction de $L^p(\Omega)$ telle que $\varphi\,(x) \in g\,(x, \beta\,(u)\,(x)\,)$,p.p. $x \in \Omega$.

Le problème fort et le problème faible :

Ces définitions et conventions étant posées, nous pouvons maintenant formuler correctement le problème aux limites : on recherche u dans $H_0^1\,(\Omega) \cap C^0(\bar{\Omega})$ solution du problème fort :

$$(1)\begin{cases} -\Delta u = g \circ \bar{\beta}\,(u) & \text{dans } \Omega, \\ u = 0 \text{ sur } \partial\Omega, \end{cases}$$

ou du problème faible :

$$(2)\begin{cases} -\Delta u \in g \circ \beta\,(u) & \text{dans } \Omega, \\ u = 0 \text{ sur } \partial\Omega. \end{cases}$$

On notera, d'après Agmon-Douglis-Niremberg [1] , qu'une telle solution est nécessairement dans $W^{2,p}(\Omega)$. Cette appellation de "problème fort" et "problème faible" se justifie, comme l'indique la proposition suivante, de démonstration immédiate.

Proposition 1 : Toute solution dans $H_0^1(\Omega)\cap C^0(\bar{\Omega})$ de (1) est une solution de (2), et réciproquement si u dans $H_0^1(\Omega)\cap C^0(\bar{\Omega})$ est une solution de (2) qui vérifie en outre

$$(3)\ \forall t \in \mathbb{R},\ \{y \mid u(y) = t\} \text{ est de mesure nulle } [3],$$

alors $\underline{\beta}(u) = \bar{\beta}(u)$ partout et u est une solution de (1).

Enfin, sous une hypothèse supplémentaire sur g, les problèmes fort et faible sont équivalents :

Proposition 2 : Si g vérifie pour presque tout x de Ω

$$\text{(iii)}\ g(x, s) \neq 0\ ,\ \forall s \in [0, |\Omega|],$$

alors u dans $H_0^1(\Omega)\cap C^0(\bar{\Omega})$ est solution de (1) si et seulement si u est solution de (2).

---

[3] La condition (3) signifie que u est "sans palier".

Démonstration : Il suffit d'après la proposition 1 de montrer que (iii) implique (3) pour toute solution u dans $H_o^1(\Omega) \cap C^o(\bar{\Omega})$ de (2). Supposons donc que (iii) soit vérifié, et que u dans $H_o^1(\Omega) \cap W^{2,p}(\Omega)$ soit une solution de (2) qui ne vérifie pas (3). Il existe alors un ensemble E de mesure non nulle où u est presque partout constante . D'après Stampacchia [12] on aura :

$$\Delta u(x) = 0 \quad \text{p.p. } x \in E,$$

et comme u est solution de (2) :

$$0 = \Delta u(x) \in g(x, \beta(u)(x)) \text{ p.p. } x \in E,$$

en contradiction avec (iii).

Nous allons maintenant nous attacher à résoudre par régularisation le problème faible (2).

Le problème régularisé :

Nous commençons par définir un opérateur univoque β_ε, "voisin" de β. Soit h la fonction d'Heaviside :

$$h(t) = 1 \text{ si } t \geq 0 , \quad 0 \text{ sinon} .$$

Pour $\varepsilon > 0$ fixé, h_ε est une régularisée de h :

$$h_\varepsilon(t) = \begin{cases} 0 & \text{si } t \leq 0, \\ \frac{t}{\varepsilon} & \text{si } 0 \leq t \leq \varepsilon, \\ 1 & \text{si } t \geq \varepsilon. \end{cases}$$

On note que pour tout t, pour tout $\varepsilon \leqslant \varepsilon_o$ quelconque mais fixé,

$$h_{\varepsilon_o}(t) \leqslant h_\varepsilon(t) \leqslant h(t),$$

et l'on introduit, pour u dans $C^o(\bar{\Omega})$, la fonction $\beta_\varepsilon(u)$ de Ω dans R définie partout par :

$$\beta_\varepsilon(u)(x) = \int_\Omega h_\varepsilon(u(x) - u(y))\, dy .$$

On remarque alors que pour tout x de Ω, pour tout $\varepsilon \leqslant \varepsilon_o$,

$$0 \leqslant \beta_{\varepsilon_o}(u)(x) \leqslant \beta_\varepsilon(u)(x) \leqslant \bar{\beta}(u)(x) \leqslant |\Omega|.$$

De plus β_ε possède la propriété ci-après :

<u>Lemme 1</u> : <u>L'opérateur β_ε envoie $C^o(\bar{\Omega})$ dans lui-même de façon continue</u>.

<u>Démonstration</u> : Tout d'abord, si φ est dans $C^o(\bar{\Omega})$, $\beta_\varepsilon(\varphi)$ est aussi dans $C^o(\bar{\Omega})$. Cela découle immédiatement du théorème de Lebesgue et du fait que h_ε est continue et bornée.

Il reste à montrer que si φ_n est une suite qui tend vers φ dans $C^o(\bar{\Omega})$, $\beta_\varepsilon(\varphi_n)$ tend vers $\beta_\varepsilon(\varphi)$ dans $C^o(\bar{\Omega})$.

Or $|\beta_\varepsilon(\varphi_n) - \beta_\varepsilon(\varphi)|_{C^o(\bar{\Omega})} = \sup_{x\in\bar{\Omega}} \left| \int_\Omega [h_\varepsilon(\varphi_n(x) - \varphi_n(y)) - h_\varepsilon(\varphi(x) - \varphi(y))]\, dy \right|$

$$\leqslant \frac{1}{\varepsilon} \sup_{x\in\bar{\Omega}} \int_\Omega [|\varphi_n(x) - \varphi(x)| + |\varphi_n(y) - \varphi(y)|] \, dy$$

(puisque h_ε est lipschitzienne de constante $\frac{1}{\varepsilon}$)

$$\leqslant \frac{2}{\varepsilon} |\Omega| \; |\varphi_n - \varphi|_{C^\circ(\bar{\Omega})},$$

ce qui démontre même que β_ε est une application lipschitzienne de $C^\circ(\bar{\Omega})$ dans lui-même.

On considère maintenant le problème suivant, qui est une forme régularisée de (2) : on recherche u_ε dans $H_0^1(\Omega) \cap C^\circ(\bar{\Omega})$ solution de :

$$(2)_\varepsilon \quad \begin{cases} - \Delta u_\varepsilon = g \circ \beta_\varepsilon(u_\varepsilon) \text{ dans } \Omega, \\ u_\varepsilon = 0 \text{ sur } \partial\Omega . \end{cases}$$

On a alors un théorème d'existence pour ce problème régularisé :

<u>Théorème 1</u> : <u>Le problème $(2)_\varepsilon$ admet au moins une solution u_ε dans $H_0^1(\Omega) \cap C^\circ(\bar{\Omega})$.</u>

<u>Démonstration</u> : Soit $\mathscr{s}_\varepsilon$ l'application qui à φ dans $C^\circ(\bar{\Omega})$ associe l'unique solution u_φ dans $H_0^1(\Omega)$ de :

$$\begin{cases} - \Delta u_\varphi = g \circ \beta_\varepsilon(\varphi) \text{ dans } \Omega, \\ u_\varphi = 0 \text{ sur } \partial\Omega . \end{cases}$$

La démonstration repose sur le théorème de Leray-Schauder applicable à $\mathscr{s}_\varepsilon$ grace aux lemmes suivants :

<u>Lemme 2 : Les solutions u_φ sont bornées dans $W^{2,p}(\Omega)$ indépendamment de φ,</u>
<u>c'est à dire</u> :

$$\exists C_o \quad , \quad \forall \varphi \in C^o(\overline{\Omega}), \ ||u_\varphi||_{W^{2,p}(\Omega)} \leq C_o.$$

<u>Lemme 3 : L'application $\mathcal{A}_\varepsilon$ est continue de $C^o(\overline{\Omega})$ dans $W^{2,p}(\Omega)$</u>.

<u>Démonstration du lemme 2</u> : D'après (ii), $g \circ \beta_\varepsilon(\varphi)$ est borné dans $L^p(\Omega)$ indépendamment de φ, et le résultat annoncé suit.

<u>Démonstration du lemme 3</u> : Il suffit de voir que $g \circ \beta_\varepsilon$ est continue de $C^o(\overline{\Omega})$ dans $L^p(\Omega)$. Or β_ε est continue de $C^o(\overline{\Omega})$ dans $C^o(\overline{\Omega}) \cap B$ par le lemme 1, et g est continue de B dans $L^p(\Omega)$ grâce à (ii), d'où la continuité de l'application composée.

Nous pouvons à présent appliquer le théorème de Leray-Schauder. Soit $C = \{v \in W^{2,p}(\Omega) \mid ||v||_{W^{2,p}(\Omega)} \leq C_o\}$, où C_o est le même qu'au lemme 2. L'ensemble C est un convexe compact de $C^o(\overline{\Omega})$ (on utilise la compacité de l'injection $W^{2,p}(\Omega) \subset C^o(\overline{\Omega})$ pour $p > \frac{N}{2}$, cf [10]), invariant par l'application continue $\mathcal{A}_\varepsilon$ ($\mathcal{A}_\varepsilon$ est continue de $C^o(\overline{\Omega})$ dans lui-même grâce au lemme 3 et à la continuité de l'injection $W^{2,p}(\Omega) \subset C^o(\overline{\Omega})$), et le théorème de Leray-Schauder donne alors l'existence d'un point fixe $u_\varepsilon = \mathcal{A}_\varepsilon u_\varepsilon$, ce qui revient à dire que u_ε est dans $H_0^1(\Omega) \cap C^o(\overline{\Omega})$ et vérifie $(2)_\varepsilon$.

Nous allons maintenant passer à la limite lorsque ε tend vers zéro.

Le passage à la limite :

Les estimations à priori sont faciles à établir : comme au lemme 2, $\beta_\varepsilon(u_\varepsilon)$ étant dans B, l'hypothèse (ii) donne que go $\beta_\varepsilon(u_\varepsilon)$ est borné dans $L^p(\Omega)$ indépendamment de ε; il suit que u_ε est borné dans $W^{2,p}(\Omega)$ et, utilisant encore la compacité de l'injection de $W^{2,p}(\Omega)$ dans $C^o(\overline{\Omega})$, on peut extraire une sous suite, notée encore u_ε, telle que

$$\begin{cases} u_\varepsilon \rightharpoonup u \text{ dans } W^{2,p}(\Omega) \text{ faible,} \\ u_\varepsilon \longrightarrow u \text{ dans } C^o(\overline{\Omega}), \\ \text{go } \beta_\varepsilon(u_\varepsilon) \rightharpoonup \chi \quad \text{dans } L^p(\Omega) \text{ faible.} \end{cases}$$

Tout le problème est maintenant de démontrer que $\chi \in$ go $\beta(u)$.

D'après une remarque faite plus haut sur β_ε, pour tout $\varepsilon \leq \varepsilon_0$ quelconque mais fixé :

$$\beta_{\varepsilon_0}(u_\varepsilon)(x) \leq \beta_\varepsilon(u_\varepsilon)(x) \leq \bar{\beta}(u_\varepsilon)(x), \qquad \forall x \in \Omega.$$

Comme u_ε converge uniformément vers u, il existe $\varepsilon_1 > 0$ tel que $|u_\varepsilon(x) - u(x)| \leq \frac{\varepsilon_0}{2}$ pour tout x, dès que $\varepsilon \leq \varepsilon_1$.

Posons $\eta = \inf(\varepsilon_0, \varepsilon_1)$. Dans la suite ε sera toujours pris inférieur à η. Il est alors clair que :

$$u_\varepsilon(y) \leq u_\varepsilon(x) \text{ implique } u(y) \leq u(x) + \varepsilon_0 .$$

Donc $\bar{\beta}(u_\varepsilon)(x) = \text{mes}\{y | u_\varepsilon(y) \leq u_\varepsilon(x)\} \leq \text{mes}\{y | u(y) \leq u(x) + \varepsilon_0\}$.

De même, $u(y) \leq u(x) - 2\varepsilon_0$ implique $u_\varepsilon(y) \leq u_\varepsilon(x) - \varepsilon_0$, et par suite :

$$\begin{aligned} \beta_{\varepsilon_0}(u_\varepsilon)(x) &= \int_\Omega h_{\varepsilon_0}(u_\varepsilon(x) - u_\varepsilon(y))\,dy \\ &\geq \text{mes}\{y | u_\varepsilon(y) \leq u_\varepsilon(x) - \varepsilon_0\} \\ &\geq \text{mes}\{y | u(y) \leq u(x) - 2\varepsilon_0\} . \end{aligned}$$

Il vient alors, pour tout x de Ω,

$$\text{mes}\{y | u(y) \leq u(x) - 2\varepsilon_0\} \leq \beta_\varepsilon(u_\varepsilon)(x) \leq \text{mes}\{y | u(y) \leq u(x) + \varepsilon_0\},$$

et, d'après l'hypothèse (i) de monotonie de g, cela entraine que $g(x, \beta_\varepsilon(u_\varepsilon)(x))$ est compris, pour presque tout x de Ω, entre

$$\varphi_1(\varepsilon_0, x) = g(x, \text{mes}\{y | u(y) \leq u(x) - 2\varepsilon_0\})$$

et

$$\varphi_2(\varepsilon_0, x) = g(x, \text{mes}\{y | u(y) \leq u(x) + \varepsilon_0\})$$

(on ne précise pas l'ordre des bornes, qui est interchangé suivant que g est croissante ou décroissante).

Maintenant l'ensemble des φ de $L^p(\Omega)$ comprises entre $\varphi_1(\varepsilon_0, .)$ et $\varphi_2(\varepsilon_0, .)$ p.p. dans Ω est un convexe fermé de $L^p(\Omega)$, donc faiblement fermé, et l'on a vu que

$g \circ \beta_\varepsilon(u_\varepsilon) \longrightarrow \chi$ dans $L^p(\Omega)$ faible.

Il suit que $\chi(x)$ est compris, pour presque tout x de Ω, entre $\varphi_1(\varepsilon_0, x)$ et $\varphi_2(\varepsilon_0, x)$.

Mais ε_0 est quelconque. On le fait tendre vers zéro en utilisant la continuité de g par rapport à s et le lemme ci-après dont on trouvera la démonstration plus loin :

<u>Lemme 4</u> : <u>Soit t un réel quelconque et soit u dans $C^0(\bar{\Omega})$. Lorsque τ tend vers zéro en décroissant, on a les convergences suivantes</u> :

$$\text{mes}\{y \mid u(y) \leq t + \tau\} \longrightarrow \text{mes}\{y \mid u(y) \leq t\},$$

$$\text{et} \quad \text{mes}\{y \mid u(y) \leq t - \tau\} \longrightarrow \text{mes}\{y \mid u(y) < t\},$$

On obtient alors, pour presque tout x de Ω,

$$\varphi_1(\varepsilon_0, x) \longrightarrow g(x, \underline{\beta}(u)(x)),$$

$$\text{et} \quad \varphi_2(\varepsilon_0, x) \longrightarrow g(x, \bar{\beta}(u)(x)),$$

$$\text{d'où} \quad \chi(x) \in g(x, \beta(u)(x)) \quad \text{p.p. } x \in \Omega,$$

$$\text{ou encore} \quad \chi \in g \circ \beta(u).$$

Le passage à la limite faible dans $(2)_\varepsilon$ donne alors :

$$(2) \quad \begin{cases} -\Delta u \in g \circ \beta(u) \text{ dans } \Omega, \\ u = 0 \text{ sur } \partial\Omega, \end{cases}$$

et permet d'énoncer le théorème d'existence :

<u>Théorème 2 : Le problème faible (2) admet au moins une solution u dans $H_0^1(\Omega) \cap C^0(\bar{\Omega})$.</u>

Nous démontrons à présent le lemme 4.

Soit $E_\tau = \{y \mid t < u(y) \leq t + \tau\}$. On a :

$\text{mes}\,\{y \mid u(y) \leq t + \tau\} - \text{mes}\,\{y \mid u(y) \leq t\} = \text{mes}\, E_\tau$.

Les E_τ forment une suite décroissante d'ensembles emboîtés dont l'intersection est vide. Par suite leur mesure tend vers zéro. La démonstration est analogue pour la deuxième assertion.

II - APPROXIMATION PAR ELEMENTS FINIS

Nous allons voir dans cette partie II que les techniques employées plus haut permettent également de traiter l'approximation interne du problème qui nous occupe.

Soit h un paramètre destiné à tendre vers zéro. On suppose donné un espace de dimension finie $V_h \subset H_0^1(\Omega) \cap C^0(\bar{\Omega})$ (V_h est par exemple l'espace des éléments finis linéaires conformes) qui approxime $H_0^1(\Omega)$ au sens suivant :

Il existe une application r_h de $H_0^1(\Omega)$ dans V_h telle que pour tout v de $H_0^1(\Omega)$, $r_h v$ tend vers v lorsque h tend vers zéro.

<u>Le problème régularisé discret</u>

Le <u>problème discret</u> que l'on considère alors est le suivant : trouver $u_{\varepsilon h}$ dans V_h solution de :

$$(2)_{\varepsilon h} \qquad a(u_{\varepsilon h}, v_h) = \int_\Omega g(x, \beta_\varepsilon(u_{\varepsilon h})(x))\, v_h(x)\, dx, \quad \forall v_h \in V_h,$$

où a est la forme bilinéaire :

$$a(u, v) = \int_\Omega \nabla u \, \nabla v \, dx$$

Désignons par A_h l'opérateur linéaire de V_h dans lui-même défini par :

$(A_h u_h, v_h) = a(u_h, v_h)$, $\forall u_h \in V_h$, $\forall v_h \in V_h$ [4],

et, pour tout φ_h dans V_h, notons $[g \circ \beta_\varepsilon(\varphi_h)]_h$ la forme linéaire sur V_h (identifiée avec un élément de V_h) telle que :

$$([g \circ \beta_\varepsilon(\varphi_h)]_h, v_h) = \int_\Omega g(x, \beta_\varepsilon(\varphi_h)(x)) \, v_h(x) \, dx, \ \forall v_h \in V_h .$$

Il est alors clair que $(2)_{\varepsilon h}$ s'écrit encore :

$$A_h \, u_{\varepsilon h} = [g \circ \beta_\varepsilon(u_{\varepsilon h})]_h .$$

Nous allons montrer que ce problème discret a au moins une solution $u_{\varepsilon h}$, cette solution étant de plus une approximation d'une solution du problème continu (2).

<u>Théorème 3</u> : <u>Le problème régularisé discret $(2)_{\varepsilon h}$ admet au moins une solution $u_{\varepsilon h}$ dans V_h.</u>

<u>Démonstration</u> : La démonstration repose ici aussi sur le théorème de Leray-Schauder. Soit $\mathcal{S}_{\varepsilon h}$ l'application qui a φ_h dans V_h associe l'unique solution $u_h(\varphi_h)$ dans V_h de :

$$A_h \, u_h(\varphi_h) = [g \circ \beta_\varepsilon(\varphi_h)]_h .$$

---

[4] On note (.,.) le produit scalaire dans V_h.

Par définition de $[g \circ \beta_\varepsilon(\varphi_h)]_h$, nous avons par l'inégalité de Hölder :

$$|([g \circ \beta_\varepsilon(\varphi_h)]_h, v_h)| \leq ||g(., \beta_\varepsilon(\varphi_h)(.))||_{L^p(\Omega)} || v_h||_{L^{p'}(\Omega)},$$

ce qui donne, en tenant compte de l'hypothèse (ii) et du fait qu'en dimension finie toutes les normes sont équivalentes :

$$||[g \circ \beta_\varepsilon(\varphi_h)]_h|| \leq C_h,$$

où C_h est indépendant de φ_h dans V_h.

Comme l'opérateur $\mathcal{A}_h$ est un isomosphisme (d'après le théorème de Lax-Milgram et la coercivité de la forme bilinéaire a) il suit que $u_h(\varphi_h)$ est borné dans V_h indépendamment de φ_h, résultat analogue à celui du lemme 2.

On montre aussi, comme ci-dessus, en notant par C_h diverses constantes, l'inégalité :

$$|| [g \circ \beta_\varepsilon(\varphi_h)]_h - g \circ \beta_\varepsilon(\Psi_h)]_h|| \leq C_h|| g \circ \beta_\varepsilon(\varphi_h) - g \circ \beta_\varepsilon(\Psi_h)||_{L^p(\Omega)},$$

et finalement

$$|| u_h(\varphi_h) - u_h(\Psi_h)|| \leq C_h|| g \circ \beta_\varepsilon(\varphi_h) - g \circ \beta_\varepsilon(\Psi_h)||_{L^p(\Omega)}.$$

Or, d'après ce qu'on a vu dans la démonstration du lemme 3, $g \circ \beta_\varepsilon$ est continu de $C^0(\bar{\Omega})$ (donc de V_h) dans $L^p(\Omega)$, et l'inégalité précédente entraine la continuité de l'application $\mathcal{J}_{\varepsilon h}$. Le théorème de Leray-Schauder en dimension finie donne alors directement le résultat.

Nous allons maintenant passer à la limite lorsque ε et h tendent vers zéro simultanément. Signalons que l'on peut traiter de la même façon les passages à la limite "en deux étapes" : ε tend vers zéro, puis h tend vers zéro, ou vis et versa, les limites intermédiaires u_h ou u_ε étant solutions respectives d'un problème faible discret facile à expliciter, ou du problème continu régularisé.

Le passage à la limite

On utilisera dans ce paragraphe $p \geq 2$, hypothèse nécessairement vérifiée si $N \geq 2$ (car $p > \frac{N}{2}$) mais que l'on rajoute si $N = 1$.

Pour les estimations à priori, l'hypothèse (ii) assure comme toujours que $g \circ \beta_\varepsilon(u_{\varepsilon h})$ est borné dans $L^p(\Omega)$ donc dans $L^2(\Omega)$ indépendamment de ε et h, et d'après $(2)_{\varepsilon h}$ appliqué à $v_h = u_{\varepsilon h}$, on tire :

$$||u_{\varepsilon h}||^2_{H^1_0(\Omega)} \leq ||g \circ \beta_\varepsilon (u_{\varepsilon h})||_{L^2(\Omega)} ||u_{\varepsilon h}||_{L^2(\Omega)},$$

d'où $\quad ||u_{\varepsilon h}||_{H^1_0(\Omega)} \leq C||g \circ \beta_\varepsilon (u_{\varepsilon h})||_{L^2(\Omega)} \leq C$, C désignant encore diverses constantes. Donc $u_{\varepsilon h}$ est borné dans $H^1_0(\Omega)$ indépendamment de ε et h, et l'on peut extraire une sous suite, notée encore $u_{\varepsilon h}$, telle que :

$$\begin{cases} u_{\varepsilon h} \longrightarrow u \text{ dans } H^1_0(\Omega) \text{ faible,} \\ u_{\varepsilon h} \longrightarrow u \text{ dans } L^2(\Omega) \text{ fort (par compacité de l'injection} \end{cases}$$

de $H^1_0(\Omega)$ dans $L^2(\Omega)$) et p.p. dans Ω,

avec $\qquad g \circ \beta_\varepsilon (u_{\varepsilon h}) \longrightarrow \chi$ dans $L^2(\Omega)$ faible.

Et le problème est encore de démontrer que $\chi \in g \circ \beta(u)$.

Remarquons qu'ici $u_{\varepsilon h}$ n'étant pas nécessairement dans $W^{2,p}(\Omega)$ (c'est le cas si V_h est l'espace des éléments finis linéaires conformes) on n'a pas la convergence uniforme de $u_{\varepsilon h}$ qui nous a été si utile. Toutefois l'application du théorème d'Egoroff permet de contourner la difficulté, mais nécessite un minimum de technique qu'il serait trop long d'exposer ici. Nous nous contentons donc d'énoncer le lemme suivant, dont on trouvera la démonstration dans [9] en même temps que le sens donné à $\beta(u)$ si u n'est pas dans $C^o(\bar{\Omega})$.

<u>Lemme</u> : <u>Soit</u> $u_{\varepsilon h}$ <u>une suite de $L^2(\Omega)$ qui vérifie</u> :

$$u_{\varepsilon h} \to u \text{ p.p. dans } \Omega,$$

$$\text{et} \quad g \circ \beta_\varepsilon (u_{\varepsilon h}) \to \chi \text{ dans } L^2(\Omega) \text{ faible,}$$

<u>alors $\chi \in g \circ \beta(u)$</u>.

Une fois ce résultat acquis, le passage à la limite dans $(2)_{\varepsilon h}$ où l'on fait $v_h = r_h v$ donne le théorème de convergence ci-après :

<u>Théorème</u> : <u>Soit $(u_{\varepsilon h})$ une suite de solutions des problèmes régularisés discrets $(2)_{\varepsilon h}$. Lorsque ε et h tendent vers zéro simultanément, on peut extraire de $(u_{\varepsilon h})$ une sous suite, qui converge dans $H_0^1(\Omega)$ faible, $L^2(\Omega)$ fort et p.p. vers une solution du problème</u>

$$(2) \begin{cases} -\Delta u \in g \circ \beta(u) \text{ dans } \Omega, \\ u = 0 \text{ sur } \partial\Omega. \end{cases}$$

On remarque que ce théorème constitue en même temps une nouvelle preuve de l'existence de solutions du problème faible (2).

En remplaçant l'opérateur multivoque $u \longrightarrow \beta(u)$ par un autre qui dépend du signe de u, nous allons maintenant pouvoir résoudre, avec les mêmes techniques, toute une classe de problèmes à frontière libre.

III - PROBLEMES A FRONTIERE LIBRE

L'ouvert Ω et la fonction g vérifient les mêmes hypothèses qu'à la partie I. Pour u dans $C^o(\overline{\Omega})$, on définit partout les fonctions univoques de Ω dans R $\underline{\gamma}(u)$ et $\overline{\gamma}(u)$, ainsi que la fonction multivoque de Ω dans 2^R $\gamma(u)$ par :

$$\underline{\gamma}(u)(x) = \text{mes } \{y \mid u(y) < - u^-(x)\} \text{ [5]},$$

$$\overline{\gamma}(u)\ (x) = \text{mes } \{y \mid u(y) \leq - u^-(x)\},$$

$$\gamma\ (u)\ (x) = [\ \underline{\gamma}(u)(x),\ \ \overline{\gamma}(u)(x)]\ .$$

Par convention on notera encore $\varphi \in g o\ \gamma(u)$ une fonction de $L^p(\Omega)$ telle que

$$\varphi(x) \in g\ (x,\ \gamma(u)\ (x))\ \text{p.p. } x \in \Omega.$$

On recherche alors u dans $H^1_0(\Omega) \cap C^o(\overline{\Omega})$ solution du <u>problème fort</u> :

$$(4)\begin{cases} - \Delta u = g o\ \overline{\gamma}\ (u) \text{ dans } \Omega, \\ u = 0 \text{ sur } \partial\Omega\ , \end{cases}$$

---

[5] Pour t réel, on pose : $t^- = \begin{cases} 0 \text{ si } t \geq 0, \\ - t \text{ sinon.} \end{cases}$

ou du <u>problème faible</u> :

$$(5) \begin{cases} -\Delta u \in g \circ \gamma(u) \text{ dans } \Omega, \\ u = 0 \text{ sur } \partial\Omega, \end{cases}$$

une telle solution étant nécessairement d'après Agmon-Douglis- Niremberg [1] dans $W^{2,P}(\Omega)$.

On note que (4) s'écrit aussi :

$$\begin{cases} -\Delta u = \begin{cases} g(., \text{mes } \{y|u(y) \leq 0\}) \text{ si } u \geq 0 \\ g(., \text{mes } \{y|\ u(y) \leq u(.)\} \text{ si } u < 0 \end{cases} \text{ dans } \Omega, \\ u = 0 \text{ sur } \partial\Omega, \end{cases}$$

et l'on reconnait alors un <u>problème à frontière libre</u>.

D'autre part le rapport entre les problèmes fort et faible est le même qu'aux propositions 1 et 2 de la partie I, au remplacement près de β par γ.

<u>Le problème régularisé</u> :

Comme nous avons fait pour l'opérateur β, nous introduisons l'opérateur univoque γ_ε défini partout dans Ω, pour u dans $C^0(\overline{\Omega})$, par :

$$\gamma_\varepsilon(u)(x) = \int_\Omega h_\varepsilon(-u(x) - u(y))\, dy,$$

h_ε étant le même qu'à la partie I; et nous remarquons encore que pour tout x de Ω, pour tout $\varepsilon \leq \varepsilon_0$,

$$0 \leq \gamma_{\varepsilon_0}(u)(x) \leq \gamma_\varepsilon(u)(x) \leq \bar{\gamma}(u)(x) \leq |\Omega|.$$

Avec une démonstration analogue à celle du lemme 1, et en utilisant le caractère lipschitzien de la fonction (.), on obtient le

<u>Lemme 1'</u> : <u>L'opérateur γ_ε envoie $C^0(\overline{\Omega})$ dans lui-même de façon continue</u>.

Le nouveau <u>problème régularisé</u> se formule alors ainsi : trouver u_ε dans $H_0^1(\Omega) \cap C^o(\bar{\Omega})$ solution de :

$$(5)_\varepsilon \begin{cases} -\Delta u_\varepsilon = g \circ \gamma_\varepsilon (u_\varepsilon) \text{ dans } \Omega, \\ u_\varepsilon = 0 \text{ sur } \partial\Omega. \end{cases}$$

Il vérifie le théorème d'existence attendu :

<u>Théorème 1' : Le problème $(5)_\varepsilon$ admet au moins une solution u_ε dans $H_0^1(\Omega) \cap C^o(\bar{\Omega})$</u>.

<u>Démonstration</u> : On note $\mathscr{L}_\varepsilon$ l'application qui à φ dans $C^o(\bar{\Omega})$ associe l'unique solution u_φ dans $H_0^1(\Omega)$ de :

$$\begin{cases} -\Delta u_\varphi = g \circ \gamma_\varepsilon (\varphi) \text{ dans } \Omega, \\ u_\varphi = 0 \text{ sur } \partial\Omega, \end{cases}$$

et au remplacement près de $\mathscr{S}_\varepsilon$ par $\mathscr{L}_\varepsilon$ la démonstration est exactement la même qu'au théorème 1.

<u>Le passage à la limite</u>

Les estimations à priori n'ont pas changé par rapport à la partie I, et il suffit de montrer (tout le reste étant identique à ce qui a été fait plus haut) que si

$$\begin{cases} u_\varepsilon \longrightarrow u \text{ dans } C^o(\bar{\Omega}), \\ \text{et } g \circ \gamma_\varepsilon (u_\varepsilon) \longrightarrow \chi \text{ dans } L^p(\Omega) \text{ faible,} \end{cases}$$

alors $\chi \in g \circ \gamma(u)$.

Il est clair que

$$u_\varepsilon^- \longrightarrow u^- \text{ dans } C^o(\bar{\Omega}),$$

puisque la fonction $(.)^-$ est lipschitzienne, et que les convergences uniformes entraînent l'existence de $\varepsilon_1 > 0$ tel que
$|u_\varepsilon(x) - u(x)| \leq \frac{\varepsilon_0}{2}$ et $|u_\varepsilon^-(x) - u^-(x)| \leq \frac{\varepsilon_0}{2}$ dès que $\varepsilon \leq \varepsilon_1$. On a maintenant pour tout $\varepsilon \leq \varepsilon_0$:

$$\gamma_{\varepsilon_0}(u_\varepsilon)(x) \leq \gamma_\varepsilon(u_\varepsilon)(x) \leq \bar{\gamma}(u_\varepsilon)(x), \quad \forall x \in \Omega,$$

et pour tout $\varepsilon \leq \varepsilon_1$ et tout x de Ω :

$$\begin{aligned}\bar{\gamma}(u_\varepsilon)(x) &= \text{mes } \{y \mid u_\varepsilon(y) \leq - u_\varepsilon^-(x)\} \\ &\leq \text{mes } \{y \mid u(y) \leq - u^-(x) + \varepsilon_0\},\end{aligned}$$

tandis que :

$$\begin{aligned}\gamma_{\varepsilon_0}(u_\varepsilon)(x) &= \int_\Omega h_{\varepsilon_0}(-u_\varepsilon^-(x) - u_\varepsilon(y))\, dy \\ &\geq \text{mes } \{y \mid u_\varepsilon(y) \leq - u_\varepsilon^-(x) - \varepsilon_0\} \\ &\geq \text{mes } \{y \mid u(y) \leq - u^-(x) - 2\varepsilon_0\}\end{aligned}$$

(Il suffit pour cela d'utiliser les arguments habituels).

Le reste de la démonstration correspondante dans la partie I se reproduit point par point, au remplacement près des anciens φ_1 et φ_2 par :

$$\varphi_1(\varepsilon_0, x) = g(x, \text{mes } \{y \mid u(y) \leq - u^-(x) - 2\varepsilon_0\})$$

et $$\varphi_2(\varepsilon_0, x) = g(x, \text{mes } \{y \mid u(y) \leq - u^-(x) + \varepsilon_0\}).$$

Et l'on obtient le nouveau théorème d'existence ci-après.

<u>Théorème 2'</u> : <u>Le problème faible (5) admet au moins une solution u dans $H_0^1(\Omega) \cap C^0(\bar{\Omega})$</u>

Notons pour terminer que les résultats d'approximation donnés à la partie II sont aussi valables avec les mêmes démonstrations, pour cette classe de problèmes à frontière libre.

La méthode de résolution employée ici est la régularisation. Signalons que l'on peut aussi utiliser pour ce genre de problèmes des théorèmes "de points fixes" d'analyse non linéaire multivoque.

D'autre part la théorie des inéquations quasi- variationnelles (ou I.Q.V.) introduite par Lions- Bensoussan [2] [5] et Tartar [13] permet d'obtenir par une preuve constructive l'existence d'une solution minimale et d'une solution maximale (cf [7] [8] où cette théorie est appliquée au cas où g est l'identité)

Pour ces deux approches différentes nous renvoyons le lecteur à [9], où l'on trouvera aussi le traitement effectif de nombreux exemples numériques.

BIBLIOGRAPHIE

[1] S. Agmon, A. Douglis, L. Niremberg ,Estimates near the boundary for solutions of elliptic partial differential equations satisfying general boundary conditions, I. Comm. Pure Appl. Math. 12 (1959), p. 623 - 727.

[2] A. Bensoussan - J.L. Lions, Comptes rendus, 276, série A, 1973, p. 1189.

[3] H. Grad, A. Kadish, D.C. Stevens, A free boundary Tokamak Equilibrium, Comm. Pure Appl. Math XXVII, p. 39 - 57 (1974)

[4] J.M. Lasry et R. Robert, Degré et théorèmes de points fixes pour les fonctions multivoques , applications ;Séminaire Goulaouic - Lions - Schwartz, mars 1975.

[5] J.L. Lions, Inéquations quasi-variationnelles, Cours au Collège de France 1974 - 1976, à paraître .

[6] C. Mercier, The magnetohydrodynamic approach to the problem of plasma confiment in closed magnetic configurations. Publication of EURATOM C.E.A., Luxembourg 1974.

[7] J. Mossino, Comptes Rendus, 282, série A, 1976, p. 187.

[8] J. Mossino, Etude d'une inéquation quasi-variationnelle apparaissant en physique, Exposé au Colloque d'Analyse Convexe et ses Applications, Murat-Le - Quaire, Mars 1976, Lecture Notes in Economics and Mathematical Systems, Springer, à paraître.

[9] J. Mossino, Thèse et article à paraître.

[10] J.Nečas , Les méthodes directes en théorie des équations elliptiques, Masson et Cie, Academia, 1967.

[11] R. Robert, Contributions à l'analyse non linéaire, Thèse, Université scientique et médicale de Grenoble, Institut National Polytechnique de Grenoble (1976).

[12] G. Stampacchia, Equations elliptiques du second ordre à coefficients discontinus , Montréal, Presses de l'Université de Montréal, 1966, Séminaire de mathématiques supérieures, Eté 1965.

[13] L. Tartar, Comptes Rendus, 278,série A, 1974, p. 1193.

[14] R. Temam, Configuration d'équilibre d'un plasma : un problème de valeur propre non linéaire, Comptes Rendus, 280, série A, 1975, p. 419.

[15] R. Temam, A non linear eigenvalue problem : the shape at equilibrium of a confined plasma, Arch. Rat. Math. Anal., 60, nb 1. 1976, p. 51.

L_∞-CONVERGENCE OF FINITE ELEMENT APPROXIMATIONS

Joachim Nitsche

The first boundary value problem for the differential equation $-\Delta u = f$ without and with a side-condition $u \leq z$ is considered. Using linear finite element spaces S_h (h mesh-size) approximations u_h are defined by means of the variational formulation. As is shown in the two-dimensional case the error is of order $h^2|\ln h|$ in both cases.

1. Formulation of the problem

In this paper we will consider finite element approximations on the solutions of the model problems:

Minimize

$$(1) \qquad I(v) = D(v,v) - 2(f,v)$$

$$= \iint_\Omega \left\{\Sigma\left(\frac{\partial v}{\partial x_i}\right)^2\right\} dx - 2 \iint_\Omega f\ v\ dx$$

in

(I) the space $\overset{o}{W}{}^1_2 = \overset{o}{W}{}^1_2(\Omega)$

respective

(II) the convex domain

(2) $$K = \{v \mid v \in \overset{o}{W}{}_2^1 \wedge v \leq z \text{ a.e. in } \Omega\} .$$

Here $\Omega \subset R^N$ is a bounded domain with boundary $\partial\Omega$ sufficiently smooth, $f \in L_2$ and $z \in W_\infty^2$ are given functions. In order to have $K \neq \emptyset$ we assume

$$\underline{z} = \inf \{z(x) \mid x \in \partial\Omega\} > 0 .$$

We will not discuss the regularity of the solutions u, u^K of (I) resp. (II) in detail but in the restricted case we refer especially to BREZIS-STAMPACCHIA [1], LIONS-STAMPACCHIA [6].

In this paper we will use only linear finite element spaces S_h for approximation spaces. They are defined by means of a regular subdivision Γ_h of Ω into simplices Δ: There is a constant $\varkappa_1$ such that for any $\Delta \in \Gamma_h$ there are two spheres $\underline{K}, \overline{K}$ of radii $\underline{r}, \overline{r}$ with $\underline{K} \subseteq \Delta \subseteq \overline{K}$ and $\varkappa_1^{-1} h \leq \underline{r} < \overline{r} \leq \varkappa_1 h$. In the case (II) of side conditions we assume moreover that no "angle" exceeds $\frac{\pi}{2} - \varkappa_2$ with a positive $\varkappa_2$ (compare CIARLET-RAVIART [2]). The simplices near $\partial\Omega$ may be curved. $\overset{o}{S}_h$ consists of all $\chi \in C^0(\Omega)$ with $\chi = 0$ on $\partial\Omega$ whose restriction to $\Delta \in \Gamma_h$ is a linear function (with isoparametric modifications in the curved simplices - see CIARLET-RAVIART [2]).

The finite element approximations u_h, u_h^K are the elements in (I) $\overset{o}{S}_h$ respective (II) $K_h = K \cap \overset{o}{S}_h$ minimizing the functional $I(.)$. We will prove for $N = 2$

<u>Theorem: If u resp. u^K is in W_∞^2 then the error estimates hold true:</u>

(3) $$\|u-u_h\|_{L_\infty} \leq c\, h^2 |\ln h|\, \|u\|_{W_\infty^2} ,$$

(4) $$\|u^K-u_h^K\|_{L_\infty} \leq c\, h^2 |\ln h|\, \{\|u\|_{W_\infty^2} + \|z\|_{W_\infty^2}\} .$$

The error estimate (3) for the unrestricted Neuman problem in two dimensions was also given by SCOTT [12], for higher order finite elements see also NITSCHE [11]

We mention with respect to the restricted case II the results of FALK [4], [5], and STRANG [13] who give under different construction of K_h error bounds in the W_2^1-norm, see also NATTERER [8] for L_2-estimates. The isoparametric modifications are superfluous if we modify the variational principle as was done in NITSCHE [9], [10], for the W_2^1-estimates in this case see FALK [5].

For general N the power of $|\ln h|$ is $(N+2)/4$. Since Green's function has a logarithmic singularity in case $N = 2$, the cases $N = 2$ and $N \geq 3$ cannot be handled in exactly the same way. We consider here only the more cumbersome case $N = 2$, the other will be given elsewhere.

2. Error estimates for the unrestricted problem in weighted norms

Let $x_o \in \overline{\Omega}$ and $\rho = \rho(h) > 0$ be fixed. With $\mu(x) = |x-x_o|^2 + \rho^2$ we introduce the norms respective seminorms

$$\|v\|_\alpha^2 = \iint_\Omega \mu^{-\alpha} v^2 \, dx \quad ,$$

(5)

$$\|\nabla v\|_\alpha^2 = \iint_\Omega \mu^{-\alpha} \left\{\Sigma\left(\frac{\partial v}{\partial x_i}\right)^2\right\} dx \quad .$$

Besides the obvious inequality

(6) $$|D(v,w)| \leq \|\nabla v\|_\alpha \, \|\nabla w\|_{-\alpha}$$

we will use a shift theorem $\left(|\nabla^2 v|^2 = \Sigma(\partial^2 v \, / \, \partial x_i \, \partial x)^2\right)$:

Lemma 1: Let $v \in \overset{\circ}{W}{}_2^1 \cap W_2^2$, then

$$\|\nabla^2 v\|_{-1}^2 \le c_o \left\{\|\Delta v\|_{-1}^2 + \|\nabla v\|_o^2\right\} \quad . \tag{7}$$

The proof follows directly applying the well-known a priori estimate $\|\nabla^2 w\| \le c\|\Delta w\|$ for $w \in \overset{\circ}{W}{}_2^1 \cap W_2^2$ to the functions ρv, xv, yv .

The properties of $\overset{\circ}{S}_h$ with respect to weighted norms needed below are summarized in

Lemma 2: For α fixed there is a $\gamma_1 = \gamma_1(\alpha, \varkappa_1)$ such that for $\rho \ge \gamma_1 h$ the statements are true with $c_1 = c_1(\varkappa_1, \gamma_1)$

(i) To any $v \in \overset{\circ}{W}{}_2^1 \cap W_2^2$ there is a $\chi \in \overset{\circ}{S}_h$ according to $\|v-\chi\|_\alpha + h\|\nabla(v-\chi)\|_\alpha \le c_1 h^2 \|\nabla^2 v\|_\alpha$.

(ii) To any $\varphi \in \overset{\circ}{S}_h$ there is a $\chi \in \overset{\circ}{S}_h$ according to $\|\nabla(\mu^{-\alpha}\varphi - \chi)\|_{-\alpha} \le c_2(h/\rho) \left\{\|\varphi\|_{\alpha+1} + \|\nabla\varphi\|_\alpha\right\}$.

For the proof see NATTERER [7], NITSCHE [11].

Now let u be the solution of problem (I) and $\Phi = u_h \in \overset{\circ}{S}_h$ minimizing $I(.)$ in $\overset{\circ}{S}_h$. Then Φ is the Ritz approximation and is characterized by

$$D(\Phi, \chi) = D(u, \chi) \qquad \text{for } \chi \in \overset{\circ}{S}_h \quad . \tag{8}$$

Theorem 1: Let $\rho = \gamma h |\ln h|^{1/2}$ with γ properly chosen, then

$$\|\Phi\|_2 + \|\nabla\Phi\|_1 \le c \left\{\|u\|_2 + \|\nabla u\|_1\right\} \tag{9}$$

with $c = c(\gamma, \varkappa_1)$.

The proof is divided into three steps:

Step 1: The identity

$$\iint_\Omega p\ |\nabla v|^2\ dx = D(v,pv) + \frac{1}{2}\iint_\Omega (\Delta p)\ v^2\ dx$$

gives with $p = \mu^{-1}$, $v = \Phi$

$$\|\nabla\Phi\|_1^2 \le D(\Phi,\mu^{-1}\Phi) + 2\ \|\Phi\|_2^2 \quad . \tag{10}$$

Using an appropriate $\chi \in \overset{\circ}{S}_h$ according to Lemma 2 we get

$$\begin{aligned} D(\Phi,\mu^{-1}\Phi) &= D(\Phi,\mu^{-1}\Phi-\chi) - D(u,\mu^{-1}\Phi-\chi) + D(u,\mu^{-1}\Phi) \\ &\le \left\{\|\nabla\Phi\|_1 + \|\nabla u\|_1\right\} c_2 \frac{h}{\rho} \left\{\|\Phi\|_2 + \|\nabla\Phi\|_1\right\} \\ &\quad + \|\nabla u\|_1\ \|\nabla(\mu^{-1}\Phi)\|_{-1} \quad . \end{aligned}$$

We may estimate

$$\|\nabla(\mu^{-1}\Phi)\|_{-1} \le c_3(\|\nabla\Phi\|_1 + \|\Phi\|_2) \quad .$$

Now let $\gamma_2 = \mathrm{Max}\ (\gamma_1, 2c_2)$. Then the factor of the quadratic term $\|\nabla\Phi\|_1^2$ on the right hand side of (10) is less than one for $\rho \ge \gamma_2\ h$. By standard arguments we come to

$$\|\nabla\Phi\|_1^2 \le c_4\ (\|\nabla u\|_1^2 + \|\Phi\|_2^2) \quad . \tag{11}$$

Step 2: Next let w be the solution of

$$\begin{aligned} -\Delta w &= \mu\ \Phi \quad \text{in } \Omega \ , \\ w &= 0 \quad \text{on } \partial\Omega \ . \end{aligned} \tag{12}$$

Then with $\chi \in \overset{\circ}{S}_h$ chosen properly

$$\begin{aligned}\|\Phi\|_2^2 &= D(\Phi,w)\\ &= D(\Phi,w-\chi) - D(u,w-\chi) + \iint_\Omega \mu^{-2}\Phi u \, dx\\ &\le (\|\nabla\Phi\|_1 + \|\nabla u\|_1)\, c_1\, h\, \|\nabla^2 w\|_{-1} + \|\Phi\|_2\, \|u\|_2 \quad .\end{aligned}$$

From this we conclude with $\delta > 0$ arbitrary

$$(13) \quad \|\Phi\|_2^2 \le \delta\, \|\nabla\Phi\|_1^2 + c_5 \left\{\|u\|_2^2 + \|\nabla u\|_1^2 + \delta^{-1} h^2 \|\nabla^2 w\|_{-1}^2\right\} .$$

(11) and (13) with $\delta < 1/c_4$ give

$$(14) \quad \|\Phi\|_2^2 + \|\nabla\Phi\|_1^2 \le c_6\left\{\|u\|_2^2 + \|\nabla u\|_1^2\right\} + c_7\, h^2\, \|\nabla^2 w\|_{-1}^2 \quad .$$

<u>Step 3:</u> Using Lemma 1 and the definition of w we get

$$(15) \qquad \|\nabla^2 w\|_{-1}^2 \le c_o \left\{\|\Phi\|_3^2 + \|\nabla w\|^2\right\} \quad .$$

The second term will be essential. the first is bounded by $\rho^{-2}\|\Phi\|_2^2$.

<u>Lemma 3:</u> <u>Let</u> w <u>be defined by (12), then</u>

$$(16) \qquad \|\nabla w\|^2 \le c_8\, \rho^{-2}\, |\ln \rho|^1\, \|\Phi\|_2^2 \quad .$$

With this lemma the inequality (14) gives directly Theorem 1 if γ is chosen properly.

The bound in (16) is given by

$$(17) \quad \lambda^{-1} = \sup\left\{\|\nabla w\|^2 \mid w \in \overset{\circ}{W}{}_2^1 \cap W_2^2 \wedge \|\Delta w\|_{-2}^2 \le 1\right\} \quad .$$

There exists (at least) one extremal function w which is the solution of the eigenvalue problem

$$(18)\qquad \begin{aligned} -\Delta w &= \lambda\mu^{-2} w \quad \text{in} \quad \Omega \; . \\ w &= 0 \quad\quad \text{on} \quad \partial\Omega \; . \end{aligned}$$

Therefore we ask for the smallest eigenvalue $\lambda_{Min}(\Omega)$ of (18). It is easy to see the monotony of this quantity with respect to Ω : If $\tilde{\Omega} \supseteq \Omega$ then $\lambda_{Min}(\tilde{\Omega}) \le \lambda_{Min}(\Omega)$. Now the sphere $K_R(x_o)$ with center x_o and radius $R = \text{diam}(\Omega)$ contains Ω and so $\lambda \ge \lambda_{Min}(K_R(x_o))$. By direct computation we get $\lambda^{-1}_{Min}(K_R(x_o)) \le c_8 \, \rho^{-2} |\ln \rho|^1$ what finishes the proof of Lemma 3.

3. L_∞ error estimates for the unrestricted problem

Let $U_h \in \overset{o}{S}_h$ be an appropriate approximation on u . Since the Ritz operator is a projection we have because of Theorem 1

$$\|\Phi - U_h\|_2^2 + \|\nabla(\Phi - U_h)\|_1^2 \le c\left\{\|u - U_h\|_2^2 + \|\nabla(u - U_h)\|_1^2\right\} .$$

Using Lemma 2 and $\rho^2 = \gamma_2 \, h^2 \, |\ln h|$ we find immediately

$$(19)\qquad \begin{aligned} \|u - U_h\|_2^2 + \|\nabla(u - U_h)\|_1^2 &\le c_9 \, h^2 \, |\ln \rho| \, \|\nabla^2 u\|_{L_\infty} \\ &\le c_{10} h^2 \, |\ln h| \, \|\nabla^2 u\|_{L_\infty} \; . \end{aligned}$$

If we choose $x_o \in \Delta_o$ such that $|\nabla(\Phi - U_h)(x_o)| = \|\nabla(\Phi - U_h)\|_{L_\infty}$ then we have since $\nabla(\Phi - U_h)$ is piecewise constant and the area of Δ_o is proportional to h^2

$$(20)\qquad c_{11}(\varkappa_1) \, \frac{h^2}{\rho^2} \, \|\nabla(\Phi - U_h)\|_{L_\infty}^2 \le \|\nabla(\Phi - U_h)\|_1^2 \quad .$$

Combining (19) and (20) and using that in addition U_h can be chosen such that $\|u - U_h\|_{L_\infty} \le c_{12}(\varkappa_1) \, h^2 \, \|\nabla^2 u\|_{L_\infty}$

we get

$$(21) \qquad \|\nabla(u-u_h)\|_{L_\infty} = \|\nabla(u-\Phi)\|_{L_\infty} \le c\, h\, |\ln h|\, \|\nabla^2 u\|_{L_\infty} .$$

The same argument applied to the first term in (19) with x_o chosen according to $|(\Phi-U_h)(x_o)| = \|\Phi-U_h\|_{L_\infty}$ gives the bound

$$\|u-u_h\|_{L_\infty} = \|u-\Phi\|_{L_\infty} \le c\, h^2\, |\ln h|^{3/2}\, \|\nabla^2 u\| .$$

By a more careful analysis of (19) we could derive from this inequality alone

$$\|u-u_h\|_{L_\infty} \le c\, h^2\, |\ln h|\, \{\ln |\ln h|\}^{1/2}\, \|\nabla^2 u\|_{L_\infty} .$$

In order to come to the estimate of the theorem we need

<u>Lemma 4:</u> <u>Let</u> $K_r(x_o)$ <u>be a sphere with center</u> x_o <u>and radius</u> $r \ge \varkappa_1^{-1} h$ <u>contained in</u> Ω . <u>For</u> $\rho = \gamma_3 h\, |\ln h|^{1/2} (\gamma_3 \ge \gamma_2)$ <u>the estimate holds true</u>

$$\frac{1}{\pi r^2} \iint_{K_r(x_o)} \Phi^2\, dx \le$$

$$\le \frac{1}{\pi r^2} \iint_{K_r(x_o)} u^2\, dx + c_{12}(\varkappa_1, \gamma_3) h^2 |\ln h| (\|\nabla\Phi\|_1^2 + \|\nabla u\|_1^2)$$

The proof is analogue to Steps 2 and 3 of Section 2 and is omitted here.

Now let $\Delta \in \Gamma_h$ be a triangle with

$$\|\Phi-U_h\|_{L_\infty(\Omega)} = \|\Phi-U_h\|_{L_\infty(\Delta)} .$$

Then - since $\Phi - U_h$ is linear in Δ - we get with $K_r(x_o) \subseteq \Delta$

$$\|\Phi-U_h\|^2_{L_\infty} \leq c_{13}(\varkappa_1)\,\frac{1}{h^2}\iint\limits_{K_r(x_o)} (\Phi-U_h)^2 \quad .$$

Using (19) and Lemma 4 we finally come to (3).

4. L_∞ error estimates for the restricted problem

In this section we will make use of the discrete 'Maximum Principle' due to CIARLET-RAVIART [3], in this connection we will need the 'angle'-condition of Section 1. Let P_i denote the nodes of Γ_h in Ω. The functions $\{\varphi_i \in \overset{\circ}{S}_h\}$ defined by

$$\varphi_i(P_k) = \delta_{ik}$$

form a basis in $\overset{\circ}{S}_h$. For any i resp. P_i we define for functions $\chi \in \overset{\circ}{S}_h$ the discrete Laplace-operator Δ_h by means of

$$-(\Delta_h\ \chi)_i = D(\chi,\varphi_i) \quad .$$

The mentioned maximum principle is

Lemma 5: Let T be a mesh-domain, i.e. the union of some of $\Delta \in \Gamma_h$. If $\chi \in \overset{\circ}{S}_h$ satisfies $(\Delta_h\chi)_i \geq 0$ for all nodes P_i in the interior of T then χ attains its maximum on the boundary ∂T.

Now let $u = u^K$ and $u_h = u^K_h$ be the minimizing elements of $I(.)$ in K resp. K_h. The following domains will be of special interest:

(i) $\Omega_o = \text{int}\left\{x \mid u(x) = z(x)\right\}$,
Ω_o is the set of points in which the side condition $u \leq z$ is attained;

(ii) $\Omega_h = \text{int}\{\cup \overline{\Delta} \mid \Delta \in \Gamma_h \wedge \Delta \cap \Omega_o \neq \emptyset\}$,
Ω_h is the smallest mesh-domain containing Ω_o , obviously

$$\text{dist}(\partial\Omega_h, \overline{\Omega}_o) \le \varkappa_1 h ;$$

(iii) $T_h = \text{int}\{\cup \overline{\Delta} \mid \exists\, x \in \overline{\Delta}$ with u_h tangent to z in $x\}$.

The solutions u, u_h can be characterized by $(f_i = (f, \varphi_i))$

$$\text{(22)} \quad \begin{aligned} &\Delta u + f \ge 0 \text{ in } \Omega \quad \text{and} \quad \Delta u + f = 0 \text{ in } \Omega - \overline{\Omega}_o , \\ &(\Delta_h u_h)_i + f_i \ge 0 \text{ in } \Omega \quad \text{and} \quad (\Delta u_h)_i + f_i = 0 \text{ in } \Omega - T_h . \end{aligned}$$

The last relation is to be understood in the sense: For all nodes in Ω respective in $\Omega - \overline{T}_h$.

Because of the assumption on z for $h \le h_o(z)$ the domains Ω_h, T_h are contained properly in Ω , i.e. there is an $\Omega_1 \subset\subset \Omega$ with

$$\text{(23)} \qquad \Omega_h, T_h \subseteq \Omega_1 ,$$

for $h \le h_o$.

In Ω_h resp. T_h the functions u resp. u_h differ from z only slightly:

<u>Lemma 6:</u>

(i) <u>Let</u> $x \in \overline{\Omega}_h$, <u>then</u>

$$0 \le z(x) - u(x) \le$$
$$\le c_{14}(\varkappa_1)\, h^2 (\|\nabla^2 u\|_{L_\infty} + \|\nabla^2 z\|_{L_\infty}) ,$$

(ii) <u>let</u> $x \in \overline{T}_h$, <u>then</u>

$$0 \le z - u_h(x) \le c_{15}(\varkappa_1)\, h^2 \, \|\nabla^2 z\|_{L_\infty} .$$

The proof is obvious, in case (i) only the continuity of the first derivatives of $z - u$ are to be noted.

In order to get inequality (4) we will compare u_h with the Ritz approximation $U_h = R_h u$ defined by

$$D(U_h, \chi) = D(u, \chi) \quad \text{for} \quad \chi \in \overset{\circ}{S}_h .$$

Since (3) is already shown it suffices to prove

$$\text{(24)} \qquad \|u_h - U_h\|_{L_\infty} \le c\, h^2\, |\ln h|\, \left\{\|\nabla^2 u\|_{L_\infty} + \|\nabla^2 z\|_{L_\infty}\right\} .$$

This will be done in two steps.

<u>Step 1:</u> We first show - with

$$d_h := h^2\, |\ln h|\, \left\{\|\nabla^2 u\|_{L_\infty} + \|\nabla^2 z\|_{L_\infty}\right\} -$$

$$\text{(25)} \qquad u_h - U_h \le c_{16}(\varkappa_1)\, d_h .$$

In $\overline{\Omega}_h$ we have using Lemma 6 and (3)

$$\text{(26)} \qquad u_h - U_h \le (z-u) + (u-U_h) \le c_{17}\, d_h .$$

Since $u_h = U_h = 0$ on $\partial\Omega$ we have therefore

$$\text{(27)} \qquad \sup\left\{(u_h-U_h)(x) \mid x \in \partial(\Omega-\overline{\Omega}_h)\right\} \le c_{17}\, d_h .$$

Now let P_i be a node in the interior of $\Omega - \overline{\Omega}_h$. Using the characterization (22) and

$$- (\Delta_h U_h)_i = D(U_h, \varphi_i) = D(u, \varphi_i)$$
$$= (f, \varphi_i) = f_i$$

we find for such nodes

$$\Delta_h (u_h - U_h)_i \ge 0 .$$

The maximum principle and (26), (27) lead to (25).

<u>Step 2:</u> In order to show

$$U_h - u_h \le c_{18}\, d_h \tag{28}$$

we construct a $v_h \in \overset{\circ}{S}_h$ with

$$v_h \ge U_h - c_{18}\, d_h \quad , \tag{29}$$

$$v_h \le u_h \quad ; \tag{30}$$

then (28) and henceforth the second statement (4) of the theorem is shown.

Let ω be defined by

$$-\Delta\omega = 1 \quad \text{in} \quad \Omega \ ,$$
$$\omega = 0 \quad \text{on} \quad \partial\Omega$$

and w_h be the Ritz approximation on ω. Since ω is positive in Ω for h small enough we have - see (23) -

$$\inf_{\Omega_1} w_h \ge \frac{1}{2} \inf_{\Omega_1} \omega \ge \underline{\omega} > 0 \quad .$$

Moreover by the maximum principle $w_h \ge 0$ holds in Ω. We take

$$v_h = U_h - \lambda\, d_h\, w_h$$

with $\lambda > 0$ properly chosen. Condition (29) is met, therefore only (30) is to be checked. In $\overline{T}_h$ we have

$$\begin{aligned} v_h - u_h &\le (z-u_h) - (z-u) - (u-U_h) - \lambda\, d_h w_h \\ &\le (c_{15} + c - \lambda\underline{\omega})\, d_h \quad . \end{aligned}$$

The choice $\lambda = \underline{\omega}^{-1}(c_{15}+c)$ guarantees (30) in $\overline{T}_h$.

Similar to Step 1 we consider the nodes P_i in the interior of $\Omega - \overline{T}_h$. We get - using (22) -

$$\begin{aligned} \Delta_h(v_h-u_h)_i &= f_i - D(U_h,\varphi_i) + \lambda\, d_h\, D(w_h,\varphi_i) \\ &= f_i - D(u,\varphi_i) + \lambda\, d_h\, D(\omega,\varphi_i) \\ &= (\varphi_i,\, f + \Delta u + \lambda\, d_h) \geq 0 \quad . \end{aligned}$$

Applying once more the maximum principle (30) is shown.

Literature

[1] BREZIS, H., STAMPACCHIA, G.: Sur la régularité de la solution d'inéquations elliptiques. Bull. Soc. Math. France, 96, MR 39 No.659, 153-180 (1968).

[2] CIARLET, P.G., RAVIART, P.-A.: Interpolation Theory over Curved Elements, with Applications to Finite Element Methods. Comput. Methods in Appl. Mech. and Eng., 1, 217-249 (1972).

[3] CIARLET, P.G., RAVIART, P.-A.: Maximum Principle and Uniform Convergence for the Finite Element Method. Comput. Methods in Appl. Mech. and Eng., 2, 17-31 (1973).

[4] FALK, R.S.: Error Estimates for the Approximation of a Class of Variational Inequalities. Math. of Comp., 28, 963-971 (1974).

[5] FALK, R.S.: Approximation of an Elliptic Boundary Value Problem with Unilateral Constraints. R.A.I.R.O. R2, 5-12 (1975).

[6] LIONS, J.L., STAMPACCHIA, G.: Variational Inequalities. Comm. Pure Appl. Math., 20, 439-519 (1967).

[7] NATTERER, F.: Über die punktweise Konvergenz finiter Elemente (to appear).

[8] NATTERER, F.: Optimale L_∞-Konvergenz finiter Elemente bei Variationsungleichungen (to appear).

[9] NITSCHE, J.: Über ein Variationsprinzip zur Lösung von Dirichlet-Problemen bei Verwendung von Teilräumen, die keinen Randbedingungen unterworfen sind. Abh. d. Hamb. Math. Sem., 36, 9-15 (1971).

[10] NITSCHE, J.: On Approximation Methods for Dirichlet-Problems Using Subspaces with 'Nearly-Zero' Boundary Conditions. Proc. of a Conference "The Mathematical Foundations of the Finite Element Method with Applications to Partial Differential Equations." A.K. Aziz editor Academic Press, 603-627 (1972).

[11] NITSCHE, J.: L_∞-Convergence of Finite Element Approximation. 2. Conf. on Finite Elements, Rennes, France (1975).

[12] SCOTT. R.: Optimal L^∞-Estimates for the Finite Element Method on Irregular Meshes (to appear).

[13] STRANG, G.: Finite Elements and Variational Inequalities. Seminaires Analyse Numérique, Paris (1973/74).

Joachim Nitsche
Institut für Angewandte Mathematik
Albert-Ludwigs-Universität
78 Freiburg, Hebelstr. 40
Federal Republic of Germany

DUAL-MIXED HYBRID FINITE ELEMENT METHOD FOR SECOND-ORDER ELLIPTIC PROBLEMS

J. T. Oden and J. K. Lee

The Texas Institute for Computational Mechanics
The University of Texas, Austin, Texas

Summary. Dual-Mixed-Hybrid finite element approximations are described for second-order boundary-value problems in which independent approximations are used for the solution and its gradient in the interior of an element and the trace of the solution on the boundary of the element. A-priori error estimates are derived with some conditions for convergence. Some numerical results are also included.

1. Introduction

Let Ω be an open bounded domain in Euclidean plane with a piecewise smooth boundary $\partial\Omega$. Consider, as a model problem,

$$\left.\begin{aligned} -\Delta u + u &= f \quad \text{in } \Omega \\ u &= 0 \quad \text{on } \partial\Omega \end{aligned}\right\} \tag{1.1}$$

where $\Delta = \nabla^2$ is the Laplacian operator and $f \in L_2(\Omega)$ is a given function. Babuska, Oden, and Lee [1] studied a mixed-hybrid method for (1.1) in which the solution and the gradient are approximated in the interior of an element and the trace of the gradient on the interelement boundaries, independently. The present study is concerned with a formulation which is dual to that in [1] and which is a generalization of the so-called equilibrium method (see, e.g., [2] and [3]). The variational principle appears to be of a modified Hellinger-Reissner type, proposed by Wolf [4]. As special cases of the present method, we obtain equilibrium models, mixed models, as well as the stress-assumed hybrid models of Pian type [5].

For some general results on mixed methods, we refer to Oden [6], Oden and Reddy [7,8], and Reddy [9]. Mixed methods for solving fourth order problems have been analyzed by Ciarlet and Raviart [10], Johnson [11], and Miyoshi [12]. See also Raviart and Thomas [13], Oden and Lee [14] for analysis of the method for Poisson's equation and plane elasticity problems, respectively.

Analysis of dual-hybrid (stress assumed) methods have been made by Brezzi [15], and Brezzi and Marini [16] for fourth order problems, and by Thomas [17] for second order problems. Primal-hybrid (displacement assumed) methods have been studied by Raviart and Thomas [18] and by Babuska, Oden and Lee [1] as a special case of the mixed-hybrid method.

The present method requires more smoothness of the solution within an element than usual. However, it eliminates the difficulty of constructing a continuous finite dimensional submanifold that satisfies the equilibrium condition which is required in equilibrium methods.

The existence of a unique solution for the mixed-hybrid method is discussed in Section 3 following some preliminary results and notations in Section 2. Section 4 is devoted to examining conditions for approximate solutions to exist, followed by a-priori error estimates with a condition for convergence in Section 5. Section 6 contains an example problem with some numerical results designed to demonstrate the validity of the error estimates and for comparisons to other well known methods.

2. Preliminaries and Notations

For an integer $m \geq 0$, $H^m(\Omega)$ denotes the Hilbert space equipped with the inner product

$$(u,v)_{m,\Omega} = \int_\Omega \sum_{|\alpha| \leq m} D^{\alpha} u \cdot D^{\alpha} v \, dx \tag{2.1}$$

with the natural norm

$$||u||_{m,\Omega} = (u,u)^{\frac{1}{2}}_{m,\Omega} \tag{2.2}$$

where $\alpha = (\alpha_1,\alpha_2)$, $\alpha_i \geq 0$ integral, $|\alpha| = \alpha_1 + \alpha_2$, and $D^{\alpha}u = \partial^{|\alpha|}u/\partial x_1^{\alpha_1}\partial x_2^{\alpha_2}$. If $\sigma = (\sigma_1,\sigma_2)$ is a vector valued function with $\sigma_i \in H^m(\Omega)$, we write $\sigma \in H^m(\Omega)$ and the natural norm is denoted by

$$||\sigma||_{m,\Omega} = (||\sigma_1||^2_{m,\Omega} + ||\sigma_2||^2_{m,\Omega})^{\frac{1}{2}} \tag{2.3}$$

When $m = 0$, we also use, interchangeably, the notation $H^o(\Omega) = L_2(\Omega)$.

It is well known that if $u \in H^1(\Omega)$, the trace $\gamma_o u$ on $\partial\Omega$ is well defined. We denote by $H^{\frac{1}{2}}(\partial\Omega)$ the space of traces of functions in $H^1(\Omega)$ furnished with the norm

$$||\phi||_{\frac{1}{2},\partial\Omega} = \inf_{u \in H^1(\Omega)} \{||u||_{1,\Omega} \; ; \; \phi = \gamma_o u\} \tag{2.4}$$

Clearly

$$||\gamma_o u||_{\frac{1}{2},\partial\Omega} \leq ||u||_{1,\Omega} \qquad \forall\, u \in H^1(\Omega) \tag{2.5}$$

Conversely, there also exists a continuous map δ of $H^{\frac{1}{2}}(\partial\Omega)$ into $H^1(\Omega)$ such that

$$||\delta\phi||_{1,\Omega} \leq ||\phi||_{\frac{1}{2},\partial\Omega} \tag{2.6}$$

Indeed, if $v \in H^1(\Omega)$ is such that

$$-\Delta v + v = 0 \quad \text{in } \Omega$$

$$v = \phi \quad \text{on } \partial\Omega$$

Then

$$||v||^2_{1,\Omega} = \oint \frac{\partial v}{\partial n} w \, ds = ||\phi||^2_{\frac{1}{2},\partial\Omega} \tag{2.7}$$

As usual, we define also

$$H_o^1(\Omega) = \{u \in H^1(\Omega) \ ; \quad \gamma_o u = 0\}$$

We now record a two part theorem which plays a fundamental role in later developments. Proof can be found in [19] (see also [20] and [21].

Theorem 2.1. Let $\mathcal{U}$ and $\mathcal{V}$ be two real Hilbert spaces and $B: \mathcal{U} \times \mathcal{V} \to \mathbb{R}$ a linear functional on $\mathcal{U} \times \mathcal{V}$ such that for every $u \in \mathcal{U}$ and $v \in \mathcal{V}$ the following hold:

$$|B(u,v)| \leq C_1 ||u||_{\mathcal{U}} ||v||_{\mathcal{V}} \tag{2.8}$$

$$\inf_{||u||_{\mathcal{U}}=1} \ \sup_{||v||_{\mathcal{V}}\leq 1} |B(u,v)| \geq C_2 \geq 0 \tag{2.9}$$

$$\sup_{u \in \mathcal{U}} |B(u,v)| > 0 , \qquad v \neq 0 , \qquad v \in \mathcal{V} \tag{2.10}$$

Here C_1 and C_2 are positive constants independent of u and v and $||\cdot||_{\mathcal{U}}$ and $||\cdot||_{\mathcal{V}}$ denote the norms on $\mathcal{U}$ and $\mathcal{V}$ respectively. In addition, let $f \in \mathcal{V}'$ be given. Then there exists a unique element $u_o \in \mathcal{U}$ such that

$$B(u_o,v) = f(v) \qquad \forall\, v \in \mathcal{V} \tag{2.11}$$

Moreover,

$$||u_o||_{\mathcal{U}} \leq \frac{1}{C_2} ||f||_{\mathcal{V}'} \tag{2.12}$$

If the same conditions hold for finite dimensionalsubspaces $\mathcal{U}_h \subset \mathcal{U}$ and $\mathcal{V}_h \subset \mathcal{V}$, then there exists a unique element $U_o \in \mathcal{U}_h$ such that

$$B(U_o,V) = f(V) \qquad \forall\, V \in \mathcal{V}_h \tag{2.13}$$

and satisfies

$$||u_o - U_o||_{\mathcal{U}} \leq (1 + C_1/C_2^h) \inf_{\tilde{U} \in \mathcal{U}_h} ||u_o - \tilde{U}||_{\mathcal{U}} \tag{2.14}$$

where C_2^h is the constant in (2.9) in the subspace $\mathcal{U}_h \times \mathcal{V}_h$.

3. Dual-Mixed-Hybrid Variational Principle

Consider a partition P of $\bar{\Omega}$ into subdomains $\bar{\Omega}_e$ such that

(i) $\bar{\Omega} = \bigcup_{e=1}^{E} \bar{\Omega}_e$

(ii) $\Omega_e \cap \Omega_f = \emptyset$ if $e \neq f$

(iii) Ω_e, $1 \leq e \leq E$ is an open subset of Ω with piecewise smooth boundary $\partial\Omega_e$.

(iv) $\Gamma = \bigcup_{e=1}^{E} \partial\Omega_e$

If v is any function defined on Ω, we denote its restriction to Ω_e and $\partial\Omega_e$ by

v_e and $\gamma_o v_e$, respectively; i.e.,

$$v_e = v|_{\Omega_e} \ , \quad \gamma_o v_e = v|_{\partial\Omega_e} \ , \quad 1 \le e \le E$$

Over the partition P just described, we introduce the following spaces:

- $$\underset{\sim}{S}(P) = \{\underset{\sim}{\sigma} \in \underset{\sim}{L}_2(\Omega) \ ; \ \underset{\sim}{\nabla} \cdot \underset{\sim}{\sigma}_e \in L_2(\Omega_e) \ , \ 1 \le e \le E\} \tag{3.1}$$

with the norm

$$||\underset{\sim}{\sigma}||_S = (||\underset{\sim}{\sigma}||^2_{0,\Omega} + \sum_{e=1}^{E} ||\underset{\sim}{\nabla} \cdot \underset{\sim}{\sigma}||^2_{0,\Omega_e})^{\frac{1}{2}} \tag{3.2}$$

- $$W(\Gamma) = \{w \in \prod_{e=1}^{E} H^{\frac{1}{2}}(\partial\Omega_e) \ ; \ \text{there exists } u \in H^1_o(\Omega) \text{ such that } w = u \text{ on } \Gamma\} \tag{3.3}$$

with the norm

$$||w||_{W(\Gamma)} = \inf_{u \in H^1_o(\Omega)} \{|u|_{1,\Omega} \ ; \ w = u \text{ on } \Gamma\} \tag{3.4}$$

where $|u|^2_{1,\Omega} = \int_\Omega \nabla u \cdot \nabla u \, dx$.

Finally, we introduce a product space

$$X = L_2(\Omega) \times \underset{\sim}{S}(P) \times W(\Gamma) \tag{3.5}$$

so that if $\underset{\sim}{\lambda} \in X$ then $\underset{\sim}{\lambda} = (u, \underset{\sim}{\sigma}, w)$. Also, we define

$$||\underset{\sim}{\lambda}||_X = (||u||^2_{0,\Omega} + ||\underset{\sim}{\sigma}||^2_S + ||w||^2_W)^{\frac{1}{2}} \tag{3.6}$$

It is also convenient to denote by $\underset{\sim}{\lambda}_e$ the triple of restrictions $(u_e, \underset{\sim}{\sigma}_e, w_e)$ to $\bar{\Omega}_e$ and to use the notation for $1 \le e \le E$,

$$||\underset{\sim}{\lambda}_e||^2_{X_e} = ||u_e||^2_{0,\Omega_e} + ||\underset{\sim}{\sigma}_e||^2_{\underset{\sim}{S}_e} + ||w_e||^2_{W_e} \tag{3.7}$$

where $||\underset{\sim}{\sigma}_e||^2_{\underset{\sim}{S}_e} = ||\underset{\sim}{\sigma}_e||^2_{0,\Omega_e} + ||\underset{\sim}{\nabla} \cdot \underset{\sim}{\sigma}_e||^2_{0,\Omega_e}$ and $||w_e||^2_{W_e} = ||w_e||^2_{\frac{1}{2},\partial\Omega}$. Then, (3.6) can also be written

$$||\underset{\sim}{\lambda}||_X = (\sum_{e=1}^{E} ||\underset{\sim}{\lambda}_e||^2_{X_e})^{\frac{1}{2}} \tag{3.8}$$

<u>Lemma 3.1</u>. Let z be a function defined over Ω such that $z|_{\Omega_e} = z_e$ for each $e = 1,2,\cdots,E$ where z_e solves, for a $w_e = w|_{\partial\Omega_e}$, $w \in W(\Gamma)$,

$$\left. \begin{aligned} -\nabla^2 z_e &= 0 \quad \text{in } \Omega_e \\ z_e &= w_e \quad \text{on } \partial\Omega_e \end{aligned} \right\} \tag{3.9}$$

Then

$$|z|^2_{H^1(\Omega)} = \sum_{e=1}^{E} \oint_{\partial\Omega_e} \frac{\partial z_e}{\partial n_e} w_e \, ds = ||w||^2_{W(\Gamma)} \tag{3.10}$$

Proof. Let Ω_e and Ω_f be two neighboring subdomains of Ω with the common boundary $\Omega_{ef} = \partial\Omega_e \cap \partial\Omega_f$. Then

$$z_e|_{\Gamma_{ef}} = w|_{\Gamma_{ef}} = z_f|_{\Gamma_{ef}}$$

This implies that $z \in H^1_0(\Omega)$ because $w = 0$ on $\partial\Omega$ and w is continuous on Γ. Now since z_e satisfies (3.9),

$$|z_e|^2_{H^1(\Omega_e)} = \oint_{\partial\Omega_e} \frac{\partial z_e}{\partial n_e} w_e \, ds$$

from which the first half of (3.10) follows by summing over all Ω_e's. The second part of (3.10) follows from (2.6) and (3.4). ∎

Let now $B(\cdot,\cdot)$ be a bilinear form on $X \times X$ given by

$$B(\underset{\sim}{\lambda},\overline{\underset{\sim}{\lambda}}) = \sum_{e=1}^{E} B_e(\underset{\sim}{\lambda}_e,\overline{\underset{\sim}{\lambda}}_e) \tag{3.11}$$

where

$$B_e(\underset{\sim}{\lambda}_e,\overline{\underset{\sim}{\lambda}}_e) = \int_{\Omega_e} (-\underset{\sim}{\nabla} \cdot \underset{\sim}{\sigma}_e \overline{u}_e - u_e \underset{\sim}{\nabla} \cdot \overline{\underset{\sim}{\sigma}}_e - \underset{\sim}{\sigma}_e \cdot \overline{\underset{\sim}{\sigma}}_e + u_e \overline{u}_e) dx$$

$$+ \int_{\partial\Omega_e} (\overline{\underset{\sim}{\sigma}}_e \cdot n w_e + \underset{\sim}{\sigma}_e \cdot n \overline{w}_e) ds \tag{3.12}$$

We also define a linear continuous functional on X by

$$F(\overline{\underset{\sim}{\lambda}}) = \int_{\Omega} f\overline{u} \, dx \tag{3.13}$$

Now we can prove:

Theorem 3.1. The dual-mixed-hybrid variational problem of finding $\underset{\sim}{\lambda} \in X$ such that

$$B(\underset{\sim}{\lambda},\lambda) = F(\overline{\underset{\sim}{\lambda}}) \qquad \forall \overline{\underset{\sim}{\lambda}} \in X \tag{3.14}$$

has a unique solution $\lambda^o \in X$. Furthermore there exists a constant $A_1 > 0$ such that

$$||\underset{\sim}{\lambda}||_X \leq A_1 ||f||_{L_2(\Omega)} \tag{3.15}$$

Proof. Proof is done by checking that the bilinear form $B(\cdot,\cdot)$ of (3.11) satisfies all the hypothesis of Theorem 2.1.

By the divergence theorem and (2.6)

$$\oint_{\partial\Omega_e} \underset{\sim}{\sigma}_e \cdot \underset{\sim}{n} w_e \, ds = \int_{\Omega_e} (\underset{\sim}{\sigma}_e \cdot \nabla z_e + \underset{\sim}{\nabla} \cdot \underset{\sim}{\sigma}_e z_e) dx \leq ||\underset{\sim}{\sigma}_e||_{\underset{\sim}{S}_e} \, ||w_e||_{\frac{1}{2},\partial\Omega_e} \tag{3.16}$$

Applying the Schwarz inequality successively with the aid of (3.16), we obtain

$$|B_e(\underset{\sim}{\lambda}_e, \underset{\sim}{\bar{\lambda}}_e)| \leq 2||\underset{\sim}{\lambda}_e||_{X_e} \, ||\underset{\sim}{\bar{\lambda}}_e||_{X_e}$$

Then, in view of (3.11) and by the Schwarz inequality, we see that (2.8) holds with the choice of $C_1 = 2$.

To show that the bilinear form also satisfies the conditions (2.9) and (2.10), we begin by selecting a special triple $\hat{\underset{\sim}{\lambda}} = (\hat{u}, \hat{\underset{\sim}{\sigma}}, \hat{w})$ where $\hat{\underset{\sim}{\lambda}}_e$ is given by

$$\left.\begin{aligned} \hat{u}_e &= u_e - \underset{\sim}{\nabla} \cdot \underset{\sim}{\sigma}_e \\ \hat{\underset{\sim}{\sigma}}_e &= -2\underset{\sim}{\sigma}_e + 2\underset{\sim}{\nabla} z_e \\ \hat{w}_e &= 2w_e \end{aligned}\right\} \tag{3.17}$$

where z_e is as in (3.9). Clearly, $\hat{\underset{\sim}{\lambda}} \in X$. Indeed, with the aid of (3.10) and (3.8), it is easy to see that

$$||\hat{\underset{\sim}{\lambda}}||_X \leq \sqrt{12}\, ||\underset{\sim}{\lambda}||_X \tag{3.18}$$

Now, a direct computation reveals that,

$$\begin{aligned} B(\underset{\sim}{\lambda}, \hat{\underset{\sim}{\lambda}}) &= \sum_{e=1}^{E} \Big\{ \int_{\Omega_e} ((\underset{\sim}{\nabla} \cdot \underset{\sim}{\sigma}_e)^2 + 2\underset{\sim}{\sigma}_e \cdot \underset{\sim}{\sigma}_e + u_e^2 - 2\underset{\sim}{\sigma}_e \cdot \nabla z_e) dx \\ &\quad + \oint_{\partial\Omega_e} 2 \frac{\partial z_e}{\partial n} w_e \, ds \Big\} \\ &\geq ||\underset{\sim}{\lambda}||_X^2 \end{aligned} \tag{3.19}$$

in view of (3.10), the Schwarz inequality, and (3.8). Hence

$$\inf_{||\underset{\sim}{\lambda}||_X = 1} \; \sup_{||\underset{\sim}{\bar{\lambda}}||_X \leq 1} |B(\underset{\sim}{\lambda}, \underset{\sim}{\bar{\lambda}})| \geq \inf_{||\underset{\sim}{\lambda}||_X = 1} |B(\underset{\sim}{\lambda}, \hat{\underset{\sim}{\lambda}})| / ||\hat{\underset{\sim}{\lambda}}||_X \geq 1/\sqrt{12} \tag{3.20}$$

That condition (2.10) also holds followed by interchanging the roles of $\underset{\sim}{\lambda}$ and $\hat{\underset{\sim}{\lambda}}$ due to the symmetry of $B(\cdot,\cdot)$.

The assertions follow from Theorem 2.1. We emphasize here that $C_1 = 2$, $C_2 = 1/\sqrt{12}$, and $A_1 = \sqrt{12}$ are independent of the partition P. ■

A remaining question is how the solution $\underset{\sim}{\lambda}^o$ of (3.14) is related to the solution of (1.1). To show the relationship, let u^* be the solution of (1.1) and let $\underset{\sim}{\lambda}^* = (u^*, \nabla u^*, u^*)$. Then, obviously, $\underset{\sim}{\lambda}^* \in X$, and a direct computation shows that

$$B(\underset{\sim}{\lambda}^*,\overline{\underset{\sim}{\lambda}}) = \int_\Omega f\bar{u}\,dx + \sum_{e=1}^{E} \oint_{\partial\Omega_e} \frac{\partial u^*}{\partial n}\bar{w}\,ds \qquad \forall\ \overline{\underset{\sim}{\lambda}} \in X \tag{3.21}$$

Since u^* is the solution of (1.1), $u^* \in H^2(\tilde{\Omega})$ where $\tilde{\Omega}$ is arbitrary but such that closure $(\Omega) \subset \tilde{\Omega}$. Then the second term in (3.21) vanishes because $\bar{w} = 0$ on $\partial\Omega$ and $\partial u^*/\partial n$ is continuous on $\partial\Omega_e \cap \tilde{\Omega}$. Thus, $\underset{\sim}{\lambda}^*$ also solves (3.14) which implies that $\underset{\sim}{\lambda}^* = \underset{\sim}{\lambda}^o$ due to the uniqueness of $\underset{\sim}{\lambda}^o$ and $\underset{\sim}{\lambda}^*$.

<u>Remark 3.1</u>. If (1.1a) is replaced by $-\Delta u = f$ in Ω, $u\bar{u}$ term in the bilinear form should be removed and one can select $\hat{\underset{\sim}{\lambda}} = (\hat{u},\hat{\underset{\sim}{\sigma}},\hat{w})$, in the proof of Theorem 3.1, such that

$$\hat{u}_e = 2u_e - \underset{\sim}{\nabla} \cdot \underset{\sim}{\hat{\sigma}}_e \;, \quad \underset{\sim}{\hat{\sigma}}_e = -2\underset{\sim}{\sigma}_e + 2\nabla z_e + \nabla y_e \;, \quad \hat{w}_e = 2w_e \tag{3.22}$$

where z_e is as in Lemma 3.1 and y_e solves

$$-\Delta y_e = u_e \quad \text{in } \Omega_e \qquad \frac{\partial y_e}{\partial n} = 0 \quad \text{on } \partial\Omega_e \tag{3.23}$$

Then

$$B(\underset{\sim}{\lambda},\hat{\underset{\sim}{\lambda}}) \geq \frac{1}{2}\,||\underset{\sim}{\lambda}||_X^2$$

and all of the subsequent arguments are the same as before.

4. Finite Element Approximations

The partition P, described at the beginning of Section 3, is now viewed as a finite element discretization of $\bar{\Omega}$ into E elements, as expected. Over each element we introduce local approximations of u_e, $\underset{\sim}{\sigma}_e$, and w_e using polynomials of possibly differing degree.

As usual, we associate to every partition P a parameter h such that

$$h = \max_{1 \leq e \leq E} h_e$$

where h_e is the diameter of Ω_e.

We now construct a collection of finite-dimensional subspaces over the partition having the following properties:

(i) $Q_k(P) = \{U \in L_2(\Omega);\ U_e \in P_{k'}(\Omega_e),\ 1 \leq e \leq E,\ 0 \leq k \leq k'\}$ (4.2)

so that for any $u_e \in H^s(\Omega_e)$, $s \geq 0$, there exists a constant $C_1 > 0$ independent of h_e, and a $\tilde{U}_e \in P_{k'}(\Omega_e)$ such that

$$||u_e - \tilde{U}_e||_{0,\Omega_e} \leq C_1 h_e^\alpha ||u_e||_{s,\Omega_e} \tag{4.3}$$

where $\alpha = \min\{k+1,s\}$, $s \geq 0$.

(ii) $\underset{\sim}{Q}_r(P) = \{\underset{\sim}{\Sigma} \in \underset{\sim}{S}(P);\ \underset{\sim}{\Sigma}_e \in \underset{\sim}{P}_{r'}(\Omega_e),\ 1 \leq e \leq E,\ 1 \leq r \leq r'\}$ (4.4) so that for any $\sigma_e \in H^q(\Omega_e)$, $q \geq 1$, there exists a constant $C_2 > 0$ independent of h_e, and a $\tilde{\Sigma}_e \in P_{r'}(\Omega_e)$ such that

$$||\sigma_e - \tilde{\Sigma}_e||_{S_e} \leq C_2 h_e^{\eta} ||\sigma_e||_{q,\Omega_e} \tag{4.5}$$

where $\eta = \min\{r, q-1\}$, $q \geq 1$.

(iii) $Q_t(\Gamma) = \{W \in W(\Gamma)$; there exists a $V_e \in P_{t'}(\Omega_e)$ such that

$$W_e = \gamma_o V_e \ , \quad 1 \leq e \leq E, \ 1 \leq t \leq t'\} \tag{4.6}$$

so that for any $u \in H^{\ell}(\Omega)$, $\ell \geq 1$, such that $w = u$ on Γ, there exists a constant $C_3 > 0$ independent of h, and a $\tilde{W} \in Q_t(\Gamma)$ such that

$$||w - \tilde{W}||_{W(\Gamma)} \leq C_3 h^{\xi} ||u||_{\ell,\Omega} \tag{4.7}$$

where $\xi = \min\{t, \ell-1\}$, $\ell \geq 1$. ■

The inequality (4.3) and (4.5) can be shown to hold if all Ω_e are quasi-uniform. For general results of this type we refer to Ciarlet and Raviart [23]. We call an element Ω_e a quasi-uniform if there exists a constant $C_o > 0$ independent of P such that $h_e/\rho_e \leq C_o$ where ρ_e is the diameter of the largest circle that can be inscribed in Ω_e. For the inequality (4.7), we note that

$$u|_{\Gamma} \in W(\Gamma) \quad \text{if} \quad u \in H_o^1(\Omega) \cap H^{\ell}(\Omega) \ , \quad \ell \geq 1$$

and that

$$\tilde{V}|_{\Gamma} \in Q_t(\Gamma) \quad \text{if} \quad \tilde{V} \quad H_o^1(\Omega) \cap P_{t'}(\Omega_e)$$

Then, from (3.4)

$$||u|_{\Gamma} - \tilde{V}|_{\Gamma}||_{W(\Gamma)} \leq ||u - \tilde{V}||_{1,\Omega}$$

and the inequality (4.7) follows from the usual result of C^o-elements.

Clearly, the product space

$$Q = Q_k(P) \times Q_r(P) \times Q_t(\Gamma) \tag{4.8}$$

is a subspace of X defined in (3.5). The dual-mixed hybrid finite element method is to seek an element $\Lambda \in Q$ such that

$$B(\Lambda, \bar{\Lambda}) = (\bar{\Lambda}) \qquad \forall \ \bar{\Lambda} \in Q \tag{4.9}$$

where $B(\cdot,\cdot)$ and $F(\cdot)$ are as in (3.11) and (3.13), respectively. The remainder of this section is devoted to a study of certain conditions under which (4.9) has a unique solution. First, a necessary condition:

Lemma 4.1. Let the dual-mixed-hybrid approximation (4.9) have a unique solution. Then the following must hold: for a $W \in Q_t(\Gamma)$

$$(C1) \begin{cases} \displaystyle\sum_{e=1}^{E} \int_{\partial\Omega_e} \overline{\underset{\sim}{\Sigma}}_e \cdot \underset{\sim}{n} W_e \, ds = 0 \qquad \forall \, \overline{\underset{\sim}{\Sigma}} \in \underset{\sim}{Q}_r(P) \qquad (4.10) \\[2mm] \text{implies that } W = 0 \end{cases}$$

Proof. Suppose there exists a unique solution but (C1) does not hold. Then there exists a $W^* \in Q_t(\Gamma)$, $W^* \neq 0$, such that (4.10) holds. Then

$$B((0,0,W^*), \overline{\underset{\sim}{\Lambda}}) = 0 \qquad \forall \, \overline{\underset{\sim}{\Lambda}} \in Q$$

which implies that $\underset{\sim}{0} = (0,0,0)$ is not a unique solution for $f = 0$, a contradiction. ∎

Remark 4.1. For the dual-hybrid model of Pian type, it can be shown that (C1) is a necessary and sufficient condition for the existence of a unique solution. Consult with Brezzi [24] and Thomas [17] for further details.

Remark 4.2. If $-\Delta u = f$ is approximated instead of (1.1a), necessary conditions are

(i) (C1) holds, and

(ii) for a $U_e \in P_{k'}(\Omega_e)$

$$(C2) \begin{cases} \displaystyle\int_{\Omega_e} U_e \underset{\sim}{\nabla} \cdot \overline{\underset{\sim}{\Sigma}}_e \, dx = 0 \qquad \forall \, \overline{\underset{\sim}{\Sigma}}_e \in \underset{\sim}{P}_{r'}(\Omega_e) \qquad (4.11) \\[2mm] \text{implies that } U_e = 0 \end{cases}$$

We also remark that a necessary condition for (C2) to hold is

$$Q_k(P) \subset \underset{\sim}{\nabla} \cdot \underset{\sim}{Q}_r(P) \qquad (4.12)$$

that is, for every $U_e \in P_{k'}(\Omega_e)$, $1 \le e \le E$, there exists a $\underset{\sim}{\Sigma}_e \in P_{r'}(\Omega_e)$ such that $U_e = \underset{\sim}{\nabla} \cdot \underset{\sim}{\Sigma}_e$ which will be the case if $k' \le r'-1$. The inclusion (4.12) is also a sufficient condition for the existence of a unique solution [22].

Turning to a sufficient condition, we now let $\underset{\sim}{\nabla} V \in \underset{\sim}{P}_{r'}(\Omega_e)$ for any $V \in P_{t'}(\Omega_e)$, symbolically,

$$\underset{\sim}{\nabla}(P_{t'}(\Omega_e)) \subset \underset{\sim}{P}_{r'}(\Omega_e) \qquad (4.13)$$

which occurs when $r' \ge t'-1$. Define a subspace of $\underset{\sim}{P}_{r'}(\Omega_e)$ by

$$\underset{\sim}{T}_e = \{ \underset{\sim}{s} \in \underset{\sim}{P}_{r'}(\Omega_e) : \underset{\sim}{\nabla} \cdot \underset{\sim}{s} = 0 \} \qquad (4.14)$$

The following lemma is similar to Lemma 3.1.

Lemma 4.2. Let (4.13) hold and let $z \in H_o^1(\Omega)$ satisfy, for each e, $1 \le e \le E$,

$$- \Delta z_e = 0 \text{ in } \Omega_e \text{ and } z_e = W_e \text{ on } \partial\Omega_e \qquad (4.15)$$

where $W_e = W|_{\partial\Omega_e}$, $W \in Q_t(\Gamma)$. Then

$$\sum_{e=1}^{E} \oint (\underset{\sim}{\Pi}_r^e \nabla z_e) \cdot \underset{\sim}{n} W_e \, ds = |z|^2_{1,\Omega} = ||W||^2_{W(\Gamma)} \tag{4.16}$$

where $\underset{\sim}{\Pi}_r^e : \underset{\sim}{L}_2(\Omega_e) \to \underset{\sim}{T}_e$ is the orthogonal projection.

Proof: Let $V_e \in P_{t'}(\Omega_e)$ be such that $\gamma_o V_e = W_e$ on $\partial\Omega_e$. Then

$$\oint_{\partial\Omega_e} \Pi_r^e(\nabla z_e) \cdot \underset{\sim}{n} W_e \, ds = \int_{\Omega_e} (\underset{\sim}{\Pi}_e^e \nabla z_e) \cdot \nabla V_e \, dx = \oint_{\partial\Omega_e} \frac{\partial z_e}{\partial n} W_e \, ds \tag{4.17}$$

due to the fact that $\underset{\sim}{\Pi}_r^e$ is self-adjoint and $\underset{\sim}{\Pi}_r^e \nabla V_e = \nabla V_e$ because of the inclusion (4.13). The equality (4.16) is clear in view of Lemma 3.1.■

Now we can prove:

Theorem 4.1. Let the inclusion (4.13) hold. Then there exists a unique solution $\hat{\underset{\sim}{\Lambda}}^* \in \mathcal{Q}$ to the dual-mixed-hybrid finite element approximation (4.9).

Proof. The proof is parallel to that of Theorem 3.1. Let $\hat{\underset{\sim}{\Lambda}}_e = (\hat{U}_e, \hat{\underset{\sim}{\Sigma}}_e, \hat{W}_e) = \hat{\underset{\sim}{\Lambda}}|_{\Omega_e}$ be given by

$$\hat{U}_e = U_e - \Pi_k^e(\underset{\sim}{\nabla} \cdot \underset{\sim}{\Sigma}_e), \quad \hat{\underset{\sim}{\Sigma}} = -2\underset{\sim}{\Sigma}_e + 2\underset{\sim}{\Pi}_r^e \nabla z_e, \quad \hat{w}_e = 2W_e \tag{4.18}$$

where Π_k^e is the orthogonal projection of $L_2(\Omega_e)$ onto $P_{k'}(\Omega_e)$, $\underset{\sim}{\Pi}_r^e$ and z_e are as in Lemma 4.2.

Then, by the continuity of projection operators,

$$||\hat{\underset{\sim}{\Lambda}}||_X \leq \sqrt{12}\, ||\underset{\sim}{\Lambda}||_X \tag{4.19}$$

By observing the orthogonality

$$(U_e, \underset{\sim}{\nabla} \cdot \underset{\sim}{\Sigma}_e - \Pi_k^e(\underset{\sim}{\nabla} \cdot \underset{\sim}{\Sigma}_e))_{0,\Omega_e} = 0 \qquad \forall\, U_e \in P_{k'}(\Omega_e)$$

we obtain

$$B(\underset{\sim}{\Lambda}, \hat{\underset{\sim}{\Lambda}}) = \sum_{e=1}^{E} \Big\{ \int_{\Omega_e} (\underset{\sim}{\nabla} \cdot \underset{\sim}{\Sigma}_e \Pi_k^e(\underset{\sim}{\nabla} \cdot \underset{\sim}{\Sigma}_e) + 2\underset{\sim}{\Sigma}_e \cdot \underset{\sim}{\Sigma}_e + U_e^2$$

$$- 2\underset{\sim}{\Sigma}_e \cdot \Pi_r^o \nabla z_e) dx + \oint_{\partial\Omega_e} 2\underset{\sim}{\Pi}_r^o \nabla z_e \cdot \underset{\sim}{n} W_e \, ds \Big\} \tag{4.20}$$

Utilizing (4.16) and the Schwarz inequality,

$$|B(\underset{\sim}{\Lambda}, \hat{\underset{\sim}{\Lambda}})| \geq \nu ||\underset{\sim}{\Lambda}||_X^2 \tag{4.21}$$

where

$$\nu = \min_{1 \leq e \leq E} \Big\{ \inf_{\underset{\sim}{\Sigma}_e \in P_{r'}(\Omega_e)} ||\Pi_k^e \underset{\sim}{\nabla} \cdot \underset{\sim}{\Sigma}_e||^2_{0,\Omega_e} / ||\underset{\sim}{\nabla} \cdot \underset{\sim}{\Sigma}_e||^2_{0,\Omega_e} \Big\} \tag{4.22}$$

Thus, with the choice of $C_2^h = \nu/\sqrt{12}$,

$$\inf_{||\underset{\sim}{\Lambda}||_X = 1} \; \sup_{||\overline{\underset{\sim}{\Lambda}}||_X \leq 1} |B(\underset{\sim}{\Lambda}, \overline{\underset{\sim}{\Lambda}})| \geq C_2^h > 0 \tag{4.23}$$

This completes the proof in view of Theorem 2.1.■

Remark 4.3. The following remarks apply [22] for approximation of either (1.1a) or $-\Delta u = f$:

(i) $\nu = 1$ if $k' \geq r' - 1$, i.e., if

$$Q_k(P) \supset \underset{\sim}{\nabla} \cdot \underset{\sim}{Q}_r(P) \tag{4.24}$$

(ii) If the inclusion (4.24) does not hold, i.e., if $k' < r' - 1$, then $\nu = 0$. For this case, using the inverse property

$$||\underset{\sim}{\Sigma}_e||^2_{0,\Omega_e} \geq Ch^2_e||\underset{\sim}{\nabla} \cdot \underset{\sim}{\Sigma}_e||^2_{0,\Omega_e} \tag{4.25}$$

C^h_2 can be chosen as

$$C^h_2 = \min_{1 \leq e \leq E} \{\tfrac{1}{2}, Ch^2_e\}/\sqrt{12} \tag{4.26}$$

Then, C^h_2 is still positive so as to guarantee a unique solution. However, in view of (2.14), the dependence of C^h_2 on h (for h small enough) will destroy convergence of the method.

5. A-Priori Error Estimates

The question of convergence can be resolved easily following results obtained thus far.

Theorem 5.1. Let $u \in H^\ell(P)$, $\ell \geq 2$ be the exact solution of (1.1) and let the inclusion (4.13) hold, i.e., $r' \geq t' - 1$. Then the error between u and the approximate solution $\Lambda = (U, \Sigma, W)$ satisfies

$$||\underset{\sim}{e}_\lambda||_X \leq C^* h^\alpha \left(\sum_{e=1}^{E} ||u_e||^2_{\ell,\Omega_e} \right)^{\frac{1}{2}} \tag{5.1}$$

where

$$\begin{aligned}
&\underset{\sim}{e}_\lambda = (e_u, \underset{\sim}{e}_\sigma, e_w) = (u - U, \nabla u - \underset{\sim}{\Sigma}, u|_\Gamma - W) \\
&\alpha = \min\{k+1, r, t, \ell-2\} \\
&C^* = A_o \max\{C_1, C_2, C_3\}, \quad A_o = 1 + 2/C^h_2
\end{aligned} \tag{5.2}$$

in which C^h_2 is as in (4.23), k, r, t, and C_i are as in (4.2)-(4.7).

Proof. Let $\underset{\sim}{\lambda} = (u, \nabla u, u|_\Gamma)$ where u is the solution of (1.1). By combining the results of Sections 3 and 4, we have, in accordance with (2.14),

$$||\underset{\sim}{e}_\lambda|| \leq A_o \inf_{\underset{\sim}{\Lambda} \in \mathcal{Q}} ||\underset{\sim}{\lambda} - \underset{\sim}{\tilde{\lambda}}||$$

$$\leq A_o \left\{ \sum_{e=1}^{E} (||u_e - \tilde{U}_e||^2_{0,\Omega_e} + ||\nabla u_e - \underset{\sim}{\tilde{\Sigma}}_e||^2_{S_e}) + ||u - \tilde{W}||^2_{W(\Gamma)} \right\}^{\frac{1}{2}}$$

where $\tilde{U}_e$, $\underset{\sim}{\tilde{\Sigma}}_e$, and $\tilde{W}$ are as in (4.3), (4.5), and (4.7), respectively. Introducing inequalities (4.3), (4.5), and (4.7) into the above, we obtain the desired result.

In view of the estimate (5.1) and Remark 4.3, sufficient conditions to guarantee a successful approximation are

(i) the inclusion (4.13), i.e., $r' \geq t' - 1$ for uniqueness

(ii) the inclusion (4.24), i.e., $k' \geq r' - 1$ for convergence. If the inclusion (4.24) does not hold, i.e., if $k' < r' - 1$, by (ii) of Remark 4.3,

$$||\underset{\sim}{e}_\lambda||_X \leq \tilde{C}h^{\alpha-2}(\cdot)$$

which means the approximation diverges at the rate of h if t = 1.

It can also be shown that if $-\Delta u = f$ is to be approximated successfully instead of (1.1a), the inclusions (4.12), (4.13), and (4.24) must be satisfied. This means that subspaces should be chosen so that $k' + 1 = r' \geq t' - 1$ to guarantee convergence for the case of $-\Delta u = f$.

6. Numerical Experiments

On a unit square $\Omega = (0,1) \times (0,1)$, we consider the following problems:

$$\left.\begin{aligned} -\Delta u &= 2x^2(1-3y)(x-1) + 2y^2(1-3x)(y-1) \quad \text{in } \Omega \\ u &= 0 \quad \text{on } \partial\Omega \end{aligned}\right\} \tag{6.1}$$

where the solution is

$$u = x^2y^2(1-x)(1-y) \tag{6.2}$$

By dividing Ω into $(N \times N)$ square elements, on each element $\bar{\Omega}_e$, we use linear approximation of the trace of u on each side of $\bar{\Omega}_e$, i.e., t = 1. Then W_e can be conveniently expressed by

$$W_e = \sum_{e=1}^{4} W_e^i N_i$$

where N_i are the usual bilinear isoparametric shape functions (see, e.g., [25]). For the choice of $Q_t(\Gamma)$ in such a way, we examine the following order of approximations:

Case	Approx. of u	Approx. of grad. u
(i)	constant $(k = k' = 0)$	linear $(r = r' = 1)$
(ii)	constant $(k = k' = 0)$	quadratic $(r = r' = 2)$
(iii)	linear $(k = k' = 1)$	quadratic $(r = r' = 2)$

The numerical results plotted in Fig. 1 confirm our theoretical estimates given in Section 5. Interestingly enough, for case (ii) (dotted lines in Figs. 1, 2, and 3), $k' < r' - 1$ and the solution is divergent (see Remark 4.3).

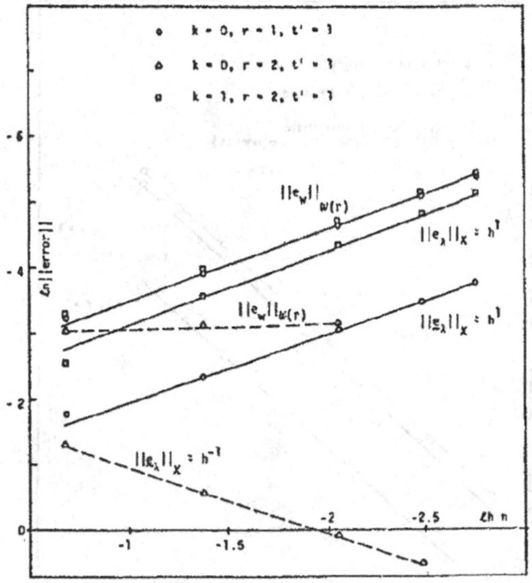

Fig. 1. Rates of Convergence

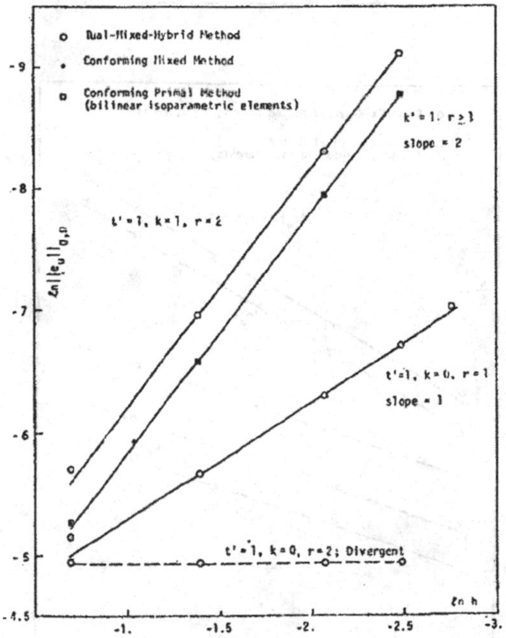

Fig. 2. L_2-Errors in approximating u.

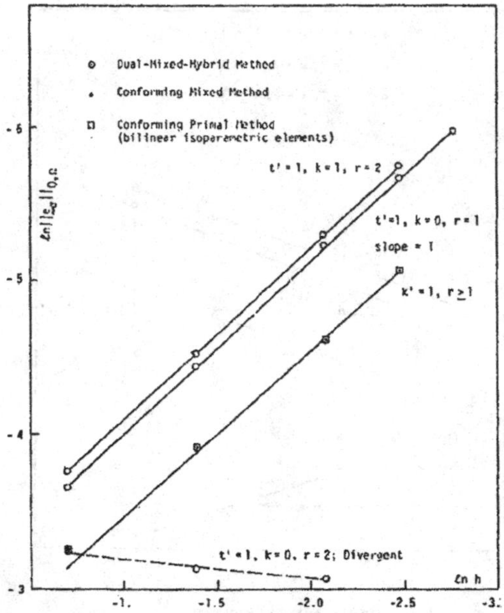

Fig. 3. L_2-Errors in approximating grad u

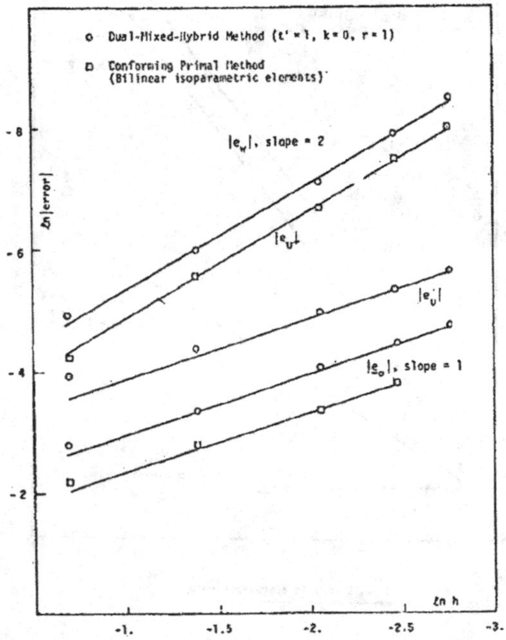

Fig. 4. Pointwise errors computed at 6 × 6 Gauss-Points

We also make some comparisons of accuracy with the well known conforming displacement method using bilinear isoparametric shape functions and the conforming mixed method (displacement connector type). Results are shown in Figs. 2, 3, and 4. Although the dual-mixed-hybrid method involves a bit more computational effort, accuracies are generally superior to other methods when it converges. The simplest convergent dual-hybrid model (case (i)), for example, provides very accurate nodal values of u (exact up to 6 digits even for 2×2 mesh) and approximation of grad. u in the interior of an element.

An interesting point, which is somewhat natural, is observed during computations. That is the stiffness matrix obtained by this method is the same as the one obtained by the displacement model when the order of approximation of the trace of u on inter-element boundaries equals the order of conforming shape functions and $r' = k' + 1$. This property may possibly be used in some applications to reduce computational efforts.

Finally, we comment that this model has an advantage over other hybrid models because:

(1) The subspaces need not satisfy any requirements such as satisfaction of equilibrium condition.

(2) The final equation is stiffness equation having the same degrees of freedom as a compatible displacement model which will provide a good mixability with the usual models. Due to these advantages and the usual characteristic of the hybrid method (relaxed continuity requirement), this method can be effectively applied to problems with singularities. For applications of other types of hybrid models, see [26], [27], and [28], for example.

Acknowledgement: We have greatly benefited from discussions of this subject with Professor Ivo Babuska and wish to record our sincere thanks to him for his comments on this work. This research was supported by the U.S. Air Force Office of Scientific Research under Grant 74-2660.

References

1. Babuska, I., Oden, J. T., and Lee, J. K., "Mixed-Hybrid Finite Element Approximations of Second Order Elliptic Boundary-Value Problems." (to appear).

2. Fraeijs de Veubeke, B., "Displacement and Equilibrium Models in the Finite Element Method," Stress Analysis, Ed. by O. C. Zienkiewicz and G. S. Holister, John Wiley & Sons, pp. 145-197 (1965).

3. Fraeijs de Veubeke, B., "Diffusion Equilibrium Models," Univ. of Calgary Lecture Notes, Int. Research Seminar on the Theory and Applications of the Finite Element Methods (1973).

4. Wolf, J. P., "Generalized Hybrid Stress Finite Element Models," AIAA Journal, Vol. 11, No. 3, pp. 386-388 (1973).

5. Pian, T. H. H., "Element Stiffness Matrices for Boundary Compatibility and for Prescribed Boundary Stresses", Proceedings of the First Conference on Matrix Methods in Structural Mechanics, Wright-Patterson Air Force Base, 1965, AFDL-TR-66-80, pp. 457-477 (1966).

6. Oden, J. T., "Some Contributions to the Mathematical Theory of Mixed Finite Element Approximations," Theory and Practice in Finite Element Structural Analysis, Ed. by Y. Yamada and R. H. Gallagher, University of Tokyo Press, pp. 3-23 (1973).

7. Oden, J. T. and Reddy, J. N., "On Mixed Finite Element Approximations," SIAM J. Num. Anal., Vol. 13, No. 3, pp. 393-404 (1976).

8. Reddy, J. N. and Oden, J. T., "Mathematical Theory of Mixed Finite Element Approximations," Q. Appl. Math., Vol. 33, pp. 255-280 (1975).

9. Reddy, J. N., "A Mathematical Theory of Complementary-Dual Variational Principles and Mixed Finite Element Approximations of Linear Boundary-Value Problems in Continuum Mechanics," Ph.D. Dissertation, The Univ. of Alabama in Huntsville (1974).

10. Ciarlet, P. G. and Raviart, P. A., "A Mixed Finite Element Method for the Biharmonic Equation," Mathematical Aspects of Finite Elements in Partial Differential Equations, Ed. by C. deBoor, Academic Press, N.Y., pp. 125-145 (1974).

11. Johnson, C., "On the Convergence of a Mixed Finite-Element Method for Plate Bending Problems," Num. Math., Vol. 21, pp. 43-62 (1973).

12. Miyoshi, T., "A Finite Element Method for the Solution of Fourth-Order Partial Differential Equations," Kumamoto J. Sci. Math., Vol. 9, pp. 87-116 (1973).

13. Raviart, P. A. and Thomas, J. M., "A Mixed Finite Element Method for 2nd Order Elliptic Problems," (to appear).

14. Oden, J. T. and Lee, J. K., "Theory of Mixed and Hybrid Finite-Element Approximations in Linear Elasticity," Proc. of IUTAM/IUM Symp. on Applications of Methods of Functional Analysis to Problems of Mechanics, September, 1975, Marseille, France (to be published by Springer-Verlag).

15. Brezzi, F., Sur la Methode des Elements Finis Hybrides pour le Probleme Biharmonique," Num. Math., Vol. 24, pp. 103-131 (1975).

16. Brezzi, F. and Marini, L. D., "On the Numerical Solution of Plate Bending Problems by Hybrid Methods," RAIRO Report (1975).

17. Thomas, J. M., "Methods des Elements Finis Hybrides Duaux Pour les Problems Elliptiques du Second-Order," Report 75006, Universite Paris VI et Centre National de la Research Scientifique (1975).

18. Raviart, P. A. and Thomas, J. M., "Primal Hybrid Finite Element Methods for 2nd Order Elliptic Equations," Report 75025, Universite Paris VI, Laboratoire Analyse Numerique.

19. Babuska, I., "Error Bounds for Finite Element Method," Numerische Mathematik, Vol. 16, pp. 322-333 (1971).

20. Babuska, I. and Aziz, A. K., "Survey Lectures on the Mathematical Foundations of the Finite Element Method," The Mathematical Foundations of the Finite Element Method with Applications to Partial Differential Equations, Ed. by A. K. Aziz, Academic Press, N.Y., pp. 322-333 (1971).

21. Oden, J. T. and Reddy, J. N., An Introduction to the Mathematical Theory of Finite Elements, John Wiley & Sons, New York (in press).

22. Lee, J. K., "Convergence of Mixed-Hybrid Finite Element Methods," Ph.D. Dissertation, Div. of Engr. Mech., The University of Texas at Austin (1976).

23. Ciarlet, P. G. and Raviart, P. A., "General Lagrange and Hermite Interpolation

in $\mathbb{R}^n$ with Applications to Finite Element Methods," Arch. Rational Mech. Anal., Vol. 46, pp. 177-199 (1972).

24. Brezzi, F., "On the Existence, Uniqueness and Approximation of Saddle-Point Problems Arising from Lagrangian Multipliers," RAIRO, R2, pp. 125-151 (1974).

25. Zienkiewicz, O. C., The Finite Element Methods in Engineering Science, McGraw-Hill, London (1971).

26. Tong, P., Pian, T. H. H., and Larry, S. J., "A Hybrid-Element Approach to Crack Problems in Plane Elasticity," IJNME, Vol. 7, pp. 297-308 (1973).

27. Lin, K. Y., Tong, P., and Orringer, O., "Effect of Shape and Size on Hybrid Crack-Containing Finite Elements," Computational Fracture Mechanics, Ed. by Rybicki, E. F. and Benzley, S. E., ASME, pp. 1-20 (1975).

28. Atluri, S. and Kathiresan, K., "An Assumed Displacement Hybrid Finite Element Model for Three-Dimensional Linear-Fracture Mechanics Analysis," Proc. 12th Annual Meeting, Soc. of Engr. Science, The University of Texas at Austin, pp. 391-402 (1975).

A MIXED FINITE ELEMENT METHOD FOR 2-nd ORDER ELLIPTIC PROBLEMS

P.A. Raviart* and J.M.Thomas**

1. INTRODUCTION

Let Ω be a bounded open subset of R^n with a Lipshitz continuous boundary Γ. We consider the 2nd order elliptic model problem

$$(1.1) \qquad \begin{cases} -\Delta u = f & \text{in } \Omega , \\ u = 0 & \text{on } \Gamma , \end{cases}$$

where f is a given function of the space $L^2(\Omega)$. A variational form of problem (1.1), known as the *complementary energy principle*, consists in finding $\underset{\sim}{p} = \underset{\sim}{\text{grad}}\, u$ which minimizes the *complementary energy functional*

$$(1.2) \qquad I(\underset{\sim}{q}) = \frac{1}{2}\int_{\Omega} |\underset{\sim}{q}|^2\, dx$$

over the affine manifold $\underset{\sim}{W}$ of vector-valued functions $\underset{\sim}{q} \in (L^2(\Omega))^n$ which satisfy the *equilibrium equation*

$$(1.3) \qquad \text{div}\, \underset{\sim}{q} + f = 0 \quad \text{in } \Omega.$$

The use of complementary energy principle for constructing finite element discretizations of elliptic problems has been first advocated by Fraeijs de Veubeke [5],[6],[7]. The so-called *equilibrium method* consists first in constructing a finite-dimensional submanifold $\underset{\sim}{W}_h$ of $\underset{\sim}{W}$ and then in finding $\underset{\sim}{p}_h \in \underset{\sim}{W}_h$ which minimizes the complementary energy functional $I(\underset{\sim}{q})$ over the affine manifold $\underset{\sim}{W}_h$. For 2nd order elliptic problems, the numerical analysis of the equilibrium method has been

---

* Centre de Mathématiques Appliquées, Ecole Polytechnique and Université de Paris VI.

** Université de Paris VI.

made by Thomas [19],[20].

Now, we note that the practical construction of the submanifold $\underset{\sim}{W}_h$ is not in general a simple ptoblem since it requires a search for explicit solutions of the equilibrium equation (1.3) in the whole domain Ω.

In order to avoid the above difficulty, we can use a more general variational principle, known in elasticity theory as the *Hellinger-Reissner principle,* in which the constraint (1.3) has been removed at the expense however of introducing a Lagrange multiplier. This paper will be devoted to the study of a finite element method based on this variational principle. In fact, this so-called mixed method has been found very useful in some practical problems and refer to [17] for an application to the numerical solution of a nonlinear problem of radia tive transfer.

For some general results concerning mixed methods, we refer to Oden [12],[13], Oden & Reddy [14], Reddy [16]. Mixed methods for solving 4th order elliptic equations have been particularly analyzed: see Brezzi & Raviart [2], Ciarlet & Raviart [4], Johnson [9],[10],and Miyoshi [11]. For related results we refer also to Haslinger & Hlåvaĉek [8].

An outline of the paper is as follows. In § 2, we derive the mixed variational formulation of problem (1.1) and we define the related discrete elements, and in § 4, the error analysis of the associated finite element method is made. Finally, in § 5, we generalize the results of §§ 3,4 to mixed methods using rectangular elements.

Let us describe some of the notations used throughout this paper. Given an integer $m \geq 0$,

$$H^m(\Omega) = \{v \in L^2(\Omega);\ \partial^\alpha v \in L^2(\Omega),\ |\alpha| \leq m\}$$

denotes the usual Sobolev space provided the norm and semi-norm

$$\|v\|_{m,\Omega} = \Big(\sum_{|\alpha| \leq m} \int_\Omega |\partial^\alpha v|^2 dx\Big)^{\frac{1}{2}},\quad |v|_{m,\Omega} = \Big(\sum_{|\alpha| = m} \int_\Omega |\partial^\alpha v|^2 dx\Big)^{\frac{1}{2}}.$$

Given a vector-valued function $\underset{\sim}{q} = (q_1, \ldots\ldots, q_n) \in (H^m(\Omega))^n$, we set:

$$\|\underset{\sim}{q}\|_{m,\Omega} = \Big(\sum_{i=1}^n \|q_i\|^2_{m,\Omega}\Big)^{\frac{1}{2}},\quad |\underset{\sim}{q}|_{m,\Omega} = \Big(\sum_{i=1}^n |q_i|^2_{m,\Omega}\Big)^{\frac{1}{2}}.$$

We denote by $H^{\frac{1}{2}}(\Gamma)$ the space of the traces $v|_\Gamma$ over Γ of the functions $v \in H^1(\Omega)$.

2. THE MIXED MODEL

In order to derive the appropriate variational form of problem (1.1), we introduce the space

$$(2.1) \qquad \underset{\sim}{H}(\mathrm{div}\ ;\ \Omega) = \{\underset{\sim}{q} \in (L^2(\Omega))^n\ ;\ \mathrm{div}\ \underset{\sim}{q} \in L^2(\Omega)\}$$

provided with the norm

$$(2.2) \qquad \|\underset{\sim}{q}\|_{\underset{\sim}{H}(\mathrm{div};\Omega)} = \left(\|\underset{\sim}{q}\|^2_{o,\Omega} + \|\mathrm{div}\ \underset{\sim}{q}\|^2_{o,\Omega}\right)^{\frac{1}{2}}.$$

Given a vector-valued function $\underset{\sim}{q} \in \underset{\sim}{H}(\mathrm{div}\ ;\ \Omega)$, we may define its normal component $\underset{\sim}{q} \cdot \underset{\sim}{\nu} \in H^{-\frac{1}{2}}(\Gamma)$ where $H^{-\frac{1}{2}}(\Gamma)$ is the dual space of $H^{\frac{1}{2}}(\Gamma)$ and $\underset{\sim}{\nu}$ is the unit outward normal along Γ. Moreover, we have Green's formula

$$(2.3) \qquad \forall v \in H^1(\Omega),\ \int_\Omega \{\underset{\sim}{\mathrm{grad}}\, v \cdot \underset{\sim}{q} + v\, \mathrm{div}\, \underset{\sim}{q}\}dx = \int_\Gamma v\, \underset{\sim}{q} \cdot \underset{\sim}{\nu}\, d\gamma$$

where the integral $\int_\Gamma$ represents the duality between the spaces $H^{-\frac{1}{2}}(\Gamma)$ and $H^{\frac{1}{2}}(\Gamma)$.

We next define *problem* (P). Find a pair of functions $(\underset{\sim}{p},u) \in \underset{\sim}{H}(\mathrm{div}\ ;\ \Omega) \times L^2(\Omega)$ such that

$$(2.4) \qquad \forall \underset{\sim}{q} \in \underset{\sim}{H}(\mathrm{div}\ ;\ \Omega)\ ,\ \int_\Omega \underset{\sim}{p} \cdot \underset{\sim}{q}\, dx + \int_\Omega u\, \mathrm{div}\, \underset{\sim}{q}\, dx = 0\ ,$$

$$(2.5) \qquad \forall v \in L^2(\Omega)\ ,\ \int_\Omega v(\mathrm{div}\ \underset{\sim}{p} + f)dx = 0.$$

Theorem 1. The problem (P) has a unique solution $(\underset{\sim}{p},u) \in \underset{\sim}{H}(\mathrm{div}\ ;\ \Omega) \times L^2(\Omega)$. In addition, u is the solution of the problem (1.1) and we have (2.6) $\underset{\sim}{p} = \underset{\sim}{\mathrm{grad}}\ u$.

Proof. Let us first check the uniqueness of the solution of problem (P). Hence, assume that $f = 0$; from (2.5), we get $\mathrm{div}\ \underset{\sim}{p} = 0$. Taking $\underset{\sim}{q} = \underset{\sim}{p}$ in (2.4), we obtain $p = 0$. Therefore, we have

$$(2.7) \qquad \forall q \in \underset{\sim}{H}(\mathrm{div}\ ;\ \Omega)\ ,\ \int_\Omega u\ \mathrm{div}\ \underset{\sim}{q}\ dx = 0$$

Now, let $w \in H^1(\Omega)$ be a function such that

$$\Delta w = u \quad \text{in} \quad \Omega.$$

Then, by choosing $\underset{\sim}{q} = \underset{\sim}{\text{grad}}\, w$ in (2.7), we get $u = 0$.

It remains only to show that the pair ($\underset{\sim}{p} = \underset{\sim}{\text{grad}}\, u, u$) is a solution of problem (P), where u is the solution of problem (1.1). On the one hand, we have

$$\text{div}\, \underset{\sim}{p} + f = \Delta u + f = 0.$$

On the other hand, since $u = 0$ on Γ, we get by using the Green's formula

$$(2.8) \qquad \int_\Omega \{\underset{\sim}{p}\cdot\underset{\sim}{q} + u\, \text{div}\, \underset{\sim}{q}\} dx = \int_\Omega u q\cdot\nu\, d\gamma = 0$$

Remark 1. One can easily check that the solution (p,u) of problem (P) may be characterized as the unique seddle-point of the quadratic functional

$$L(\underset{\sim}{q},v) = I(\underset{\sim}{q}) + \int_\Omega v(\text{div}\, \underset{\sim}{q} + f)\, dx$$

over the space $\underset{\sim}{H}(\text{div}\,;\Omega) \times L^2(\Omega)$, i.e.,

$$\forall \underset{\sim}{q} \in \underset{\sim}{H}(\text{div}\,;\Omega),\ \forall v \in L^2(\Omega),\ L(\underset{\sim}{p},v) \leq L(\underset{\sim}{p},u) \leq L(\underset{\sim}{q},u)$$

Hence, the function u is the Lagrange multiplier associated with the constraint $\text{div}\, \underset{\sim}{p} + f = 0$.

Let us now introduce a general method of discretization of problem (1.1) based on the mixed variational formulation (2.4),(2.5). We are given two finite-dimensional spaces $\underset{\sim}{Q}_h$ and V_h such that

$$(2.8) \qquad \underset{\sim}{Q}_h \subset \underset{\sim}{H}(\text{div}\,;\Omega) \quad ; \quad V_h \subset L^2(\Omega).$$

Then we define *problem* (P_h) : Find a pair of functions ($\underset{\sim}{p}_h$, u_h) $\in \underset{\sim}{Q}_h \times V_h$ such that

$$(2.9) \qquad \forall\, \underset{\sim}{q}_h \in \underset{\sim}{Q}_h,\ \int_\Omega \underset{\sim}{p}_h\cdot\underset{\sim}{q}_h dx + \int_\Omega u_h\, \text{div}\, \underset{\sim}{q}_h\, dx = 0\,,$$

$$(2.10) \qquad \forall\, v_h \in V_h,\ \int_\Omega v_h(\text{div}\, \underset{\sim}{p}_h + f)\, dx = 0\,.$$

Using a general result of Brezzi [1 , Theorem 2.1] concerning the approximation of variational problems, we get the following

Theorem 2. Assume that

$$(2.11)\quad \left\{\begin{array}{l} \underset{\sim}{q}_h \in \underset{\sim}{Q}_h \\ \\ \forall\, v_h \in V_h\,,\ \int v_h \operatorname{div} \underset{\sim}{q}_h \, dx = 0 \end{array}\right. \Rightarrow \operatorname{div} \underset{\sim}{q}_h = 0$$

and that there exists a constant $\alpha > 0$ such that

$$\forall\, v_h \in V_h\,,\quad \sup_{\underset{\sim}{q}_h \in \underset{\sim}{Q}_h} \frac{\int_\Omega v_h \operatorname{div} \underset{\sim}{q}_h \, dx}{\|\underset{\sim}{q}_h\|_{\underset{\sim}{H}(\operatorname{div}\,;\,\Omega)}} \geq \alpha \|v_h\|_{o,\Omega} \,.$$

Then the problem (P_h) has a unique solution $(\underset{\sim}{p}_h\,, u_h) \in \underset{\sim}{Q}_h \times V_h$ and there exists a constant $\tau > 0$ which dependes only on α such that

$$(2.13)\quad \left\{\begin{array}{l} \|\underset{\sim}{p}-\underset{\sim}{p}_h\|_{\underset{\sim}{H}(\operatorname{div};\Omega)} + \|u-u_h\|_{o,\Omega} \leq \\ \\ \leq \tau \left\{ \inf\limits_{\underset{\sim}{q}_h \in \underset{\sim}{Q}_h} \|\underset{\sim}{p}-\underset{\sim}{q}_h\|_{\underset{\sim}{H}(\operatorname{div}\,;\,\Omega)} + \inf\limits_{v_h \in V_h} \|u-v_h\|_{o,\Omega} \right\}. \end{array}\right.$$

Remark 2. Define the operator $\underset{\sim}{\nabla}_h \in L(V_h\,;\,\underset{\sim}{Q}_h)$ by

$$(2.14)\quad \forall\, v_h \in V_h\,,\ \forall\, \underset{\sim}{q}_h \in \underset{\sim}{Q}_h\,,\ \int_\Omega \underset{\sim}{\nabla}_h v_h \cdot \underset{\sim}{q}_h \, dx = -\int_\Omega v_h \operatorname{div} \underset{\sim}{q}_h \, dx \,.$$

Clearly, $\underset{\sim}{\nabla}_h$ can be viewed as an approximation of the operator grad. Now, the function u_h may be characterized as the unique solution of the following problem : Find $u_h \in V_h$ such that

$$(2.15)\quad \forall\, v_h \in V_h\,,\ \int_\Omega \underset{\sim}{\nabla}_h u_h \cdot \underset{\sim}{\nabla}_h v_h \, dx = \int_\Omega f v_h \, dx \,.$$

In fact, from the assumption (2.11) and (2.12), it follows that problem (2.15) has a unique solution $u_h \in V_h$. Moreover, it is readily seen that the pair $(\underset{\sim}{\nabla}_h u_h,\, u_h)$ is the solution of problem (P_h).

Since in general $V_h \not\subset H^1_o(\Omega)$, (2.15) is *non-conforming displacement* model for solving problem (1.1). For other non-conforming methods based on hybrid models, we refer to [15].

It remains to construct the finite-dimensional subspaces $\underset{\sim}{Q}_h$ and V_h of the spaces $\underset{\sim}{H}(\text{div} ; \Omega)$ $L^2(\Omega)$ respectively so that they satisfy "good" approximation properties and the compatibility conditions (2.11) and (2.12) with a constant α independent of the parameter h.

For convenience, we shall assume in the sequel that $\bar{\Omega}$ is a bounded *polygon* of $\mathbb{R}^2$. We then establish a triangulation $\mathcal{K}_h$ of $\bar{\Omega}$ made up with triangles and parallelograms K whose diameters are $\leq h$. We begin by consrtuction finite-dimensional $\underset{\sim}{Q}_h$ of the space $\underset{\sim}{H}(\text{div};\Omega)$. Given a finite element $K \in \mathcal{K}_h$, we denote by $\underset{\sim}{\nu}_K$ the unit outward normal along the boundary ∂K of K. Using the Green's formula (2.3) in each $K \subset \mathcal{K}_h$, one can easily prove that a function $\underset{\sim}{q} \in (L^2(\Omega))^2$ belongs to the space $\underset{\sim}{H}(\text{div} ; \Omega)$ if and only if the two following conditions hold :

(i) for all $K \in \mathcal{K}_h$, the restriction $\underset{\sim}{q}_{|K}$ of $\underset{\sim}{q}$ to the set K belongs to the space $\underset{\sim}{H}(\text{div} ; K)$;

(ii) for any pair of adjacent elements $K_1, K_2 \in \mathcal{K}_h$, we have the reciprocity relation

$$(2.16) \qquad \underset{\sim}{q}_1 \cdot \underset{\sim}{\nu}_{K_1} + \underset{\sim}{q}_2 \cdot \underset{\sim}{\nu}_{K_2} = 0 \text{ on } K' = K_1 \subset K_2 ,$$

where $\underset{\sim}{q}_i$ stands for $\underset{\sim}{q}_{|K_i}$, $i = 1,2$.

Hence the functions of $\underset{\sim}{Q}_h$ will be assumed to be smooth in each element $K \in \mathcal{K}_h$ and to satisfy the reciprocity conditions.

3. MIXED TRIANGULAR ELEMENTS

In this § , we shall assume that K is a triangle. With K and for any integer $k \geq 0$, we shall associate a space $\underset{\sim}{Q}_K$ of vector-valued functions $\underset{\sim}{q} \in \underset{\sim}{H}(\mathrm{div}\ ; K)$ such that :

(i) div $\underset{\sim}{q}$ is a polynomial of degree $\leq k$;

(ii) the restriction of $\underset{\sim}{q} \cdot \underset{\sim}{\nu}_K$ to any side K' of K is a polynomial of degree $\leq k$.

We begin by introducing the space $\hat{\underset{\sim}{Q}}$ associated with the unit right triangle $\hat{K}$ in the (ξ,η)-plane whose vertices are $\hat{a}_1 = (1,0)$, $\hat{a}_2 = (0,1)$, $\hat{a}_3 = (0,0)$. Let us first give some notations. We denote by P_k the space of all polynomials of degree $\leq k$ in the two variables ξ,η and by $\hat{S}_k$ the space of all functions defined over $\partial\hat{K}$ whose restrictions to any side $\hat{K}'$ of $\hat{K}$ are polynomials of degree $\leq k$. Given a point $\hat{x} = (\xi,\eta)$ of R^2, we denote by $\lambda_i = \lambda_i(\hat{x})$, $1 \leq i \leq 3$, the barycentric coordinates of $\hat{x}$ with respect to the vertices $\hat{a}_i$ of $\hat{K}$.

Now, the space $\hat{\underset{\sim}{Q}}$ is required to satisfy the following properties:

(3.1) $$(P_k)^2 \subset \hat{\underset{\sim}{Q}}\ ;$$

(3.2) $$\dim(\hat{\underset{\sim}{Q}}) = (k+1)(k+3)\ ;$$

(3.3) $$\forall \hat{\underset{\sim}{q}} \in \hat{\underset{\sim}{Q}}\ ,\ \mathrm{div}\ \hat{\underset{\sim}{q}} = \frac{\partial \hat{q}_1}{\partial \xi} + \frac{\partial \hat{q}_2}{\partial \eta} \in P_k\ ;$$

(3.4) $$\forall \hat{\underset{\sim}{q}} \in \hat{\underset{\sim}{Q}}\ ,\ \hat{\underset{\sim}{q}} \cdot \hat{\underset{\sim}{\nu}} \in \hat{S}_k \ \text{(where } \hat{\underset{\sim}{\nu}} \text{ stands for } \underset{\sim}{\nu}_{\hat{K}})\ ;$$

(3.5) $$\hat{\underset{\sim}{Q}}_o = \{\hat{\underset{\sim}{q}} \in \hat{\underset{\sim}{Q}}\ ;\ \mathrm{div}\ \hat{\underset{\sim}{q}} = 0\} \subset (P_k)^2\ .$$

Lemma 1. *Assume that the conditions* (3.2)-(3.5) *hold. Then a function* $\hat{\underset{\sim}{q}} \in \hat{\underset{\sim}{Q}}$ *is uniquely determined by:*

(a) *the values of* $\hat{\underset{\sim}{q}} \cdot \underset{\sim}{\nu}$ *at* (k+1) *distinct points of each side* $\hat{K}'$ *of* $\hat{K}$;

(b) *the moments of order* $\leq k-1$ *of* $\hat{\underset{\sim}{q}}$, i.e.,

$$\int_{\hat{K}} \hat{q}_i \lambda_1^{\alpha_1}\ \lambda_2^{\alpha_2}\ \lambda_3^{\alpha_3}\ d\hat{x}\ ,\ i = 1,2,\ a_1 + \alpha_2 + \alpha_3 = k-1\ .$$

Proof. Since by (3.2) the number of degrees of freedom (a),(b) is equal to the dimension of the space $\hat{\underset{\sim}{Q}}$, it is sufficient to prove that a function $\hat{\underset{\sim}{q}} \in \hat{\underset{\sim}{Q}}$ which satisfies the two conditions:

(3.6) $\tilde{q} \cdot \underset{\sim}{\nu} = 0$ at (k+1) distinct points of each side $\hat{K}'$ of $\hat{K}$,

(3.7) $\int_{\hat{K}} \hat{q}_i \lambda_1^{\alpha_1} \lambda_2^{\alpha_2} \lambda_3^{\alpha_3} \, dx = 0$, $i = 1, 2$, $\alpha_1 + \alpha_2 + \alpha_3 = k-1$

must vanish identically. In fact, conditions (3.4) and (3.6) imply $\tilde{q} \cdot \underset{\sim}{\hat{\nu}} = 0$ on $\partial \hat{K}$. Hence, using (3.7) and applying the Green's formula (2.3) in $\hat{K}$, we obtain for all $\hat{\varphi} \in P_k$

$$\int_{\hat{K}} \hat{\varphi} \, \text{div} \, \hat{\underset{\sim}{q}} \, d\hat{x} = - \int_{\hat{K}} \text{grad} \, \hat{\varphi} \cdot \hat{\underset{\sim}{q}} \, d\hat{x} + \int_{\partial \hat{K}} \hat{\varphi} \, \hat{\underset{\sim}{q}} \cdot \hat{\underset{\sim}{\nu}} \, d\hat{\gamma} = 0 \, .$$

Since, by (3.3), $\text{div} \, \hat{\underset{\sim}{q}} \in P_k$, we get $\text{div} \, \hat{\underset{\sim}{q}} = 0$ so that $\hat{\underset{\sim}{q}} \in \hat{\underset{\sim}{Q}}_o$.

Now, it follows from (3.5) that there exists a polynomial $\hat{w} \in P_{k+1}$ uniquely determined up to an additive constant such that

$$\hat{\underset{\sim}{q}} = \text{curl} \, \hat{w} = \left(\frac{\partial \hat{w}}{\partial \eta} \, , \, - \frac{\partial \hat{w}}{\partial \xi} \right) \, .$$

Note that $\hat{\underset{\sim}{q}} \cdot \hat{\nu} = \frac{\partial \hat{w}}{\partial \tau} = 0$ on $\partial \hat{K}$, where $\frac{\partial}{\partial \hat{\tau}}$ stands for the tangential derivative along $\partial \hat{K}$. Thus we may assume that $\hat{w} = 0$ on $\partial \hat{K}$ and we may write

$$\hat{w} = \lambda_1 \lambda_2 \lambda_3 \hat{z}, \quad \hat{z} \in P_{k-2} \quad (\hat{z} = 0 \quad \text{for } k = 0, 1) \, .$$

Using again (3.7), we obtain for any $\hat{\underset{\sim}{r}} \in (P_{k-1})^2$

$$0 = \int_{\hat{K}} \hat{\underset{\sim}{q}} \cdot \hat{\underset{\sim}{r}} \, d\hat{x} = \int_{\hat{K}} \text{curl} \, \hat{w} \cdot \hat{\underset{\sim}{r}} \, d\hat{x} = \int_{\hat{K}} \hat{w} \, \text{curl} \, \hat{\underset{\sim}{r}} \, d\hat{x} = \int_{\hat{K}} \lambda_1 \lambda_2 \lambda_3 \hat{z} \, \text{curl} \, \hat{\underset{\sim}{r}} \, d\hat{x},$$

where $\text{curl} \, \hat{\underset{\sim}{r}} = \frac{\partial \hat{r}_2}{\partial \xi} - \frac{\partial \hat{r}_1}{\partial \eta} \in P_{k-2}$. Clearly, we can choose $\hat{\underset{\sim}{r}}$ so that $\hat{z} = \text{curl} \, \hat{\underset{\sim}{r}}$ and then

$$\int_{\hat{K}} \lambda_1 \lambda_2 \lambda_3 \hat{z}^2 \, d\hat{x} = 0 \, .$$

Therefore, we get $\hat{z} = 0$ so that $\hat{w} = 0$ and $\hat{\underset{\sim}{q}} = \text{curl} \, \hat{w} = \underset{\sim}{0}$.

Remark 3. As regards the degrees of freedom of a function $\hat{\underset{\sim}{q}} \in \hat{\underset{\sim}{Q}}$, one could have equivalently specified the moments of order $\leq k$

$$\int_{\hat{K}'} \hat{\varphi} \, \hat{\underset{\sim}{q}} \cdot \hat{\underset{\sim}{\nu}} \, d\hat{\gamma} \, , \quad \hat{\varphi} \in P_k$$

of $\hat{\underset{\sim}{q}} \cdot \hat{\underset{\sim}{\nu}}$ on the side $\hat{\mathcal{R}}'$ instead of its values at (k+1) distinct points of $\hat{\mathcal{R}}'$.

Let us give some examples of spaces $\hat{\underset{\sim}{Q}}$.

Example 1. Let k $\geq$ 0 be an *even* integer; we define $\hat{\underset{\sim}{Q}}$ to be the space of all functions $\hat{\underset{\sim}{q}}$ of the form

(3.8)
$$\begin{cases} \hat{q}_1 = \text{pol}_k(\xi,\eta) + \alpha_0 \xi^{k+1} + \alpha_1 \xi^k + \ldots + \alpha_{\frac{k}{2}} \xi^{\frac{k}{2}+1} \eta^{\frac{k}{2}} \\[2ex] \hat{q}_2 = \text{pol}_k(\xi,\eta) + \beta_0 \eta^{k+1} + (\beta_1 \xi \eta^k + \ldots + \beta_{\frac{k}{2}} \xi^{\frac{k}{2}} \eta^{\frac{k}{2}+1} \end{cases}$$

with

(3.9)
$$\sum_{i=0}^{\frac{k}{2}} (-1)^i (\alpha_i - \beta_i) = 0$$

In (3.8), $\text{pol}_k(\xi,\eta)$ denotes any polynomial of degree k in the two variable ξ,η. Clearly, conditions (3.1),(3.2) hold. Next $\hat{\underset{\sim}{q}} \cdot \hat{\underset{\sim}{\nu}}$ is obviously a polynomial of degree $\leq$ k on each side $\xi = 0$ and $\eta = 0$ of $\hat{\mathcal{R}}$. On the other hand, it follows from (3.9) that $\hat{\underset{\sim}{q}} \cdot \hat{\underset{\sim}{\nu}}$ is also a polynomial of degree $\leq$ k on the side $\xi + \eta = 1$. Finally, we have

$$\text{div } \hat{\underset{\sim}{q}} = \text{pol}_{k-1}(\xi,\eta) + \sum_{i=0}^{\frac{k}{2}} (k+1-i)(\alpha_i \xi^{k-i}\eta^i + \beta_i \xi^i \eta^{k-i}) \in P_k$$

so that div $\hat{\underset{\sim}{q}}$ = 0 implies

$$\begin{cases} \alpha_i = \beta_i = 0 , \ 0 \leq i \leq \frac{k}{2} - 1 , \\[2ex] \alpha_{\frac{k}{2}} + \beta_{\frac{k}{2}} = 0 \end{cases}$$

and, by the condition (3.9)

$$\alpha_i = \beta_i = 0 , \quad 0 \leq i \leq \frac{k}{2} ,$$

Hence, hypotheses (3.1)-(3.5) hold.

Consider for instance the case k = 0. Then a function $\hat{\underset{\sim}{q}} \in \hat{\underset{\sim}{Q}}$ is of the form

(3.10)
$$\begin{cases} \hat{q}_1 = a_0 + a_1 \xi \\ \hat{q}_2 = b_0 + b_1 \eta \end{cases} , \quad a_1 = b_1 ,$$

and by Lemma 1, the degrees of freedom of $\hat{\underset{\sim}{q}}$ may be chosen as the values of $\hat{\underset{\sim}{q}}\cdot\hat{\underset{\sim}{\nu}}$ at the midpoints of the sides of the triangle $\hat{K}$.

Example 2. Now, let $k \geq 1$ be an *odd* integer; we then define $\hat{\underset{\sim}{Q}}$ to be the space of all functions $\hat{\underset{\sim}{q}}$ of the form

$$(3.11)\qquad \begin{cases} \hat{q}_1 = \mathrm{pol}_k(\xi,\eta) + \alpha_o\xi^{k+1} + \alpha_1\xi^k\eta + \ldots + \alpha_{\frac{k+1}{2}}\xi^{\frac{k+1}{2}}\eta^{\frac{k+1}{2}} , \\ \\ \hat{q}_2 = \mathrm{pol}_k(\xi,\eta) + \beta_o\eta^{k+1} + \beta_1\xi\eta^k + \ldots + \beta_{\frac{k+1}{2}}\xi^{\frac{k+1}{2}}\eta^{\frac{k+1}{2}} , \end{cases}$$

with

$$(3.12)\qquad \sum_{i=0}^{\frac{k+1}{2}} (-1)^i\alpha_i = \sum_{i=0}^{\frac{k+1}{2}} (-1)^i\beta_i = 0 .$$

Here again, one can easily check that conditions (3.1)-(3.5) hold. For $k = 1$, a function $\hat{\underset{\sim}{q}} \in \hat{\underset{\sim}{Q}}$ is of the form

$$(3.13)\qquad \begin{cases} \hat{q}_1 = a_o + a_1\xi + a_2\eta + a_3\xi(\xi+\eta) , \\ \\ \hat{q}_2 = b_o + b_1\xi + b_2\eta + b_3\eta(\xi+\eta) , \end{cases}$$

and, by Lemma 1, the degrees of freedom of $\hat{\underset{\sim}{q}}$ may be chosen as the values of $\hat{\underset{\sim}{q}}\cdot\hat{\underset{\sim}{\nu}}$ at two distinct points of each side of $\hat{K}$ (for the Gauss-Legendre points) and as the mean value

$$\frac{1}{\mathrm{mes}(\hat{K})} \int_{\hat{K}} \hat{\underset{\sim}{q}}\, d\hat{x} = \frac{1}{2} \int_{\hat{K}} \hat{\underset{\sim}{q}}\, d\hat{x}$$

of $\hat{\underset{\sim}{q}}$ over $\hat{K}$.

Next, consider any triangle K in the (x_1, x_2)-plane whose vertices are denoted by a_i, $1 \leq i \leq 3$. We set :

$$(3.14)\qquad h_K = \text{diameter of } K,$$

$$(3.15)\qquad \rho_K = \text{diameter of the inscribed circle in } K.$$

Let $F_K : \hat{x} \to F_K(\hat{x}) = B_K\hat{x} + b_K$, $B_K \in \mathcal{L}(\mathbb{R}^2)$, $b_K \in \mathbb{R}^2$, be the unique affine invertible mapping such that

$$F_K(\hat{a}_i) = a_i \ , \ 1 \leq i \leq 3 .$$

With any *scalar* function $\hat{\varphi}$ defined on $\hat{K}$ (resp. on $\partial\hat{K}$), we associate the function φ defined on K (resp. on ∂K) by

$$(3.16) \qquad \varphi = \hat{\varphi} \circ F_K^{-1} \quad (\hat{\varphi} = \varphi \circ F_K) \; .$$

On the other hand, with any *vector-valued* function $\hat{\underset{\sim}{q}} = (\hat{q}_1 , \hat{q}_2)$ defined on $\hat{K}$, we associate the function $\underset{\sim}{q}$ defined on K by

$$(3.17) \qquad \underset{\sim}{q} = \frac{1}{J_K} B_K \hat{\underset{\sim}{q}} \circ F_K^{-1} \quad (\hat{\underset{\sim}{q}} = J_K B_K^{-1} \underset{\sim}{q} \circ F_K) \; ,$$

where $J_K = \det(B_K)$. We shall *constantly* use in the sequel the one-to-one *correspondences* $\hat{\varphi} \longleftrightarrow \varphi$ $\hat{\underset{\sim}{q}} \leftrightarrow \underset{\sim}{q}$

The choice of the transformation (3.17) is based on the followong standard result.

Lemma 2. *For any function* $\hat{\underset{\sim}{q}} \in (H^1(\hat{K}))^2$, *we have:*

$$(3.18) \qquad \forall\, \hat{\varphi} \in L^2(\hat{K}) \; , \int_{\hat{K}} \hat{\varphi} \operatorname{div} \hat{\underset{\sim}{q}} \, d\hat{x} = \int_K \varphi \operatorname{div} \underset{\sim}{q} \, dx \; ,$$

$$(3.19) \qquad \forall\, \hat{\varphi} \in L^2(\partial\hat{K}), \int_{\partial\hat{K}} \hat{\varphi}\, \hat{\underset{\sim}{q}} \cdot \hat{\underset{\sim}{\nu}} \, d\hat{\gamma} = \int_{\partial K} \varphi \, \underset{\sim}{q} \cdot \underset{\sim}{\nu}_K \, d\gamma \; .$$

For the proof, see [18] for instance. We shall also need,

Lemma 3. *We have for any integer* $\ell \geq 0$:

$$(3.20) \qquad \forall\, \hat{\varphi} \in H^\ell(\hat{K}), \; |\hat{\varphi}|_{\ell,\hat{K}} \leq \|B_K\|^\ell \, |J_K|^{-\frac{1}{2}} \, |\varphi|_{\ell,K} \; ,$$

$$(3.21) \qquad \forall\, \hat{\underset{\sim}{q}} \in (H^\ell(\hat{K}))^2 \; , \; |\hat{\underset{\sim}{q}}|_{\ell,\hat{K}} \leq \|B_K\|^2 \, \|B_K^{-1}\| \, |J_K|^{\frac{1}{2}} \, |\underset{\sim}{q}|_{\ell,K}$$

where $\|B_K\|$ (resp. $\|B_K^{-1}\|$) *denotes the spectral norm of* B_K (resp. B_K^{-1}).

Proof. The inequality (3.20) has been derived in [3 , inequality (4.15)]. By using (3.17), the inequality (3.21) can be obtained in a very similar way.

Now, with the triangle K, we associate the space

$$(3.22) \qquad \underset{\sim}{Q}_K = \{\underset{\sim}{q} \in \underset{\sim}{H}(\operatorname{div} \; ; K) \; ; \; \hat{\underset{\sim}{q}} \in \hat{\underset{\sim}{Q}}\}$$

Assume that conditions (3.3) and (3.4) hold. Then, by Lemma 2, the functions $\underset{\sim}{q}$ of the space $\underset{\sim}{Q}_K$ satisfy the desired properties (i) and (ii).

Concerning the approximation of smooth vector-valued functions $\underset{\sim}{q}$ by functions of the space $\underset{\sim}{Q}_K$, we have

Theorem 3. Assume that the conditions (3.1)-(3.5) hold and let the space $\underset{\sim}{Q}_K$ be defined as in (3.22). Then there exist an operator $\underset{\sim}{\pi}_K \in \mathcal{L}((H^1(K))^2;$ and a constant $C > 0$ independent of K such that:

(i) for each side K' of K and for all $\varphi \in P_k$,

$$(3.23) \qquad \int_{K'} (\underset{\sim}{\pi}_K \underset{\sim}{q} - \underset{\sim}{q}) \cdot \underset{\sim}{\nu}_K \, \varphi \, d\gamma = 0 \; ,$$

(ii) for all function $\underset{\sim}{q} \in (H^{k+1}(K))^2$ with $\operatorname{div} \underset{\sim}{q} \in H^{k+1}(K)$,

$$(3.24) \qquad \| \underset{\sim}{\pi}_K \underset{\sim}{q} - \underset{\sim}{q} \|_{\underset{\sim}{H}(\operatorname{div};K)} \leq C \frac{h_K^{k+2}}{\rho_K} \left(|\underset{\sim}{q}|_{k+1,K} + |\operatorname{div} \underset{\sim}{q}|_{k+1,K} \right)$$

Proof. Given a function $\hat{\underset{\sim}{q}} \in (H^1(\hat{K}))^2$, there exists by Lemma 1 and Remark ~~1~~ 3 a unique function $\hat{\underset{\sim}{\pi}} \hat{\underset{\sim}{q}} \in \hat{\underset{\sim}{Q}}$ such that

$$(3.25) \qquad \forall \hat{\varphi} \in P_k \; , \int_{\hat{K}'} (\hat{\underset{\sim}{\pi}} \hat{\underset{\sim}{q}} - \hat{\underset{\sim}{q}}) \cdot \hat{\underset{\sim}{\nu}} \, \hat{\varphi} \, d\hat{\gamma} = 0 \text{ for each side } \hat{K}' \text{ of } \hat{K} \; ,$$

$$(3.26) \qquad \forall \underset{\sim}{\hat{r}} \in (P_{K-1})^2, \int_{\hat{K}} (\hat{\underset{\sim}{\pi}} \hat{\underset{\sim}{q}} - \hat{\underset{\sim}{q}}) \cdot \underset{\sim}{\hat{r}} \, d\hat{x}$$

It follows from (3.1) that $\hat{\underset{\sim}{\pi}} \hat{\underset{\sim}{q}} = \hat{\underset{\sim}{q}}$ for all $\hat{\underset{\sim}{q}} \in (P_K)^2$. Then, by applying Lemma 7 of [3] in vector form, we get for all $\hat{\underset{\sim}{q}} \in (H^{k+1}(\hat{K}))^2$

$$(3.27) \qquad \| \hat{\underset{\sim}{\pi}} \hat{\underset{\sim}{q}} - \hat{\underset{\sim}{q}} \|_{0,\hat{K}} \leq c_1 |\hat{\underset{\sim}{q}}|_{k+1,\hat{K}}$$

for some constant $c_1 = c_1(\hat{K})$. On the other hand, using (3.25),(3.26) and the Green's formula, we obtain for all $\hat{\varphi} \in P_k$

$$\int_{\hat{K}} \operatorname{div}(\hat{\underset{\sim}{\pi}} \hat{\underset{\sim}{q}} - \hat{\underset{\sim}{q}}) \hat{\varphi} \, d\hat{x} = - \int_{\hat{K}} (\hat{\underset{\sim}{\pi}} \hat{\underset{\sim}{q}} - \hat{\underset{\sim}{q}}) \cdot \underset{\sim}{\operatorname{grad}} \, \hat{\varphi} \, dx + \int_{\partial \hat{K}} (\hat{\underset{\sim}{q}} - \hat{\underset{\sim}{\pi}} \hat{\underset{\sim}{q}}) \cdot \hat{\underset{\sim}{\nu}} \, \hat{\varphi} \, d\hat{\gamma} = 0.$$

Hence $\operatorname{div}(\hat{\underset{\sim}{\pi}} \hat{\underset{\sim}{q}})$ is the orthogonal projection in $L^2(\hat{K})$ of $\operatorname{div} \hat{\underset{\sim}{q}}$ upon P_K. Then, assuming that $\operatorname{div} \hat{\underset{\sim}{q}} \in H^{k+1}(\hat{K})$ and applying again [3, Lemma 7], we obtain for some constant $c_2 = c_2(\hat{K})$

$$(3.28) \qquad \| \operatorname{div}(\hat{\underset{\sim}{\pi}} \hat{\underset{\sim}{q}} - \hat{\underset{\sim}{q}}) \|_{0,\hat{K}} \leq c_2 \; |\operatorname{div} \hat{\underset{\sim}{q}}|_{k+1,\hat{K}} \; .$$

Define now the operator π_K by

$$\forall\, q \in (H^1(K))^2,\quad \widehat{\pi_K q} = \hat{\pi}\,\hat{q}\ .$$

Clearly, (3.23) follows from (3.25) and Lemma 2. Since

$$\pi_K q - q = \frac{1}{J_K} B_K(\hat{\pi}\,\hat{q} - \hat{q}) \circ F_K^{-1}\ ,$$

we have

$$\|\pi_K q - q\|_{o,K} \leq \|B_K\|\,|J_K|^{-\frac{1}{2}}\ \|\hat{\pi}\,\hat{q} - \hat{q}\|_{o,\hat{K}}\ .$$

Thus, by using inequalities (3.27) and (3.21) for $\ell = k+1$, we get for all $q \in (H^{k+1}(K))^2$

$$\|\pi_K q - q\|_{o,K} \leq c_1 \|B_K\|^{k+2} \|B_K^{-1}\|\,|q|_{k+1,K}\ . \tag{3.29}$$

Finally, from (3.18) we have

$$\operatorname{div}(\pi_K q - q) = \frac{1}{J_K}(\operatorname{div}(\hat{\pi}\,\hat{q} - \hat{q})) \circ F_K^{-1}$$

so that

$$\|\operatorname{div}(\pi_K q - q)\|_{o,K} = |J_K|^{-\frac{1}{2}}\ \|\operatorname{div}(\hat{\pi}\,\hat{q} - \hat{q})\|_{o,\hat{K}}\ .$$

Therefore, noticing that

$$\operatorname{div} \hat{q} = J_K(\widehat{\operatorname{div} q})$$

and applying the inequalities (3.28) and (3.20) (with $\ell = k+1$ and $\varphi = \operatorname{div} q$), we obtain when $\operatorname{div} q \in H^{k+1}(K)$

$$\|\operatorname{div}(\pi_K q - q)\|_{o,K} \leq c_2 \|B_K\|^{k+1} |\operatorname{div} q|_{k+1,K}\ . \tag{3.30}$$

Since, by [15 , Lemma 2] , we have

$$\|B_K\| \leq \frac{h_K}{\rho_{\hat{K}}}\ ,\quad \|B_K^{-1}\| \leq \frac{h_{\hat{K}}}{\rho_K}\ , \tag{3.31}$$

the desired inequality (3.24) follows from (3.29) and (3.30).

4. ERROR BOUNDS

Assume that $\mathcal{K}_h$ is a triangulation of $\bar{\Omega}$ made up with triangles K whose diameters are $\leq$ h. We now introduce the space

(4.1) $$\underset{\sim}{Q}_h = \{\underset{\sim}{q}_h \in \underset{\sim}{H}(\text{div}\,;\Omega)\ ;\ \forall\, K \in \mathcal{K}_h,\ \underset{\sim}{q}_h|_K \in \underset{\sim}{Q}_K\}$$

where, for all $K \in \mathcal{K}_h$, the space $\underset{\sim}{Q}_K$ is defined as in (3.22).

The degrees of freedom of a function $\underset{\sim}{q}_h \in \underset{\sim}{Q}_h$ are easily determined; they can be chosen as

(i) the values of $\underset{\sim}{q}_h \cdot \nu_K$, at (k+1) distinct points of each side K' of the triangulation $\mathcal{K}_h$;

(ii) the moments of order $\leq$ k-1 of $\underset{\sim}{q}_h$ over each triangle $K \in \mathcal{K}_h$.

On the other hand, for any $\underset{\sim}{q}_h \in \underset{\sim}{Q}_h$ and any $K \in \mathcal{K}_h$, we have $(\text{div}\, \underset{\sim}{q}_h)|_K \in P_k$. Hence, a natural choice for the space V_h is given by

(4.2) $$V_h = \{v_h \in L^2(\Omega)\ ;\ \forall\, K \in \mathcal{K}_h,\ v_{h|K} \in P_k\},$$

so that condition (2.11) is automatically satisfied.

Note that the function $v_h \in V_h$ do not satisfy any continuity constraint at the interelement boundaries.

Now, in order to apply Theorem 2, the essential step consists in proving that the compatibility condition (2.12) holds with a constant α independent of h. In fact, we want to show that, for any function $v_h \in V_h$, there exists a function $\underset{\sim}{q}_h \in \underset{\sim}{Q}_h$ such that

(4.3) $$\text{div}\, \underset{\sim}{q}_h = v_h \text{ in } \Omega$$

and

(4.4) $$\|\underset{\sim}{q}_h\|_{\underset{\sim}{H}(\text{div}\,;\Omega)} \leq C\|v_h\|_{o,\Omega},$$

where the constant C is independent of h. For the proof, we need some technical preliminary results.

Let K a triangle of $\mathcal{K}_h$; we denote by $S_{k,\partial K}$ the space of all functions defined over ∂K whose restrictions to any side K' of K are polynomials of degree $\leq$ k.

Lemma 4. Let there be given functions $v \in P_k$ *and* $\mu \in S_{k,\partial K}$ *such that*

(4.5) $$\int_K v\, dx = \int_{\partial K} \mu\, d\gamma .$$

Assume that conditions (3.2)-(3.5) *hold. Then there exists a function* $\underset{\sim}{q} \in \underset{\sim}{Q}_K$ *such that*

$$(4.6) \qquad \begin{cases} \operatorname{div} \underset{\sim}{q} = v & \textit{in } K, \\ \underset{\sim}{q} \cdot \underset{\sim}{\nu}_K = \mu & \textit{on } \partial K, \end{cases}$$

and

$$(4.7) \qquad \|\underset{\sim}{q}\|_{\underset{\sim}{H}(\operatorname{div}; K)} \leq C\left(\|v\|^2_{o,K} + \frac{h_K^2}{\rho_K}\|u\|^2_{o,\partial K}\right)^{\frac{1}{2}}$$

where the constant C *is independent of* K.

Proof. Let $\hat{v}_1 \in P_k$ and $\hat{\mu}_1 \in \hat{S}_k$ be functions such that

$$(4.8) \qquad \int_{\hat{K}} \hat{v}_1 \, d\hat{x} = \int_{\partial\hat{K}} \hat{\mu}_1 \, d\hat{\gamma}$$

Then the Neumann problem

$$(4.9) \qquad \begin{cases} \Delta\hat{w} = \hat{v}_1 & \text{in } \hat{K}, \\ \dfrac{\partial\hat{w}}{\partial\hat{\nu}} = \hat{\mu}_1 & \text{on } \partial\hat{K}, \end{cases}$$

has a solution $\hat{w} \in H^1(\hat{K})$ which is unique up to an additive constant. Moreover, there exists a constant $c_1 = c_1(\hat{K}) > 0$ such that

$$(4.10) \qquad |\hat{w}|_{1,\hat{K}} \leq c_1\left(\|\hat{v}_1\|^2_{o,\hat{K}} + \|\hat{\mu}_1\|^2_{o,\partial\hat{K}}\right)^{\frac{1}{2}}$$

Now, by Lemma 1, there exists a unique function $\hat{\underset{\sim}{q}} \in \hat{\underset{\sim}{Q}}$ such that

$$\begin{cases} \forall\, \hat{f} \in (P_{k-1})^2, & \displaystyle\int_{\hat{K}} (\hat{\underset{\sim}{q}} - \underset{\sim}{\operatorname{grad}}\, \hat{w}) \cdot \hat{f} \, d\hat{x} = 0, \\ \hat{\underset{\sim}{q}} \cdot \hat{\underset{\sim}{\nu}} = \hat{\mu}_1 & \text{on } \partial\hat{K}. \end{cases}$$

From (4.9) and the Green's formula, it follows that

$$\forall\, \hat{\varphi} \in P_k, \quad \int_{\hat{K}} \hat{\varphi} \operatorname{div} \hat{\underset{\sim}{q}} \, d\hat{x} = -\int_{\hat{K}} \hat{\underset{\sim}{q}} \cdot \underset{\sim}{\operatorname{grad}}\, \hat{\varphi} \, d\hat{x} + \int_{\partial\hat{K}} \hat{\varphi}\, \hat{\underset{\sim}{q}} \cdot \hat{\underset{\sim}{\nu}} \, d\hat{\gamma} =$$

$$= - \int_{\hat K} \underset{\sim}{\text{grad}}\, \hat w \; \underset{\sim}{\text{grad}}\, \hat\varphi \; d\hat x + \int_{\partial\hat K} \hat\varphi \frac{\partial \hat w}{\partial \hat\nu} d\hat\gamma = \int_{\hat K} \hat\varphi \; \Delta\hat w \; d\hat x \; ,$$

so that div $\hat{\underset{\sim}{q}}$ is the orthogonal projection in $L^2(\hat K)$ of $\Delta\hat w$ upon P_k. Hence, we get

$$\text{(4.11)} \quad \begin{cases} \operatorname{div} \hat{\underset{\sim}{q}} = \hat v_1 \text{ in } \hat K_1 \; , \\ \hat{\underset{\sim}{q}} \cdot \hat{\underset{\sim}{\nu}} = \hat\mu_1 \text{ on } \partial\hat K. \end{cases}$$

On the other hand, it follows from (4.10) that

$$\text{(4.12)} \qquad \|\hat{\underset{\sim}{q}}\|_{o,\hat K} \le c_2 \left(\|\hat v_1\|^2_{o,\hat K} + \|\hat\mu_1\|^2_{o,\partial\hat K} \right)^{\frac{1}{2}}$$

for some constant $c_2 = c_2(\hat K) > 0$.

Now, let $K \in \mathcal{K}_h$; with the functions $v \in P_k$ and $\mu \in S_{k,\partial K}$ such that (4.5) holds, we associate the functions $\hat v_1 \in P_k$ and $\hat\mu_1 \in \hat S_k$ defined by

$$\text{(4.13)} \quad \begin{cases} \forall\, \hat\varphi \in P_k \; , \; \int_{\hat K} \hat v_1 \hat\varphi \; d\hat x = \int_K v \; \varphi \; dx \; , \\ \forall\, \hat\varphi \in \hat S_k \; , \; \int_{\partial\hat K} \hat\mu_1 \hat\varphi \; d\hat\gamma = \int_{\partial K} \mu \; \varphi \; d\gamma \; . \end{cases}$$

Clearly we have (4.8) and there exists a function $\hat{\underset{\sim}{q}} \in \hat{\underset{\sim}{Q}}$ such that (4.11) and (4.12) hold. We next define $\underset{\sim}{q} \in \underset{\sim}{Q}_K$ by

$$\text{(4.14)} \qquad \underset{\sim}{q} = \frac{1}{J_K} B_K \hat{\underset{\sim}{q}} \circ F_K^{-1} \; ,$$

so that, by (4.11) and Lemma 2, we get (4.6).

It remains only to show the estimate (4.7). We get from (4.12) and (4.14)

$$\text{(4.15)} \qquad \|\underset{\sim}{q}\|^2_{o,K} \le c_2^2 \|B_K\|^2 |J_K|^{-1} \left(\|\hat v_1\|^2_{o,\hat K} + \|\mu_1\|^2_{o,\partial\hat K} \right) .$$

Since $\hat v_1 = |J_K| v \circ F_K$, we obtain

$$\text{(4.16)} \qquad \|\hat v_1\|^2_{o,\hat K} = |J_K| \|v\|^2_{o,K} \; .$$

On the other hand, let $\hat K'$ be a side of $\hat K$ and let $K' = F_K(\hat K')$. Since the superficial measures of $\hat{K}'$ and K' are remated by

$$\text{meas}(K') \leq \|B_K^{-1}\| |J_K| \text{meas}(\hat{K}')$$

we obtain

$$(4.17) \qquad \|\hat{\mu}_1\|^2_{o,\hat{K}'} \leq \|B_K^{-1}\| |J_K| \|\mu\|^2_{o,K'} \; .$$

By combining the inequalities (4.15) - (4.17), we get

$$(4.18) \qquad \|\underset{\sim}{q}\|^2_{o,K} \leq c_2^2 \|B_K\|^2 \left(\|v\|^2_{o,K} + \|B_K^{-1}\| \|\mu\|^2_{o,\partial K} \right) .$$

Therefore, the desired inequality follows from (4.18) and (3.31).

Let us next introduce the space

$$(4.19) \qquad M_h = \{\mu_h \in \prod_{K \in \mathcal{K}_h} S_{k,\partial K} \; ; \; \mu_h|_{\partial K_1} + \mu_h|_{\partial K_2} = 0 \quad \text{on } K_1 \cap K_2$$

for every pair of adjacent triangles K_1, $K_2 \in \mathcal{K}_h\}$.

We consider a *regular family* $(\mathcal{K}_h)$ of triangulations of $\bar{\Omega}$ in the sense of [3], in that there exists a constant $\sigma > 0$ independent of h such that

$$(4.20) \qquad \max_{K \in \mathcal{K}_h} \frac{h_K}{\rho_K} \leq \sigma \; .$$

Lemma 5. *Let there be given spaces* V_h *and* M_h *defined as in* (4.2) *and* (4.19) *which are associated with a regular family of triangulations. Then, with any function* $v_h \in V_h$ *, we can associate a function* $\mu_h \in M_h$ *such that for all* $K \in \mathcal{K}_h$

$$(4.21) \qquad \int_K v_h \, dx = \int_{\partial K} \mu_h \, d\gamma$$

and

$$(4.22) \qquad \Big(\sum_{K \in \mathcal{K}_h} h_K \|\mu_h\|^2_{o,\partial K} \Big)^{\frac{1}{2}} \leq C \|v_h\|_{o,\Omega} \; ,$$

where the constant $C > 0$ *is independent of* h .

Proof. We shall construct the function μ_h by using a hybrid finite element method as it has been described and studied in [15]. Hence, the proof of the Lemma will depend heavily upon the results of [15]. We first define the space

$$X_h = \{\varphi_h \in L^2(\Omega) \; ; \; \forall\, K \in \mathcal{K}_h \; , \; \varphi_{h|K} \in P_{k+2}\}$$

provided with the norm

$$\|\varphi_h\|_{X_h} = \left\{ \sum_{K \in \mathcal{K}_h} \left(|\varphi_h|^2_{1,K} + h_K^{-2} \|\varphi_h\|^2_{0,K} \right) \right\}^{\frac{1}{2}} .$$

Next, we set :

$$a(\varphi_h, \psi_h) = \sum_{K \in \mathcal{K}_h} \int_K \underset{\sim}{\text{grad}}\, \varphi_h \cdot \underset{\sim}{\text{grad}}\, \psi_h dx, \qquad \varphi_h, \psi_h \in X_h \; ,$$

$$b(\varphi_h, \mu_h) = - \sum_{K \in \mathcal{K}_h} \int_{\partial K} \varphi_h \mu_h d\gamma \; , \qquad \varphi_h \in X_h, \mu_h \in M_h \; .$$

Then, by using [15, Theorem 2 and Lemmas 2,3,4], there exists a unique pair of functions $(\varphi_h \, , \, \mu_h) \in X_h \times M_h$ such that

$$(4.23) \quad \forall\, \psi_h \in X_h \; , \; a(\varphi_h \, , \, \psi_h) + b(\psi_h \, , \, \mu_h) = \int_\Omega v_h \psi_h dx \; ,$$

$$(4.24) \quad \forall\, \rho_h \in M_h \; , \; b(\varphi_h \, , \, \rho_h) = 0.$$

By choosing in (4.23)

$$\psi_h = \text{characteristic function of the set } K \in \mathcal{K}_h$$

we get (4.21) for all $K \in \mathcal{K}_h$.

Now, in order to prove the inequality (4.22), we introduce the following subspace of the space X_h

$$Y_h = \left\{ \psi_h \in X_h \; ; \; \forall \rho_h \in M_h \; , \; b(\psi_h \, , \, \rho_h) = 0 \right\}$$

Clearly, the function $\varphi_h \in Y_h$ may be characterized as the solution of

$$\forall\, \psi_h \in Y_h \; , \; a(\varphi_h \, , \, \psi_h) = \int_\Omega v_h \psi_h dx \; .$$

Therefore, we get

$$a(\varphi_h , \varphi_h) \le \|v_h\|_{o,\Omega} \|\varphi_h\|_{o,\Omega} .$$

By, [15, inequality (6.18)], we have the discrete analogue of the Poincaré-Friedrichs inequality :

$$\forall \psi_h \in Y_h , \|\psi_h\|_{o,\Omega} \le c_1 a(\psi_h , \psi_h)^{\frac{1}{2}} ,$$

where the constant c_1 is independent of h. Hence we obtain

$$(4.25) \qquad a(\varphi_h , \varphi_h)^{\frac{1}{2}} \le c_1 \|v_h\|_{o,\Omega} .$$

Next, it follows from (4.23) and (4.25) that

$$(4.26) \qquad b(\psi_h , \mu_h) \le (\|\psi_h\|_{o,\Omega} + c_1 a(\psi_h , \psi_h)^{\frac{1}{2}}) \|v_h\|_{o,\Omega} \le$$

$$\le c_2 \|\psi_h\|_{X_h} \|v_h\|_{o,\Omega} ,$$

where c_2 is a constant independent of h. Thus, the inequality (4.22) follows from (4.26) and the following inequality

$$\left(\sum_{K \in \mathcal{H}_h} h_K \|\mu_h\|^2_{o,\partial K} \right)^{\frac{1}{2}} \le c_3 \sup_{\psi_h \in X_h} \frac{b(\psi_h , \mu_h)}{\|\psi_h\|_{X_h}} , \quad c_3 = c_3(\Omega) .$$

(cfr. [15 , inequality (6.29)]).

We are now able to state

Theorem 4. Let there be given spaces $\underset{\sim}{Q}_h$ and V_h defined as in (4.1) and (4.2), which are associated with a regular family of triangulations. Assume in addition that the conditions (3.2)-(3.5) hold.Then, with any function $v_h \in V_h$, we can associate a function $\underset{\sim}{q}_h \in \underset{\sim}{Q}_h$ which satisfies the conditions (4.3), (4.4) with a constant C > 0 independent of h.

Proof. Let v_h be a function in V_h. By Lemma 5, we construct a function $\mu_h \in M_h$ such that the conditions (4.21) and (4.22) hold.
Next, using Lemma 4, there exists a function $\underset{\sim}{q}_h \in (L^2(\Omega))^2$ such that for all $K \in \mathcal{H}_h$, we have :

$$\begin{cases} \underset{\sim}{q}_{h|K} \in \underset{\sim}{Q}_K , \\ \operatorname{div}(\underset{\sim}{q}_{h|K}) = v_{h|K} , \\ (\underset{\sim}{q}_{h|K}) \cdot \underset{\sim}{\nu}_K = \mu_{h|\partial K} . \end{cases}$$

Since $\mu_h \in M_h$, the reciprocity conditions (2.15) hold so that $\underset{\sim}{q}_h \in \underset{\sim}{Q}_h$ and div $\underset{\sim}{q}_h = v_h$ in Ω. Moreover, it follows from (4.7) and (4.20) that

$$(4.27) \qquad \|\underset{\sim}{q}_h\|^2_{\underset{\sim}{H}(\mathrm{div};\Omega)} \leq C^2 \left(\|v_h\|^2_{o,\Omega} + \sigma \sum_{K\in\ _h} h_K \|\mu_h\|^2_{o,\partial K}\right).$$

Combining the inequalities (4.22) and (4.27), we obtain the inequality (4.4).

We now have our main result.

Theorem 5. We assume that $u \in H^{k+2}(\Omega)$ and $\Delta u \in H^{k+1}(\Omega)$ for some integer $k \geq 0$. Let there be given spaces $\underset{\sim}{Q}_h$ and V_h defined as in (4.1)-(4.2), which are associated with a regular family of triangulations. We assume in addition that the conditions (3.1)-(3.5) hold. Then problem (P_h) has a unique solution and there exists a constant $\mathfrak{K}$ independent of h such that

$$(4.28) \qquad \|\underset{\sim}{p}-p_h\|_{\underset{\sim}{H}(\mathrm{div};\Omega)} + \|u-u_h\|_{o,\Omega} \leq \mathfrak{K}\, h^{k+1} \left(|u|_{k+1,\Omega} + |u|_{k+2,\Omega} + |\Delta u|_{k+1,\Omega}\right)$$

.

Proof. Let $v_h \in V_h$; by the previous theorem, we have

$$\sup_{\underset{\sim}{q}_h \in \underset{\sim}{Q}_h} \frac{\int_\Omega v_h \,\mathrm{div}\, \underset{\sim}{q}_h \,dx}{\|\underset{\sim}{q}_h\|_{\underset{\sim}{H}(\mathrm{div}\,;\,\Omega)}} \geq \frac{1}{C} \|v_h\|_{o,\Omega}$$

so that the hypothesis (2.12) holds with $\alpha = \frac{1}{C}$. Thus, by Theorem 2, it remains only to evaluate the quantities

$$\inf_{\underset{\sim}{q}_h \in \underset{\sim}{Q}_h} \|\underset{\sim}{p}-\underset{\sim}{q}_h\|_{\underset{\sim}{H}(\mathrm{div};\Omega)} \quad \text{and} \quad \inf_{v_h \in V_h} \|u-v_h\|_{o,\Omega} .$$

On the one hand, by using Theorem 3, we define Π_h $\underset{\sim}{p} \in (L^2(\Omega))^2$ by

$$\forall K \in \mathcal{K}_h , \quad \underset{\sim}{\pi}_h p|_K = \underset{\sim}{\pi}_K(\underset{\sim}{p}|_K) .$$

It follows from (3.23) that the reciprocity relations (2.15) hold so that $\underset{\sim}{\pi}_h \in \underset{\sim}{Q}_h$. Next, we deduce from (3.24) and (4.20) that for some constant c_1 independent of h

$$(4.29) \qquad \|\underset{\sim}{p}-\underset{\sim}{\pi}_h p\|_{\underset{\sim}{H}(\mathrm{div};\Omega)} \leq c_1 h^{k+1} \left(|u|_{k+2,\Omega} + |\Delta u|_{k+1,\Omega}\right).$$

On the other hand, a straightforward application of [3, Theorem 5] gives for some constant c_2 independent of h

$$(4.30)\qquad \inf_{v_h \in V_h} \|u-v_h\|_{o,\Omega} \leq c_2 h^{k+1} |u|_{k+1,\Omega} .$$

Then, inequality (4.28) follows from inequalities (2.11),(4.29) and (4.30).

5. MIXED QUADRILATERAL ELEMENTS

We shall briefly discuss the case of quadrilateral elements. As for triangular elements, we begin by introducing the space $\hat{\underset{\sim}{Q}}$ associated with the unit square $\hat{K} = [0,1]^2$ in the (ξ,η)-plane. Given two integers $k, \ell \geq 0$, let us denote by $P_{k,\ell}$ the space of all polynomials in the two variables ξ,η of the form

$$(5.1)\qquad P(\xi,\eta) = \sum_{i=0}^{k} \sum_{j=0}^{\ell} c_{ij}\, \xi^i \eta^j \quad , \quad c_{ij} \in \mathbb{R} .$$

Now we define the space $\hat{\underset{\sim}{Q}}$ by

$$(5.2)\qquad \hat{\underset{\sim}{Q}} = \left\{ \hat{\underset{\sim}{q}} = (\hat{q}_1 , \hat{q}_2) ;\ \hat{q}_1 \in P_{k+1,k} \ ,\ \hat{q}_2 \in P_{k,k+1} \right\}$$

Note that, for $\hat{\underset{\sim}{q}} \in \hat{\underset{\sim}{Q}}$, we have :

(i) $\quad \operatorname{div} \hat{\underset{\sim}{q}} = \frac{\partial \hat{q}_1}{\partial \xi} + \frac{\partial \hat{q}_2}{\partial \eta} \in P_{k,k}$;

(ii) the restriction of $\hat{\underset{\sim}{q}} \cdot \hat{\underset{\sim}{\nu}}$ to any side $\hat{K}'$ of $\hat{K}$ is a polynomial of degree $\leq k$.

One can prove

Lemma 6. A function $\hat{\underset{\sim}{q}} \in \hat{\underset{\sim}{Q}}$ *is uniquely determined by:*

(a) *the values of* $\hat{\underset{\sim}{q}} \cdot \hat{\underset{\sim}{\nu}}$ *at* (k+1) *distinct points of each side* $\hat{K}'$ *of* $\hat{K}$;

(b) *the quantities*

$$\int_{\hat{K}} \hat{q}_1 \xi^i \eta^j d\hat{x} \ , \quad 0 \leq i \leq k-1 \ , \quad 0 \leq j \leq k \ ,$$

$$\int_{\hat{K}} \hat{q}_2 \xi^i \eta^j d\hat{x} \ , \quad 0 \leq i \leq k \ , \quad 0 \leq j \leq k-1$$

The proof goes along the same lines of that of Lemma 1.

Consider for instance the case $k = 0$. A Function $\hat{\underset{\sim}{q}} \in \hat{\underset{\sim}{Q}}$ is of the form

$$(5.3) \qquad \begin{cases} \hat{q}_1 = a_0 + a_1 \xi \ , \\ \\ \hat{q}_2 = b_0 + b_1 \eta \ , \end{cases}$$

and by Lemma 6, the degrees of freedom of $\hat{\underset{\sim}{q}}$ may be choosen as the values of $\hat{\underset{\sim}{q}} \cdot \underset{\sim}{\hat{\nu}}$ at the midpoints of the sides of the square $\hat{K}$.

Next, let K be a parallelogram in the (x_1 , x_2)-plane. There exists an affine invertible mapping $F_K : \hat{K} \to F_K(\hat{x}) = B_K \hat{x} + b_K$, such that $K = F_K(\hat{K})$. With K, we associate the space

$$(5.4) \qquad \underset{\sim}{Q}_K = \left\{ \underset{\sim}{q} : K \to \mathbb{R}^2 \ ; \ \underset{\sim}{q} = \frac{1}{J_K} B_K \ \hat{\underset{\sim}{q}} \circ F_K^{-1} , \ \hat{\underset{\sim}{q}} \in \hat{\underset{\sim}{Q}} \right\} .$$

Let $\underset{\sim}{q} \in \underset{\sim}{Q}_K$; the restriction of $\underset{\sim}{q} \cdot \underset{\sim}{\nu}_K$ to any side K' of the quadrilateral K is a polynomial of degree $\leq k$.

Assume now that $\mathcal{H}_h$ is a triangulation of $\bar{\Omega}$ made up with parallelograms K whose diameters are $\leq h$. We set :

$$(5.5) \qquad \underset{\sim}{Q}_h = \left\{ \underset{\sim}{q}_h \in \underset{\sim}{H}(\text{div} \ ; \ \Omega) \ ; \ \forall \ K \in \mathcal{H}_h \ , \ \underset{\sim}{q}_h|_K \in \underset{\sim}{Q} \right\} .$$

Note that, for any $\underset{\sim}{q}_h \in \underset{\sim}{Q}_h$ and any $K \in \mathcal{H}_h$, we have

$$(\text{div} \ \underset{\sim}{q}_h)|_K \circ F_K \in P_{k,k} .$$

So we set

$$(5.6) \qquad V_h = \left\{ v_h \in L^2(\Omega) \ ; \ \forall \ K \in \mathcal{H}_h \ , \ v_h|_K \circ F_K \in P_{k,k} \right\} .$$

By using the techniques of §§ 3,4, one can similarly prove that problem (P_h) has a unique solution $(\underset{\sim}{p}_h \ , \ u_h) \in \underset{\sim}{Q}_h \times V_h$ and that the error bound (4.28) still holds.

REFERENCES

[1] F.Brezzi : *On the existence, uniqueness and approximation of saddle-point problems arising from lagrangian multipliers.* R.A.I.R.O., R 2 Août 1974, 129-151.

[2] F.Brezzi, P.A.Raviart : *Mixed finite element methods for 4 th order problems.* To appear.

[3] P.G.Ciarlet, P.A.Raviart : *General Lagrange and Hermite interpolation in R^n with applications to finite element methods.* Arch. Rat. Mech. Anal. 46 (1972), 177-199.

[4] P.G.Ciarlet, P.A.Raviart : *A mixed finite element method for the biharmonic equation, Mathematical Aspects of Finite Elements in Partial Differential Equations.* pp. 125-145. Academic Press, New-York, 1974.

[5] B.Fraeijs De Veubeke : *Displacement and equilibrium models in the finite element method, Stress Analysis* (O.C. Zienkiewicz and G.S.Holister, Editors), ch. 9, pp. 145-197. Wiley, 1965.

[6] B.Fraeijs De Veubeke : *Diffusive equilibrium models Lecture notes.* University of Calgary, 1973.

[7] B.Fraeijs De Veubeke, M.A.Hogge : *Dual analysis for heat conductions problems by finite elements.* Int. J. for Num. Methods in Eng., 5 (1972), 65-82.

[8] J.Haslinger, I.Hlaváček : *A mixed finite element method close to the equilibrium model, I.Dirichlet problem for one equation, II. Plane elasticity.* To appear.

[9] C.Johnson : *On the convergence of a mixed finite element method for plate bending problems.* Numer. Math., 21 (1973), 43-62.

[10] C.Johnson : *Convergence of another mixed finite element method for plate bending problems.* To appear.

[11] T.Miyoshi : *A finite element method for the solutions of fourth order partial differential equations.* Kumamoto J., (Math.), 9 , (1973), 87-116.

[12] J.T.Oden : *Generalized conjugate functions for mixed finite element approximations of boundary-value problems, The Mathematical Foundations of the Finite Element Method* (A.K. Aziz Editor), pp. 629-670, Academic Press, New-York, 1973.

[13] J.T.Oden : *Some contributions to the mathematical theory of mixed finite element approximations.* Tokyo Seminar on Finite Elements, Tokyo, 1973.

[14] J.T.Oden, J.N.Reddy : *On mixed element approximations*, *Texas Institute for Computational Mechanics.* The University of Texas at Austin, 1974.

[15] P.A.Raviart, J.M.Thomas : *Primal hybrid finite element methods for 2 nd order elliptic equations.* To appear.

[16] J.N.Reddy : *A Mathematical Theory of Complementary-Dual Variational Principles and Mixed Finite Element Approximations of Linear Boundary-Value Problems in Continuous Mechanics.* Ph. D. Dissertation, The University of Alabama in Huntsville, 1973.

[17] B.Scheurer : To appear.

[18] J.M.Thomas : *Methode des éléments finis hybrides duaux pour les problèmes elliptiques du 2 nd ordre.* To appear in R.A.I.R.O., Série Analyse Numérique.

[19] J.M.Thomas : *Méthode des éléments finis équilibre pour les problèmes elliptiques du 2 nd ordre.* To appear.

[20] J.M.Thomas : *Thesis.* To appear.

THE INFLUENCE OF THE CHOICE OF CONNECTORS IN THE FINITE ELEMENT METHOD

by G. SANDER (Associated Professor)
P. BECKERS (Senior Assistant)
University of Liège, Belgium

Summary :

The patch test is interpreted from a variational point of view. This leads to conclude that non conforming (or non stress diffusive) displacement (stress) models can be considered as special hybrid models. A simple choice of mixed connectors is proposed that satisfy a priori the patch test for any assumed displacement or stress field. Finally some numerical examples are given of finite elements that do not pass the patch test but still converge.

Presented at the "Meeting on Mathematical Aspects of Finite Element Methods" December 10-12, 1975 in Rome.

1. INTRODUCTION.

The patch test can be interpreted from a variational point of view [1] as a condition that the non conforming modes in displacement models have zero virtual work along the interfaces for a constant state of stress, or, in equilibrium elements, that the non stress diffusing modes have zero virtual work for a constant state of strain.

It is shown in the following that non conforming (or non stress diffusing) models that pass the patch test are equivalent to special hybrid models. The variational interpretation of the patch test allows to define in a rational way connection modes between elements that are at the same time very simple and that insure a priori the satisfaction of the patch test. An element that has the connectors of the simplest displacement model and those of the simplest equilibrium model will pass the patch test for any assumed displacement or stress field. This concept opens new possibilities for deriving more efficient finite elements especially for shell analysis [11].

A numerical investigation of the influence of internal modes shows their possible importance in displacement models. The comparison of seven membrane elements having the same numbers of connectors reveals that convergence can occur in elements that do not pass the patch test. In elements that do not converge, a change in the definition of the connectors is sufficient to yield convergence.

2. CONNECTORS IN CONFORMING DISPLACEMENT MODELS.

The conforming or kinematically admissible finite element models of elastic continua can be derived from the principle of minimum total potential energy which is written

$$\min \pi(u) = \min \{ \frac{1}{2} \int_R \varepsilon^T H \varepsilon \, dR - \int_{\partial R_t} \bar{t}^T u \, d\partial R_t \} \tag{1}$$

where
- ε is the column matrix of strain components
- H is a positive definite matrix of elastic moduli
- t is the column matrix of surface traction components
- u is the column matrix of displacement components.

The bar denotes prescribed quantities (not subjected to variation). In the present discussion the integration domain R is restricted to the volume of one finite element. The contour ∂R of the element consists of the interfaces with the adjacent elements. It is subdivised in two parts: ∂R_t where surface tractions are prescribed and ∂R_u where displacements are prescribed. Note that in the derivation of displacement finite element all the interfaces are supposed to pertain to ∂R_t indicating simply that we want to leave the variation of the displacements free along the contour in order to let the definition of generalized forces appear automaticaly from the minimization process.

The admissible displacement field u must satisfy a priori the compatibility conditions

$$\varepsilon = \partial u \qquad \text{in } R \tag{2}$$

$$u = \bar{u} \qquad \text{along } \partial R_u \tag{3}$$

where ∂ denotes the appropriate differential operator for elastic problem. In the following we restrict the discussion to two or three dimensional linear elasticity. The operator ∂ is of first degree and therefore u has to be C_o continious. The extension of the discussion to other operators like those used in plate or shell theory does not present any additional difficulty.

The displacement field is discretized by polynomials in the form

$$u = M\,(x_i)\,\alpha \tag{4}$$

where M is a matrix of shaping functions and α a column matrix of intensities of these modes. The discretization contains at least the rigid body modes and constant strain modes. As the boundary ∂R_u does not appear in (1) it is necessary to define the displacements along the boundary by a set interface displacement modes of intensities q which are such that the simple equality of these modes across an interface between two elements insures the required C_o continuity of the displacements. The boundary degrees of freedom necessarily contain the local or nodal values of the displacements of the vertices of the element. In addition other degrees of freedom can be defined along the interfaces with more flexibility. Local values of derivatives or integrals of boundary displacement modes can be used as degrees of freedom to enforce the unique definition of u along the interfaces.

The boundary degrees of freedom q are called strong connectors [1]. They are called strong because they define in a unique way the displacements along the boundary of the element. They can be expressed in terms of the assumed displacement field (4) by a matrix of constants, called the kinematic connection matrix

$$q = C\,\alpha \tag{5}$$

If we assume for simplicity that the homogeneous adjoint problem $C^T y = o$ has only the trivial solution, the solution of (5) can be written as

$$\alpha = Qq + Aa \tag{6}$$

where

$$CQ = E \qquad \text{and } CA = o \tag{7}$$

and E denotes the identity matrix. The column matrix a contains the intensities of displacement modes that do not influence the boundary modes. They have been named "bubble" modes.

In practical situations the matrices Q and A can be obtained numerically by a Gauss-Jordan algorithm [2].

Substitution of (6) in (4) yields

$$u = W_q\ (x_i)\ q + W_a\ (x_i)\ a \tag{8}$$

with

$$W_q = M\,Q \quad \text{and} \quad W_a = M\,A$$

where the distinction between "connected modes" and bubble "unconnected modes" is now apparent in the shaping functions.

It is clear that in conforming models, unconnected modes must be such that $u \equiv o$ along the interfaces.

In practice such unconnected modes are eliminated at the element generation by a condensation process [3].

The boundary generalized forces conjugate of the displacement connectors turn out to be

$$\tilde{g} = \int_{\partial R_t} W_q^T\ \bar{t}\ d\partial R_t \tag{9}$$

weighted averages of the surface tractions. The equilibrium or stress diffusivity from an element to the other is only insured by the reciprocity (in the absence of external loading) of the average surface tractions (9). This property is called "weak diffusivity" and therefore the forces $\tilde{g}$ are called weak connectors.

More generally a strong connector has the nature of the discretized field while a weak connector is the average conjugate quantity. In a conforming displacement model all the Q connectors are strong and hence all the $\tilde{g}$ connectors are weak. The sign $\sim$ denotes a weak variable.

3. CONNECTORS IN HYBRID DISPLACEMENT MODELS.

If the boundary constraints (3) are introduced in the principle by a Lagrange multiplier technique, a new functional is obtained :

$$\pi\,(u,t) = \qquad (10)$$

$$\frac{1}{2}\int_R \varepsilon^T H\, \varepsilon\, dR - \int_{\partial R_t} \bar{t}^T u\, d\partial R_t - \int_{\partial R_u} t^T (u - \bar{u})\, d\partial R_u$$

The Lagrange multipliers t are easily identified as the surface tractions associated to the allowed discontinuity of displacements. They can be discretized independently of the u field, along the boundary ∂R_u.

The new functional (10) is the most general form from which hybrid displacement models can be derived [3].

Note that ∂R_t and ∂R_u do not necessarily represent a physical subdivision of the contour. Indeed the boundary integral over ∂R_u simply introduces a certain relaxation of the continuity required for the displacement field. The following situations can be distinguished:

- ∂R_u extends over all the interfaces. Hence ∂R_t disappears. This situation is characteristic of the hybrid displacement models as introduced by PIAN [6].
- ∂R_u extends over certain interfaces only. Hence ∂R_t covers the other interfaces. This can be useful for deriving special purpose models where one insists in satisfying in the best possible way different types of boundary conditions on different interfaces.
- ∂R_u concerns only certain displacement modes. Hence both ∂R_u and ∂R_t extend over all interfaces but for the modes appearing under ∂R_t one insists on exact codeformability or displacement continuity, while for the modes appearing under ∂R_u relaxed continuity conditions are considered as sufficient or even are preferred. In the following we concentrate the discussion on this last case.

If s denotes a coordinate current along the contour, the discretization of the surface tractions relaxing the compatibility requirements can be written

$$t\ (s) = W_g(s)\ g \qquad \text{along } \partial R_u \tag{11}$$

where $W_g(s)$ is a matrix of polynomial shaping functions, and g a column of intensities of these modes. They are often taken as local values or resultants of the surface tractions.
These degrees of freedom are, by our definition, strong force connectors and the conjugate displacements, defined by virtual work as

$$\tilde{q} = \int_{\partial R_u} W_g^T\ \bar{u}\ d\partial R_u \tag{12}$$

are weak displacement connectors. Indeed they are of exactly the name nature as those appearing in equilibrium models [3, 7, 8].

The discretization that has to be introduced in the modified principle (10) consists in two parts

$$\text{in R and along } \partial R_t \qquad u = M\ \alpha \tag{4}$$

$$\text{along } \partial R_u \qquad t = W_g\ g \tag{11}$$

Along ∂R_t we still maintain the continuity requirements for certain displacement modes described by a set of strong q connectors for which we write a kinematic connection matrix

$$q = C\ \alpha \tag{13}$$

Those connectors are however eventually unable to determine completely the internal displacement field.
Introducing (4), (11) and (13) in the principle (10) yields its discretized form

$$\{ \frac{1}{2} \alpha^T\ I\ \alpha\ - \tilde{g}^T\ C\ \alpha - g^T\ S\ \alpha + g^T \tilde{q}\ \} \tag{14}$$

where $\quad I = \int_R \partial M^T\ H\ \partial M\ dR$

$\qquad\quad S = \int_{\partial R_u} W_g^T\ M\ d\partial R_u$

Minimization with respect to α and g yields

$$I\,\alpha = C^T \tilde{g} + S^T g \qquad (15)$$

$$S\,\alpha = \tilde{q}$$

The finite element has a set of connectors of mixed nature, some are strong, some are weak for the displacements as well as for the forces. If we define

$$\hat{q}^T = \{q^T, \tilde{q}^T\} \qquad \hat{g}^T = \{\tilde{g}^T, g^T\} \qquad (16)$$

$$\hat{C}^T = \{C^T\ S^T\}$$

then (15) (16) and (13) can be written as

$$I\,\alpha = \hat{C}^T \hat{g} \qquad (17)$$

$$\hat{C}\,\alpha = \hat{q} \qquad (18)$$

If $\hat{C}$ is invertible, it turns out that

$$\hat{g} = \hat{C}^{-1T}\, I\, \hat{C}^{-1}\, \hat{q} = K\,\hat{q} \qquad (19)$$

where K is a stiffness matrix. If we assume that $\hat{C}^T y = o$ has only the trivial solution, a more general solution of (18) is in the form

$$\alpha = Q\,\hat{q} + A\,a \qquad (20)$$

with

$$\hat{C}\,Q = E \quad \text{and} \quad \hat{C}\,A = o$$

and substitution into (4) yields

$$u = M\,Q\,\hat{q} + M\,A\,a$$

$$= W_{\tilde{q}}\,\tilde{q} + W_q\,q + W_a\,a \qquad (21)$$

This allows to separate in the assumed displacement field, the modes that are exactly or strongly connected W_q , the modes that are weakly connected $W_{\tilde{q}}$ and the unconnected modes W_a, which again can be eliminated by standard condensation at the element generation. Note that now the unconnected modes are no longer those for which $u \equiv o$ along the contour, but those that do not influence the set of mixed connectors $\hat{q}$. The connectors appear therefore as filters allowing the transmission of certain internal modes (in a strong or weak sense) but stopping some others. The definition of "bubble" modes in such hybrid elements can only be done with respect to a given definition of connectors.

4. HYBRID MODELS AND THE PATCH TEST.

The patch test as imagined by IRONS [9] and justified mathematically by STRANG and FIX [10] has received a variational interpretation by FRAEIJS de VEUBEKE [1]. This interpretation can be extended as follows.

In a displacement model of non conforming type, the assumed field is split into two parts

$$u = u_1 + u_2 \tag{22}$$

where u_1 represents the modes that are correctly transmitted by a given set of strong displacement connectors q and where u_2 represents the non conforming modes only. The patch test requires that [10]

$$\int_R \sigma_o^T \varepsilon_2 \, dR = o \tag{23}$$

where σ_o is a constant state of stress and ε_2 is the strain field derived from the non conforming displacement field u_2. An equivalent but perhaps more convenient statement of (23) is obtained after integrating by parts

$$\int_{\partial R_t} n_i \, \sigma_{ij} \, u_{2_j} \, d \, \partial R_t = o \tag{24}$$

where $n_i \, \sigma_{ij}$ are the surface traction given along the interfaces by a constant stress field. If the interfaces are straight lines it turns out that

$$\sum_{\text{interfaces}} n_i \, \sigma_{ij} \int_{\ell} u_{2_j} \, d\ell = o \tag{25}$$

In some cases the integral vanishes along each interface independently while in other cases more than one interface has to be considered. Most non conforming elements that pratice had justified have been proved to pass successfully the patch test.

Suppose we want to derive a hybrid displacement element with mixed connectors (q and $\tilde{q}$) and for this purpose we split the assumed displacement field as in (22).

Here clearly u_1 must be described uniquely by the q strong connectors. The continuity of the modes u_2 is relaxed by the technique described above and consequently the boundary integral over ∂R_u in (10) concerns only these modes. We now insist for the simplest possible discretization of the surface tractions playing the role of Lagrange multipliers, that is, we use only constant modes in (11) for W_g (s).

The last term in the principle (10) becomes

$$\int_{\partial R_u} t^T (u_2 - \bar{u}_2) \, d\, \partial R_u = g^T \int_{\partial R_u} u_2 \, d\, \partial R_u - g^T \tilde{q} \tag{26}$$

As the boundary surface tractions are discretized independently along each interface, the first term in (26) vanishes only if

$$\int_{\partial R_u} u_2 \, d\, \partial R_u = o \tag{27}$$

on each interface. If this happens it indicates that the Lagrange multipliers are useless : it means that the lack of compatibility introduced by the non conforming modes u_2 does not involve any loss of energy across an interface or, in other words, that these displacement modes do not work against the simplest possible surface traction modes.

The corresponding hybrid displacement model has only strong q connectors (the $\tilde{q}$ being useless) and is identical to a non conforming model passing the patch test as (27) implies that (25) is satisfied.

However the reverse is not true as (27) must hold on each individual interface while (25) allows more freedom because the surface tractions are suppose to derive from a constant internal stress field. This implies that the corresponding generalized forces satisfy a priori the global equilibrium equation. Indeed the hybrid model could have been derived with the constraint that the constant Lagrange multipliers satisfy a priori the global equilibrium in which case (27) and (25) become identical.

This allows to formulate the following equivalence :

" A non conforming displacement element that passes successfully the patch test is identical to a mixed connector hybrid displacement element having the same set of strong q connectors and where the simplest possible choice of surface tractions relaxing the non conforming modes reveals useless. The simplest possible choice is defined as a set of constant surface tractions constrained to satisfy a priori the global equilibrium equations. "

5. EQUILIBRIUM AND HYBRID STRESS MODELS.

The principle of minimum complementary energy can be written

$$\min \pi(\sigma) = \min \left\{ \frac{1}{2} \int_R \sigma^T H^{-1} \sigma \, dR - \int_{\partial R_u} t^T \bar{u} \, d\,\partial R_u \right. \tag{28}$$

The assumed stress field σ has to satisfy the a priori equilibrium conditions

$$\partial^T \sigma + X = 0 \qquad \text{in the volume } R \tag{29}$$

$$t = \bar{t} \qquad \text{along } \partial R_t \tag{30}$$

where X is a column of body forces and ∂ still denotes the differential operator of linear two or three dimensional elasticity.
The natural connectors in such models are generalized surface traction g which are strong connectors while the conjugate displacements $\tilde{q}$ defined by the boundary term in (28) are weak connectors. In practice these elements are used in a stiffness method, that means that the solution is obtained using the weak $\tilde{q}$ connectors as unknowns.

In the same way as for displacement models, the boundary equilibrium conditions (30) can be relaxed by a Lagrange multipliers technique, yielding the modified functional :

$$\pi\ (\sigma,u) = \frac{1}{2}\int_R \sigma^T H^{-1} \sigma\ dR - \int_{\partial R_u} t^T \bar{u}\ d\ \partial R_u - \int_{\partial R_t} u^T (t-\bar{t})\ d\ \partial R_t \tag{31}$$

The Lagrange multipliers are obviously the displacements conjugate of the boundary surface tractions. If all the equilibrium conditions (30) are relaxed, then ∂R_u disappear in (31) and we are left with the principle leading to pure hybrid stress models as introduced by PIAN [6].

In this presentation however we consider the case where the equilibrium relaxation is applied only for the surface traction modes that do not satisfy the stress diffusivity for a given set of strong force connectors g. The boundary integral over ∂R_t concerns only such modes.

The displacement field u (s) can be discretized in the simplest way but has to be piece-wise differentiable along the contour. If the interfaces are straight lines, this allows to assume simply a linear variation of u (s) between the vertices where u must remain single valued.

Just as for displacement elements, if for the simplest admissible discretisation of u (s) along ∂R_t, the last boundary term in (31) vanishes, it proves that the corresponding displacement connectors are useless and that the hybrid model is in fact a non stress diffusive equilibrium model that passes the patch test. The patch test for equilibrium models can be written

$$\int_R {\sigma_2}^T \varepsilon_o\ dR = o \tag{32}$$

where σ_2 contains only the stress modes that yield non stress diffusing surface tractions for the given connectors g and where ε_o is a constant strain field satisfying the integrability condition

$$\varepsilon_o = \partial u_o \tag{33}$$

Integrating (32) by parts yields the equivalent form

$$\int_{\partial R_t} n_i\ \sigma_{2_{ij}}\ u_{o_j}\ d\ \partial R_t = o \tag{34}$$

in which u_o is a boundary displacement field integrated from (33) and consequently linear along the interfaces.

6 IMPLICATION OF THE PATCH TEST ON THE CHOICE OF CONNECTORS.

The preceding equivalence between hybrid elements and non conforming or non stress diffusive elements passing the patch test has immediate application in the choice of the connectors when designing a new element.

Let us concentrate first on displacement models. What the patch test requires, is the exact transmission or continuity of the interface displacement modes conjugate of a constant surface traction mode. Such displacement modes are exactly determined if we take as connector the simple average of displacement along each edge. In fact if the degrees of freedom appearing along an interface determine uniquely the average of the assumed displacements along the edge, the patch test will be satisfied.

Instead of checking this after having derived the element, it is much more easy to take such an average displacement among the set of displacement connectors chosen for the element. This is precisely what is done automatically in the hybrid displacement elements when the surface tractions are discretized in the simplest possible way.

$$g = t = \text{constant} \leftrightarrow \tilde{q} = \int_i^j u\ (s)\ ds \tag{35}$$

In other words, what is essential in displacement models is to be able to transmit exactly the displacement connectors that appear in the simplest corresponding equilibrium models. All additional displacement connectors including the corner displacements are not strictly necessary for satisfying the patch test. They only improve the compatibility over the required minimum.

If we turn now to equilibrium and hybrid stress models, the dual requirement is as follows : What is essential in such assumed stress models is that the chosen connectors (which are naturally weak displacement connectors) are able to transmit exactly the simplest possible interface displacement modes arising in the corresponding displacement models, that is, a linear variation of displacement between the vertices. Again this can be done automatically by taking such degrees of freedom as a priori necessary in the stress hybrid models.

This dual statement leads to the conclusion that there is a set of connectors that insures the satisfaction of the patch test for any assumed internal field, be it a displacement or stress field of any degree. This set of connectors contains :

- the local displacements at the vertices,
- the simple average of displacements along each interface.

The only restriction is that the internal field must be "rich" enough so that these connectors are linearly independent.

In practice the above conclusions are very appealing for the engineer for the following reasons :

- the local displacements at the vertices are economic to use and necessary for an easy interpretation of displacement results,
- the interface displacements are limited in number to a minimum which again is interesting for an economical point of view, especially for sophisticated assumed fields.
- the interface average displacements have, as conjugate force connectors, the constant surface traction modes which are in practice the stress results required most often by the engineer. In many cases the computation of stresses will not be necessary inside the element.
- the connectors becoming identical for assumed stress and assumed displacement models, it becomes possible to join such models in the same analysis. That means to adopt locally the assumptions that is best adapted to the problem. This has already allowed to justify such mixed assumptions in shell models [11, 12].

7. EXAMPLE OF MIXED CONNECTOR MEMBRANE ELEMENTS.

The following examples illustrate the application of the mixed connector concept. Consider the problem of deriving a quadrilateral membrane element from an assumed displacement field

$$u = \alpha_1 + \alpha_2\, x + \alpha_3\, y + \alpha_4\, x^2 + \alpha_5\, xy + \alpha_6\, y^2 + \alpha_7\, x^2y + \alpha_8\, xy^2$$
$$v = \beta_1 + \beta_2\, x + \beta_3\, y + \beta_4\, x^2 + \beta_5\, xy + \beta_6\, y^2 + \beta_7\, x^2y + \beta_8\, xy^2 \qquad (36)$$

Depending of the choice of the connectors, different elements can be derived from the same field and are schematised on figure 1. The first (figure 1.A) is based on the classical choice of local connectors at the vertices and at mid edges. It is non conforming for the general quadrilateral shape but becomes conforming for a rectangular geometry.

In the non conforming case, the patch test must be applied. If s denotes a coordinate current along the edge jk, variing from + 1 to -1 the connectors are u_j (+1) , u_k (-1) and u_{jk} (o) while the displacement is of the form

$$u\ (s) = \overset{\circ}{\alpha}_1 + \overset{\circ}{\alpha}_2\, s + \overset{\circ}{\alpha}_3\, s^2 + \overset{\circ}{\alpha}_4\, s^3 \qquad s \in (-1,\ +1) \qquad (37)$$

or equivalently

$$u\ (s) = u_{jk}\ (1 - 2s^2) + u_j\ (s^2 + \frac{s}{2}) + u_k\ (s^2 - \frac{s}{2}) + \overset{\circ}{\alpha}_4\ (s^3 - s) \qquad (38)$$

The non conforming modes u_2 defined by (22) consist only in the last term and as

$$\int_{-1}^{+1} (s^3 - s)\ ds = o \qquad (39)$$

the patch test is successfully passed.
Unless we consider isoparametric transformation, what we do not here, such quadrilateral membrane element will always be non conforming. We could allow indeed more non conformity by deciding to use as connectors only the 4 single average displacements

$$u_{jk} = \int_j^k u\ (s)\ ds \qquad (40)$$

along each edge yielding the element schematised on figure 1.B. The boundary displacement field (37) bcomes

$$u\ (s) = u_{jk} + \overset{\circ}{\alpha}_2\ s + \overset{\circ}{\alpha}_3\ (s^2 - \frac{1}{3}) + \overset{\circ}{\alpha}_4\ s^3 \qquad (41)$$

The non conforming modes are now $\overset{\circ}{\alpha}_2$, $\overset{\circ}{\alpha}_3$, $\overset{\circ}{\alpha}_4$ but the element still pass the patch test. This is obvious if it is considered as a hybrid element with constant surface traction. The non conforming or unconnected modes $\overset{\circ}{\alpha}_2$ to $\overset{\circ}{\alpha}_4$ are "bubble" modes for this set of connectors. Such an element looks "externally" like the well known equilibrium element with constant stresses [7,8] .It has the name connectors and also presents the same disadvantages of possible kinematical modes appearing in a group of such elements for certain boundary conditions.

Instead of going to this extreme situation, we could modify the element 1.A by adopting along the interfaces the average displacements (40) as connectors, in place of the local connectors $u_{jk}(o)$. This yields the element of figure 1.C. If we consider the element as a non conforming displacement model, the patch test has to be checked.

The boundary displacement field (37) becomes

$$u(s) = \frac{1}{4} u_j (-1 + 2s + 3s^2) + \frac{1}{4} u_k (-1 - 2s + 3s^2) + \frac{3}{2} u_{jk} (1 - s^2) + \overset{\circ}{\alpha}_4 (s^3 - s) \qquad (42)$$

which still satisfy the patch test. This is obvious if the element is considered as a hybrid model. Note that, even though we have adopted a weak connector along the edge, the element still becomes strictly conforming for a rectangular shape.

The advantage of the change of connector is more obvious if we refine the discretization of the displacements by adding extra terms to (36) like

$$\begin{aligned} u &= (36) \quad + \alpha_9 x^3 + \alpha_{10} y^3 + \alpha_{11} x^2 y^2 \\ v &= (36) \quad + \beta_9 x^3 + \beta_{10} y^3 + \beta_{11} x^2 y^2 \end{aligned} \qquad (43)$$

With such a field the element of figure 1.A (local connectors only) pass the patch test in the rectangular case but not in the quadrilateral one. With the connectors of figure 1.B or 1.C it remains obvious that the patch test is satisfied. If more terms are added to (43), local connectors can no longer insure the satisfaction of the patch test even for a rectangular shape.

8. SOME NUMERICAL RESULTS.

The convergence properties of various quadrilateral membrane elements have been numerically investigated. All the elements have in common the same number of connectors : 16. They all use the local values of (u.v) at the 4 vertices. They differ from each other

- by the choice of the 8 interface connectors,
- by the assumed internal fields.

From an economical point of view the elements are identical because they not only involve the same number of degrees of freedom but also the same topological connections and hence the same bandwidth, with as a consequence identical solution time on the computer. Only the generation time can be different but this is usually not significant for practical problems.

When local values of the displacement components u.v are used at the mid edges of the element, it will be referred to as Local Connector Rectangle or Quadrilateral (LCR or LCQ). When the average value of displacement are used as unknowns along the interfaces, it will be referred to as Mixed Connectors elements (MCR or MCQ).

Influence of internal or bubble modes

The influence of internal modes has been investigated on a LCR (local connector rectangle) using first the simple displacement field defined by (36). Then additional internal modes giving $u \equiv o$ on the boundary have been added in ther form

$$u = (x^2 - a^2)\ (y^2 - b^2)\ P_1\ (x.y)$$
$$v = (x^2 - a^2)\ (y^2 - b^2)\ P_1\ (x.y) \qquad (44)$$

where 2a and 2b are the dimensions of the rectangle, and P_1 (x.y) is a complete polynominal of degree n_p. The element remains obviously strictly compatible. The results used for comparison are the minimum and maximum non zero eigen values of the stiffness matrix which is always of dimension 16 X 16 as the internal modes are always condensed. In addition the trace of the stiffness matrix is also considered as giving a global indication of the evolution of the element properties. The results are

n_p	λ_{min}	λ_{max}	trace
-	33.610	943.88	4114.3
0	33.610	467.15	2604.5
1	31.435	403.23	2313.7
2	31.412	399.53	2273.0
3	31.410	399.19	2265.4
4	31.408	398.97	2263.5

(45)

with a = b = 1. Young's modulus = 10^4 Poisson's ratio = 0.3

Thickness = 1

The first result corresponds to P_1 = 0.

Using the same LCR geometry and connectors, a pure hybrid stress element in the sense of PIAN has been derived using a stress field derived from an Airy stress function represented by a complete polynomial of degree 5 up to 8. The stresses satisfy the homogeneous equilibrium and are of degree n_σ variing from 3 to 6. The stiffness matrix is still 16 X 16 and the element is "externally" identical to the preceding one. The same quantities have been computed

n_σ	λ_{min}	λ_{max}	trace
3	31.058	395.87	2180.3
4	31.401	396.86	2236.9
5	31.401	396.86	2254.0
6	31.408	398.74	2256.9

(46)

From these results, it appears that a few internal modes improve very significantly the displacement element and, as expected, yield at the limit a stiffness matrix very similar to that obtained by an internal stress field. The hybrid stress element converges from the opposite side but is much less sensitive to additional modes. Note that it is impossible to obtain a hybrid stress element without spurious kinematical modes from a stress field of degree less than 3, using quadratic boundary displacements [3, 12].

These conclusions have been confirmed by additional investigations not reported here [3, 12] .

Influence of the choice of connectors

The trapezoidal membrane problem illustrated in figure 1.D has been solved using various conforming and non conforming (or hybrid displacement) elements of quadrilateral shape and of LCQ (figure 1.A) or MCQ (figure 1.C) types. The convergence of the strain energy functional (1) or (10) has been plotted in terms of the mesh sizes. The domain has been discretized by 1 X 1, 2 X 2, 4 X 4 and 8 X 8 elements by dividing the edges of quadrilateral domain in 1, 2, 4, 8 equal segments.

Seven different displacements elements have been compared that differ only by the assumed internal displacement field and by the choice of local or average interface connectors. The eventual internal modes have been condensed and hence all the stiffness matrices used are of dimensions 16 X 16.

The seven elements can be described as follows :

LCQ - A is a strictly conforming element obtained as the assemblage of 4 triangles in each of which a complete second degree polynomial is used to discretize the displacement field. Local connectors are used. The 10 internal degrees of freedom (local connectors joining the 4 triangles) are condensed. This element is the only strictly conforming element used in this investigation.

MCQ - B
LCQ - B are two non conforming elements based on the displacement field (36). The LCQ or MCQ versions yield identical results because the boundary displacements are of third degree and the three local connectors along each edge define uniquely the mean displacement of such a function. Therefore they are referred to as a single element in the following. Note that such an element becomes conforming for a rectangular shape. This property seems to be important as will be shown later.

LCQ - C is a non conforming element based on the displacement field (36) to which the terms

$$u = (x^2 - a^2)(y^2 - b^2)\{\gamma_1 + \gamma_2 x + \gamma_3 y + \gamma_4 xy\} \qquad (47)$$
$$v = \text{similar}$$

are added. 2a and 2b are the lengths of the 2 medians of the quadrilateral.

The additional terms are pure bubble modes, that is, they vanish on the boundary of a rectangular element for which the patch test is satisfied, but they introduce additional non conformity for a quadrilateral element for which the patch test is not satisfied.

LCQ - D is a non conforming element with the displacement field

$$u = \text{full 10 terms cubic} + \alpha_{11}\, x^3 y + \alpha_{12}\, xy^3 \qquad (48)$$

$$v = \text{similar}$$

Such a field has the properties of giving a cubic displacement along the boundary of a rectangle and of avoiding any mode that vanishes along the boundary of the rectangle. Local connectors are used. The element pass the patch test when rectangular but does not pass it when quadrilateral.

MCQ - E is a non conforming element similar of LCQ - D but where the average displacements are used along the interfaces. Therefore the patch test is satisfied for any shape.

LCQ - F is a non conforming element with the displacement field

$$u = \text{full 10 terms cubic} + \alpha_{11}\, x^4 + \alpha_{12}\, y^4 + \alpha_{12}\, x^3 y + \alpha_{14}\, xy^3 + \alpha_{15}\, x^4 y + \alpha_{16}\, xy^4 \qquad (49)$$

$$v = \text{similar}$$

Such a field gives quartic displacements along the boundary of a rectangle and avoids any mode vanishing along the same boundary. This explains why the $x^2 y^2$ terms is omitted. Local connectors are used. The element does not pass the patch test for a rectangular nor for a quadrilateral shape.

MCQ - G is similar to LCQ - F but with the choice of the average connectors along the interfaces. Therefore the patch test is satisfied for any shape.

The strain energy obtained for the various meshes using these 7 elements is as follows :

elements	1 X 1	2 X 2	4 X 4	8 X 8
LCQ - A	305.503	334.642	338.882	339.548
MCQ - B	316.710	334.645	338.700	-
LCQ - C	326.424	339.832	341.929	340.959
LCQ - D	-	349.076	342.090	340.606
MCQ - E	430.971	348.293	342.543	340.688
LCQ - F	-	386.827	385.578	384.976
MCQ - G	433.890	350.543	343.681	-

Figure 2 presents the results in graphical form. They rise the following remarks.

- Only one element (LCQ - F) does not converge. The other elements that do not pass the patch test (LCQ - C, LCQ - D) converge just as well as the non conforming elements for which the choice of average connectors has been made to enforce a priori the satisfaction of the patch test (MCQ - B, MCQ - E).

- Additional investigations are certainly required to understand this behavior but it appears to be related to the fact that all the elements that converge without passing the patch test in a quadrilateral, do pass the patch test in a rectangle. As the mesh is refined, the patch test is necessarily less and less violated.

- The element LCQ - F is the only one that does not pass the patch test on a rectangle and (as a consequence ?) does not converge.

- In the same LCQ - F element, a change of definition of the interface connectors brings the expected improvement and leads to convergence.

- Going from element LCQ - A to element LCQ - G can be viewed as increasing the number of unconnected or internal modes for the given set of connectors. Note that none of the additional unconnected modes are bubble modes in the sense of the preceding section (44).
  They have precisely been chosen to add a non zero contribution to the interfaces. Their influence on the convergence behavior is especially important. Starting from a convergence from below, the increase of the number of such internal modes leads to a convergence from above.

- From a practical point of view, the engineer is still unable to decide from a sound theoretical point of view which element is the best for a given problem. From the results of figure 2 the best element is to be chosen between LCQ - C and LCQ - D or MCQ - E.
  Two of these "best" elements violate the patch test.

- If the error E in strain energy is expressed with respect to a mesh size h in a form like
  $E = k\,h^n$
  it is clear that the "best" elements are LCQ - G and LCQ - A, which shows that such criteria can not be accepted by an engineer as stand alone criteria. This is especially questionable if one recalls the well known fact that convergence curves like those of figure 2 are strongly influenced by the type of boundary conditions applied.

REFERENCES

1. FRAEIJS de VEUBEKE, B.
"Variational principles and the patch test"
Int. Jnl for Num. Meth. in Eng., vol. 8, 783-801, 1974

2. ROBINSON, J.
"Integrated theory of finite element methods"
J. Wiley, 1973.

3. FRAEIJS de VEUBEKE, B., SANDER, G. and BECKERS, P.
"Dual analysis by finite elements : linear and non linear applications"
USAF Technical Report, AFFDL-TR-72-93, 1972

4. SANDER, G.
"Application de la méthode des éléments finis à la flexion des plaques"
Coll. Publ. Fac. Sc. Appl., vol. 15, 1969, Univ. of Liège

5. FRAEIJS de VEUBEKE, B and ZIENKIEWICZ, O.C.
"Strain energy bounds in finite element analysis by slab analogy"
Jl of Strain Analysis, vol. 2, n° 4, 1967

6. PIAN, T.H.H.
"Formulations of finite element methods for Solid Continua"
in "Recent Advances in Matrix Methods of Structural Analysis and Design"
ed. Gallagher, Yamada, Oden, Univ. of Alabama Press, 1971

7. SANDER, G.
"Application of the dual analysis principle"
Proc. IUTAM Sump. "High Speed Computing of Elastic Structures",
Univ. Liège, Belgium, 1970

8. FRAEIJS de VEUBEKE, B.
"Displacements and equilibrium models"
Chap. 9 in "Stress Analysis", John Wiley, 1965

9. IRONS, B. and RAZZAQUE, A.
"Experience with the patch test"
in "Mathematical Foundations of the F.E.M." ed. by AZIZ
Academic Press, 1972

10. STRANG, G. and FIX, G.
"An analysis of the finite element method"
Prentice Hall, 1973

11. SANDER, G. and BECKERS, P.
"Delinquent finite elements for shell idealization"
Proc. World Congress on F.E.M. in Struct. Mech. (ed Robinson)
Bournemouth, 1975

12. SANDER, G. and BECKERS, P.
"Improvements of finite element solutions for structural and non structural applications"
Proc. III Dayton Conf., USAF Report AFFDL-TR-71-160, 1973

13. IDELSOHN, S.
"Analyse statique et dynamique des coques par la méthode des éléments finis"
Ph. D. Thesis, Univ. of Liège, Aerospace Lab., Report SF-25, 1974.

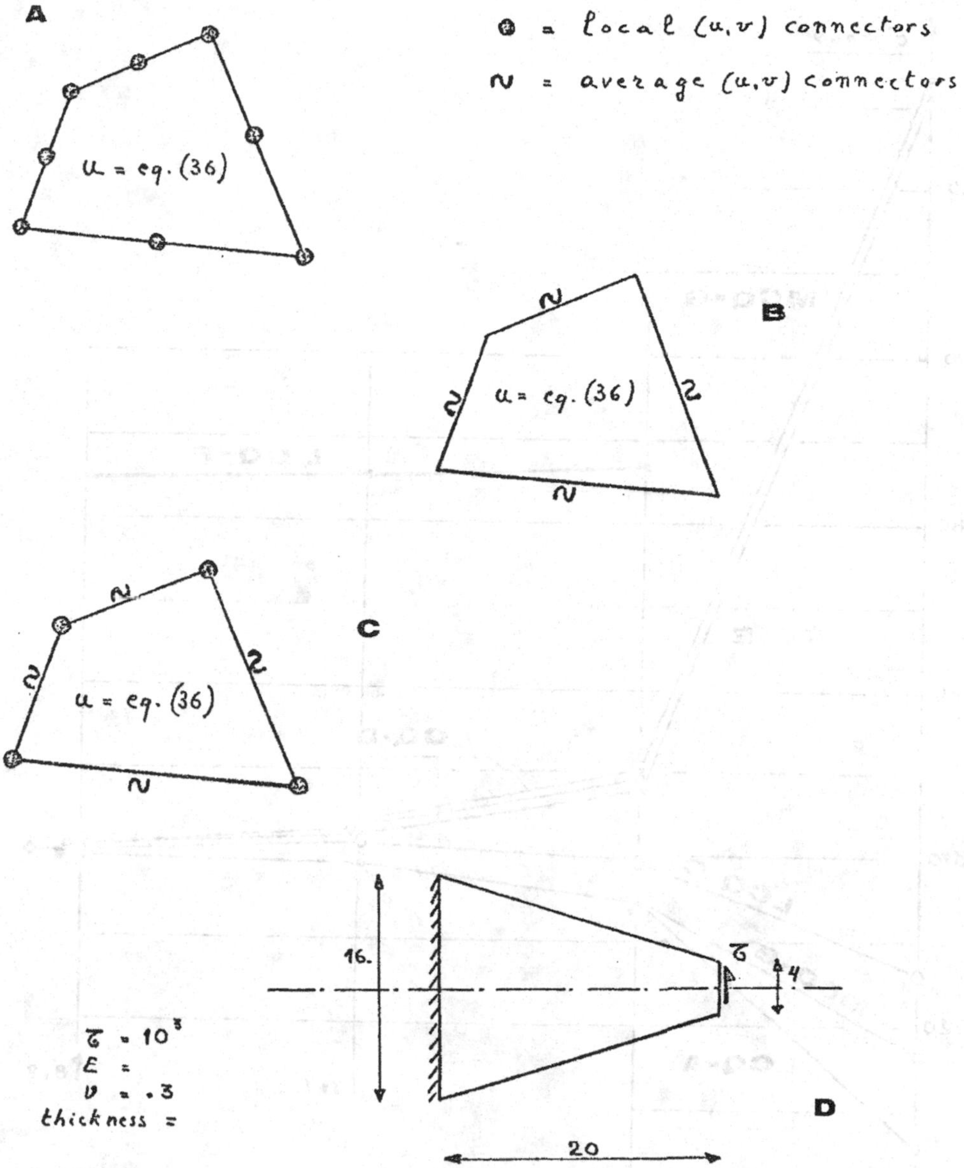
A
= local (u,v) connectors
= average (u,v) connectors
u = eq. (36)
B
u = eq. (36)
C
u = eq. (36)
16.
τ
4
20
D
τ = 10³
E =
ν = .3
thickness =

FIGURE 1

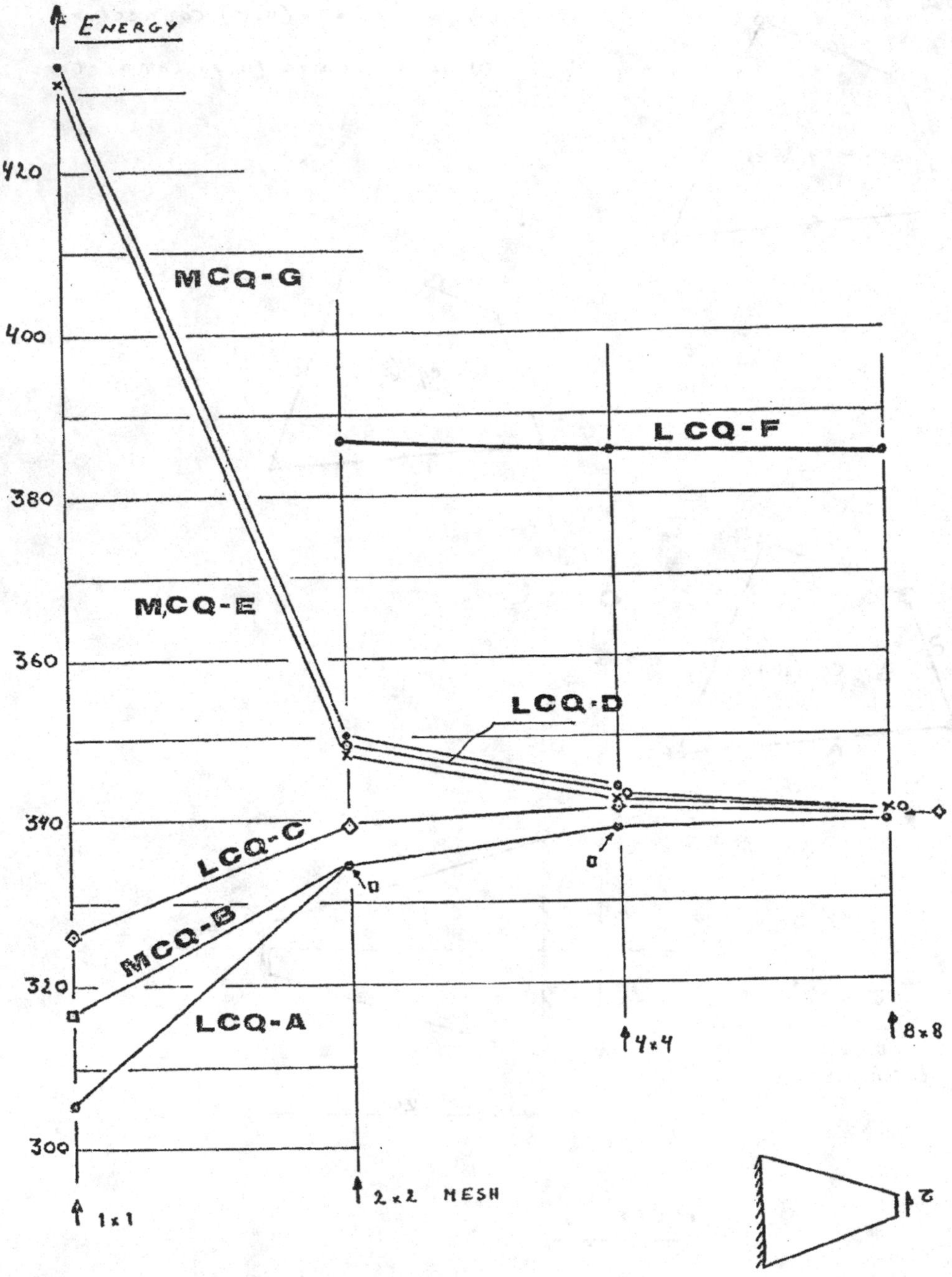

Energy
420
400
380
360
340
320
300
MCQ-G
M,CQ-E
LCQ-F
LCQ-D
LCQ-C
MCQ-B
LCQ-A
1x1
2x2 MESH
4x4
8x8

Figure 2

SOME ERROR ESTIMATES IN GALERKIN METHODS FOR PARABOLIC EQUATIONS

Vidar THOMEE
Chalmers Institute of Technology
Göteborg

1. INTRODUCTION.

Consider the initial-boundary value problem

$$(1.1)\quad \begin{cases} \dfrac{\partial u}{\partial t} = -Lu \equiv \displaystyle\sum_{j,k} \frac{\partial}{\partial x_j}\left(a_{jk}\frac{\partial u}{\partial x_k}\right) - a_o u \text{ in } \Omega\times(0,T_0], \\ u(x,t) = 0 \text{ on } \partial\Omega\times(0,T_0], \\ u(x,0) = v(x). \end{cases}$$

Here Ω is a sufficiently smooth bounded domain in R^N, a_{jk} and a_0 are sufficiently smooth functions which do not depend on t, the matrix (a_{jk}) is symmetric and uniformly positive definite and a_0 nonnegative on $\bar{\Omega}$.

For the purpose of approximating the solution of this problem, let $\{S_h\}$ (h small positive) denote a family of finite dimensional subspaces of H^1 with elements vanishing on $\partial\Omega$, such that with r an integer $\geqslant 2$ and C a constant independent of h and v ($H^s = W_2^s(\Omega)$),

$$(1.2)\quad \inf_{\chi\in S_h} \{\|v-\chi\| + h\|v-\chi\|_{H^1}\} \leqslant Ch^s\|v\|_{H^s}, \text{ for } 1\leqslant s\leqslant r.$$

Introducing the bilinear form

$$(1.3)\quad A(\phi,\psi) = \int_\Omega \left(\sum_{j,k} a_{jk}\frac{\partial\phi}{\partial x_j}\frac{\partial\psi}{\partial x_k} + a_0\phi\psi\right)dx,$$

we may then pose the semidiscrete problem : Find $u_h(t)\in S_h$ for $t>0$ such that, with $u_h(0) = v_h$ suitably chosen, $(.,.)$ the inner product in $L_2 = L_2(\Omega)$, and $D_t = \partial/\partial t$,

$$(1.4)\quad (D_t u_h(t),\chi) + A(u_h(t),\chi) = 0 \quad \forall \chi\in S_h,\ t>0.$$

Error estimates for this problem have been given by energy methods in e.g. Price and Varga [10], Douglas and Dupont [5], Dupont [7] (cf. also Bramble and Thomée [4] for the completely discrete case). In order to show $O(h^r)$ error bounds, these results require at least r derivates il L_2 of the initial data, and in addition certain compatibility assumptions for v on $\partial\Omega$. Using spectral representations of the solutions, Helfrich [8] was able to show that for t bounded away from zero the convergence is $O(h^r)$ even if v is only in L_2 (cf. also Thomée [12] for an even stronger result in one dimension).

Since the results quoted are of optimal order one might have thought that they were essentially final. However, in [11] it was shown that, in the case of the heat equation in one dimension, with S_h consisting of smooth splines of order r on a uniform mesh, then with a suitable choice of v_h (in fact, the interpolant of v) the convergence rate at mesh-points is $O(h^{2r-2})$. Douglas, Dupont and Wheeler [6] who then proved a similar result with S_h consisting of continuous functions which are piecewise polynomials of degree $<r$ on a quasi-uniform mesh, termed this phenomenon superconvergence.

For elliptic problems in N dimensions it was demonstrated by Bramble and Schatz [2], with $\{S_h\}$ assumed to satisfy certain interior regularity requirements, that if a specific averaging operator is applied to the solution of certain semi-discrete Galerkin problems, then in the interior of the domain the results approximate the exact solution to superconvergent order $O(h^{2r-2})$.

Our purpose here is to describe work by Bramble, Schatz, Thomée and Wahlbin [3] showing that results similar to these hold for N-dimensional parabolic problems. In addition we show here that under the appropriate assumptions also any derivative of the solution can be approximated to order $O(h^{2r-2})$.

Since for a smooth domain it is difficult to construct subspaces S_h which satisfy the boundary conditions, many methods have been proposed in the elliptic case to allow more general S_h. In our presentation we shall pose the semidiscrete parabolic problem in a form which permits such choices of S_h.

The proofs of our higher results depend on error estimates in norms of negative order. These are derived by spectral representation and demand only low regularity of the initial data. The analysis of the application of the averaging operator by Bramble and Schatz then requires that interior estimates are known for difference quotients of the error. Such estimates are obtained for fixed positive time by arguments similar to those of Nitsche and Schatz [9].

2. ERROR ESTIMATES BY SPECTRAL REPRESENTATION.

The exact solution of (1.1) can be represented as

$$(2.1) \quad u(x,t) = \sum_{j=1}^{\infty} (v,\phi_j)\, e^{-\lambda_j t}\, \phi_j(x),$$

where $\{\lambda_j\}_1^\infty$ and $\{\phi_j\}_1^\infty$ are the eigenvalues and normalized eigenfunctions of the operator L with vanishing boundary values.

For $s\geq 0$ let $\dot{H}^s$ be the space of w in L_2 for which

$$\|w\|_s = (\sum_j \lambda_j^s (w,\phi_j)^2)^{1/2} < \infty.$$

It can be shown that for s a nonnegative integer, $\dot{H}^s$ consists of the functions $w \in H^s$ with $L^j w = 0$ on $\partial\Omega$ for $j<s/2$, and that $\|\cdot\|_s$ is equivalent to the usual norm in H^s. For $s>0$ and $w \in L_2$ we also define

(2.2) $$\|w\|_{-s} = (\sum_j \lambda_j^{-s}(w,\phi_j)^2)^{1/2} = \sup_{\phi \in \dot{H}^s} \frac{(w,\phi)}{\|\phi\|_s}.$$

For the norm in $L_2 = \dot{H}^0$ we simply write $\|\cdot\|$.

As a consequence of the presence of the exponential factors in (2.1) it follows that the solution of (1.1) is in $\dot{H}^p$ for any p, even with v only in L_2 and that for $p \geq q$,

(2.3) $$\|u(t)\|_p \leq Ct^{-\frac{p-q}{2}} \|v\|_q.$$

We now introduce the solution operator T of the Dirichlet problem

$$Lw = f \text{ in } \Omega, \quad w = 0 \text{ on } \partial\Omega,$$

by $w = Tf$. From the eigenfunction expansion

$$Tf = \sum_j \mu_j (f,\phi_j)\phi_j \text{ with } \mu_j = \lambda_j^{-1},$$

we find easily that T is a selfadjoint positive definite operator on L_2 and is bounded as an operator $\dot{H}^s$ into $\dot{H}^{s+2}$. In terms of T we may write the parabolic problem as

(2.4) $$D_t Tu + u = 0, \quad u(0) = v.$$

Let now $\{S_h\}$ be a family of finite dimensional subspaces of L_2 and assume that we are given a corresponding family of operators $T_h : L_2 \to S_h$ which approximate T. We may then pose the semidiscrete analogue of (2.4),

(2.5) $$D_t T_h u_h + u_h = 0, \quad u_h(0) = v_h,$$

where $u_h = u_h(t) \in S_h$ for $t \geq 0$ and v_h is a suitable approximation to v.

We shall make the following assumptions about the family $\{T_h\}$:

(i) T_h is selfadjoint, positive semidefinite on L_2 and positive definite on S_h.

(ii) There is a constant C such that

$$\|(T_h - T)v\|_{-p} \leq Ch^{p+q+2} \|v\|_q, \text{ for } 0 \leq p,q \leq r-2.$$

In the case when the elements of S_h vanish on $\partial\Omega$ we may define T_h by

$$A(T_h f,\chi) = (f,\chi), \quad \forall \chi \in S_h,$$

and the semidiscrete problem (2.5) now reduces to (1.4). The property (i) is here obvious and (ii) is a well-known consequence of (1.2). The assumptions (i) and (ii) are also satisfied for methods for which the elements do not necessarily vanish on $\partial\Omega$, such as is described for certain methods of Nitsche and Babuška in [3]. Notice that in in the sequel (ii) replaces (1.2).

In order to analyze the error in the semidiscrete problem (2.5) we introduce the resolvent $R_z(T) = (z-T)^{-1}$ of the solution operator T and notice that the solution of (2.4) can then be exppressed as

$$u(t) = \frac{1}{2\pi i} \int_\Gamma e^{-t/z} R_z(T)v dz,$$

where Γ can be taken to be the positively oriented curve defined by $\arg z = \pm \pi/4$ and $\operatorname{Re} z = M$ with $M>\mu_1$.

Taking for the initial-values of the semidiscrete problem the L_2--projection $v_h = P_0 v$ onto S_h, we find similarly,

$$u_h(t) = \frac{1}{2\pi i} \int_\Gamma e^{-t/z} R_z(T_h)v dz, \text{ with } R_z(T_h) = (z-T_h)^{-1}.$$

The error is therefore

$$(2.6) \quad e_h(t) = u_h(t)-u(t) = \frac{1}{2\pi i} \int_\Gamma e^{-t/z}[R_z(T_h)-R_z(T)]v\, dz.$$

The analysis of the error is based on the following estimate:

Lemma 2.1. We have for $0 \le p,q \le r-2$,

$$\|(R_z(T_h)-R_z(T)v\|_{-p} \le C \frac{h^{p+q+2}}{|z|^2} \|v\|_q \text{ for } z \in \Gamma.$$

Using this we easily conclude:

Theorem 2.1. With $v_h = P_0 v$ we have for $0 \le p,q \le r-2$,

$$\|D_t^j e_h(t)\|_{-p} \le Ch^{p+q+2}\, t^{-1-j} \|v\|_q.$$

Proof. We obtain at once from (2.6),

$$D_t^j e_h(t) = \frac{(-1)^j}{2\pi i} \int_\Gamma z^{-j} e^{-t/z} (R_z(T_h)-R_z(T))v dz,$$

and hence, using Lemma 2.1,

$$\|D_t^j e_h(t)\|_{-p} \le Ch^{p+q+2} \|v\|_q \int_\Gamma |z|^{-j-2} e^{-ct/|z|} d|z|$$

$$= Ch^{p+q+2} t^{-j-1} \|v\|_q.$$

For $p = q = 0$ this contains an error estimate in L_2 for data in L_2:

$$\|D_t^j e_h(t)\| \le Ch^2 t^{-1-j} \|v\|.$$

Using a simple iteration argument by Helfrich this may be improved to:

Theorem 2.2. With $v_h = P_0 v$ we have

$$\|D_t^j e_h(t)\| \leq Ch^r t^{-r/2-j} \|v\| .$$

This generalizes to the present context the result by Helfrich mentioned in the introduction.

We shall consider briefly other choices of initial data than $v_h = P_0 v$. For arbitrary $v_h \in S_h$ one easily proves, using the above result for $P_0 v$:

Theorem 2.3. For any k there is a constant C such that

$$\|D_t^j e_h(t)\| \leq Ch^r t^{-r/2-j} \|v_h\| + Ct^{-k/2-j} \|v-v_h\|_{-k}.$$

In particular, if v_h is chosen to be bounded in L_2 and so that it approximates v to order $O(h^r)$ in some negative norm, then the error is $O(h^r)$ in L_2 for t positive. This is satisfied, for instance for the "elliptic projection" $P_1 v = T_h Lv$, if $v \in \dot{H}^2$, since then

$$\|v_h - v\|_{-(r-2)} = \|(T_h - T)\ Lv\|_{-(r-2)} \leq Ch^r \|Lv\| \leq Ch^r \|v\|_2 ,$$

and

$$\|v_h\| \leq \|TLv\| + \|(T_h - T)Lv\| \leq \|v\| + Ch^2 \|Lv\| \leq C \|v\|_2 .$$

For p = q = r-2 the result of Theorem 2.1 reads

$$\|D_t^j e_h(t)\|_{-(r-2)} \leq Ch^{2r-2t-1-j} \|v\|_{r-2} .$$

In the next section we shall see how such estimates in norms of negative order can be used to derive certain high order pointwise estimates. We shall conclude this section by deriving negative norm estimates also when the initial data are chosen as the elliptic projection.

Theorem 2.4. With $v_h = P_1 v$ we have for $0 \leq p,q \leq r-2$,

$$\|e_h(t)\|_{-p} \leq Ch^{p+q+2} \log(t^{-1}) \|v\|_{q+2},$$

and for j > 0,

$$\|D_t^j e_h(t)\|_{-p} \leq Ch^{p+q+2} t^{-j} \|v\|_{q+2} .$$

Proof. We obtain now

$$u_h(t) = \frac{1}{2\pi i} \int_\Gamma e^{-t/z} R_z(T_h) T_h Lvdz,$$

and we may hence write

$$e_h(t) = \frac{1}{2\pi i} \int_\Gamma e^{-t/z} (R_z(T_h)T_h - R_z(T)T)Lvdz.$$

We find at once

$$R_z(T_h)T_h - R_z(T)T = z(R_z(T_h) - R_z(T)),$$

so that

$$D_t^j e_h(t) = \frac{(-1)^j}{2\pi i} \int_\Gamma z^{-(j-1)} e^{-t/z} (R_z(T_h) - R(T) Lv dz.$$

For $j > 0$ we conclude as before, by Lemma 2.1,

$$\|D_t^j e_h(t)\|_{-p} \leq Ch^{p+q+2} \|Lv\|_q \int_\Gamma |z|^{-j-1} e^{-ct/|z|} d|z| =$$

$$= Ch^{p+q+2} t^{-j} \|v\|_{q+2}.$$

For $j = 0$, finally, the result follows from the fact that since z is bounded on Γ, we have for some positive c_0,

$$\int_\Gamma |z|^{-1} e^{-ct/|z|} d|z| \leq \int_{c_0}^\infty \sigma^{-1} e^{-ct\sigma} d\sigma \leqslant C(\log(t^{-1}) \text{ for } t \leq T_0.$$

3. INTERIOR SUPERCONVERGENCE RESULTS.

In this section we shall work in interior subdomains $\Omega_0 \subset\subset \Omega$ and we shall assume that the semidiscrete equation (2.5) is such that its solutions satisfy the interior equation

(3.1) $\quad (D_t u_h(t), \chi) + A(u_h(t), \chi) = 0, \ \forall \chi \in S_h, \text{ supp } \chi \subset \Omega_0,$

where A is the bilinear form in (1.3).

We shall now denote the norms in $L_2(\Omega_0)$ and $C(\bar{\Omega}_0)$ by $\|\cdot\|_{\Omega_0}$ and $|\cdot|_{\Omega_0}$, respectively. For the norm in $H^s = W_2^s(\Omega_0)$ and $C^s(\bar{\Omega}_0)$ we write similarly $\|\cdot\|_{s,\Omega_0}$ and $|\cdot|_{s,\Omega_0}$. For $s>0$ we also write

$$\|w\|_{-s,\Omega_0} = \sup_{\phi \in C_0^\infty(\Omega_0)} \frac{(w,\phi)}{\|\phi\|_{s,\Omega_0}} .$$

Notice that this norm is majorized by the norm defined in (2.2).

We shall first state two technical lemmas about convolutions with splines. For ℓ a positive integer, let $\psi_{(\ell)}$ denote the B-spline of order ℓ, that is the convolution $\chi * \ldots * \chi$ with ℓ factors where χ is the characteristic function of $[-\frac{1}{2}, \frac{1}{2}]$. For $\alpha = (\alpha_1, \ldots, \alpha_N)$ let $\psi_{(\alpha)}(x) = \psi_{(\alpha_1)}(x_1) \ldots \psi_{(\alpha_N)}(x_N)$ denote the B-spline of order α in R^N. The following lemma is then essentially proved in [2]:

Lemma 3.1. Given α and $r\geq 2$, there exists a function K_h of the form

$$K_h(x) = h^{-N} \sum_{\gamma} k_\gamma \psi_{(\alpha)}(h^{-1}x-\gamma),$$

with $k_j = 0$ when $|\gamma_j|>r-2$, such that for $\Omega_2 \subset\subset \Omega_1$,

$$|v-K_h * v|_{\Omega_2} \leq Ch^{2r-2}|v|_{2r-2,\Omega_1}.$$

The coefficients k_γ may in fact be determined in the following simple way: We first choose $k_j(\sigma)$ as a trigonometric polynominal of order $r-2$ such that (recall $\hat{\chi}(\sigma) = \sin \frac{1}{2}\sigma / (\frac{1}{2}\sigma)$)

$$(3.2) \quad k_j(\sigma)\hat{\chi}(\sigma)^{\alpha_j} = 1+O(\sigma^{2r-2}) \quad \text{as } \sigma\to 0,$$

and then set

$$k(\xi) = \sum_{\gamma} k_\gamma e^{-i<\xi,\gamma>} = \prod_{j=1}^{N} k_j(\xi_j).$$

In order to find k_j in (3.2) we set $\tau = \sin \frac{1}{2}\sigma$ and obtain for small τ,

$$\hat{\chi}(\sigma)^{-\alpha_j} = (\frac{\text{arc sin } \tau}{\tau})^{\alpha_j} = \sum_{j=0}^{\infty} \gamma_{j\ell}\tau^{2\ell}.$$

We may now choose

$$k_j(\sigma) = \sum_{j=0}^{r-2} \gamma_{j\ell}(\sin \tfrac{1}{2}\sigma)^{2\ell} = \sum_{j=0}^{r-2} \gamma_{j\ell} 2^{-\ell}(1-\cos\sigma)^{\ell}.$$

Denoting by ∂_h^α the finite difference quotient

$$\partial_h^\alpha = \partial_{h,1}^{\alpha_1} \cdots \partial_{h,N}^{\alpha_N} \quad \text{with } \partial_{h,j}v(x) = \frac{v(x+\frac{1}{2}he_j)-v(x-\frac{1}{2}he_j)}{h},$$

we also quotr the following lemma from [2] (Lemma 6.1). Here ε denotes the vector $(1,\ldots,1) \in R^N$ and $N_0 = [N/2]+1$.

Lemma 3.2. Let ψ denote the B-spline $\psi_{((r-2)\varepsilon)}$ in R^N and set $\psi_h(x) = \psi(h^{-1}x)$. Then for $\Omega_2 \subset\subset \Omega_1$ there is a constant C such that

$$|\psi_h * v|_{\Omega_2} \leq C\{\sum_{|\alpha|\leq N_0+r-2} \|\partial_h^\alpha v\|_{-(r-2),\Omega_1} + h^{r-2}\sum_{|\alpha|\leq r-2}|\partial_h^\alpha v|_{\Omega_2}\}.$$

Let now u and u_h be the solutions of the continuous and semi-discrete parabolic problems (2.4) and (2.5) where we assume that (3.1) holds. Set $e_h = u_h - u$. We shall see that in certain situations we shall be able to approximate u well in the interior of Ω by $K_h * u_h$ with K_h chosen as in Lemma 3.1. For this purpose it will be necessary to estimate $K_h * e_h$ which is a linear combination of translates of $\psi_h * e_h$ with ψ_h as in Lemma 3.2. In order to accomplish this we shall need interior estimates for difference quotients of the error. For the elliptic case such estimates were obtained by Nitsche and Schatz [9] and Bramble, Nitsche and Schatz [1] under the assumption that the elements of S_h are translation invariant in the interior of Ω in a specific sense. The precise conditions used in [9], [1] and also in [3] are rather lengthy to describe. We shall simply say here that $\{S_h\}$ is regular on $\Omega_0 \subset\subset \Omega$ when they are satisfied, and refer to [3] for details. Examples of families of subspaces which are regular in $\Omega_0 \subset\subset \Omega$ are furnished by the restrictions to Ω_0 of the tensor products of onedimensional splines on a uniform mesh, the plane triangular elements of Bramble-Zlámal (provided the triangulation is uniform in Ω_0) and the restrictions to Ω_0 of continuous piecewise linear functions defined on a uniform partition by N-dimensional simplices.

The basic L_2 estimate for finite difference quotients of the error which are contained in the next lemma are obtained in [3] (Lemma 6.4) using the results of [9], with $D_t u$ and $D_t u_h$ acting as inhomogeneous terms in elliptic problems.

Lemma 3.3. Assume that $\{S_h\}$ is regular on $\Omega_0 \subset\subset \Omega$ and let $j \geq 0$ and α be arbitrary. Then for any $\Omega_2 \subset\subset \Omega_1 \subset\subset \Omega_0$ there is a constant C such that

$$\| \partial_h D_t^j e_h(t) \|_{\Omega_2} \leq C\{h^r \|u\|_{r+|\alpha|+2j,\Omega_1} + \sum_{\ell \leq j+1+|\alpha|/2} \|D_t^\ell e_h(t)\|_{\Omega_1}\}.$$

For our application of Lemma 3.2 we shall need the following estimates in negative order norms and in the maximum-norm, which are obtained in [3] (Lemmas 7.3 and 6.5):

Lemma 3.4. Under the assumptions of Lemma 3.3, there is a constant C such that

$$\| \partial_h^\alpha D_t^j e_h(t) \|_{-(r-2),\Omega_2} \leq C\{h^{2r-2} \| u(t) \|_{r+|\alpha|+2j,\Omega_1} +$$

$$+ \sum_{\ell \leq j+1+|\alpha|/2} \{h^{r-2} \|D_t^\ell e_h(t)\|_{\Omega_1} + \|D_t^\ell e_h(t)\|_{-(r-2),\Omega_1}\}\},$$

$$|\partial_h^\alpha D_t^j e_h(t)|_{\Omega_2} \leq C\{h^r \|u(t)\|_{r+|\alpha|+2j+N_0,\Omega_1} +$$

$$+ \sum_{\ell \leq j+1+|\alpha|/2} \|D_t^\ell e_h(t)\|_{\Omega_1}\}.$$

The following is now the main result in this section:

Theorem 3.1. Let α be a given multi-index and let $\tilde{K}_h$ be the function in Lemma 3.1 corresponding to the B-spline of order $\alpha+(r-2)\varepsilon$. Then if $v_h = P_0 v$ we have for $\Omega_1 \subset\subset \Omega_0$,

$$|D_x^\alpha D_t^j u(t) - D_x^\alpha \tilde{K}_h * D_t^j u_h(t)|_{\Omega_1} \leq Ch^{2r-2} \|v\|_{r-2} \text{ for } 0 < t_0 \leq t \leq T_0.$$

Proof. We have with $D^{\alpha,j} = D_x^\alpha D_t^j$,

$$|D^{\alpha,j}u-D_x^\alpha\tilde{K}_h * D_t^j u|_{\Omega_1} \le |D^{\alpha,j}u-\tilde{K}_h * D^{\alpha,j}u|_{\Omega_1} +$$

$$+|D_x^\alpha\tilde{K}_h * D_t^j e_h|_{\Omega_1} = I_1 + I_2.$$

By Lemma 3.1, Sobolev's inequality and the regularity of the solution (2.3) we obtain for the first term, for $t_0 \le t \le T_0$,

$$I_1 \le Ch^{2r-2}|u|_{2r-2+|\alpha|+2j,\Omega_0} \le Ch^{2r-2}\,\|u\|_{2r-2+|\alpha|+2j+N_0}$$

$$\le C(t_0)h^{2r-2}\,\|v\|.$$

In order to estimate the second term we notice that derivatives of splines of higher order are difference quotients of splines of lower order. More precisely (with $\psi_{(\beta),h}(x) = \psi_{(\beta)}(h^{-1}x)$),

$$D_x^\alpha\psi_{(\alpha+(r-2)\varepsilon),h} = \partial_h^\alpha\psi_{((r-2)\varepsilon),h},$$

so that

(3.2) $\quad D_x^\alpha\tilde{K}_h = \partial_h^\alpha K_h$ with $K_h = h^{-N}\sum_\gamma k_\gamma\psi_{((r-2)\varepsilon)}(h^{-1}x-\gamma)$,

where k_γ are the coefficients occurring in $\tilde{K}_h$. Hence, using Lemma 3.2 we obtain with $\Omega_1 \subset\subset \Omega_2 \subset\subset \Omega_3 \subset\subset \Omega_0$,

$$I_2 \le C|\partial_h^\alpha\psi_{((r-2)\varepsilon),h} * D_t^j e_h|_{\Omega_2} = C|\psi_{((r-2)\varepsilon),h} * \partial_h^\alpha D_t^j e_h|_{\Omega_2}$$

$$\le C\{\sum_{|\beta|\le r-2+|\alpha|+N_0}\|\partial_h^\alpha D_t^j e_h\|_{-(r-2),\Omega_3} + h^{r-2}\sum_{|\beta|\le r-2+|\alpha|}|\partial_h^\beta D_t^j e_h|_{\Omega_3}\},$$

and hence using Lemma 3.4, the error estimates of Theorem 2.1 and the regularity (2.3) of the solution, for $t \ge t_0$,

$$I_2 \le Ch^{2r-2}\,\|v\|_{r-2}.$$

This completes the proof of the theorem.

In exactly the same way, using Theorem 2.4 instead of Theorem 2.1 we obtain with the elliptic projection as initial data:

<u>Theorem 3.2.</u> Under the assumptions of Theorem 3.1, with $v_h = P_1 v$, we have

$$|D_x^\alpha D_t^j u(t) - D_x^\alpha\tilde{K}_h * D_t^j u_h(t)|_{\Omega_1} \le Ch^{2r-2}\,\|v\|_r \quad \text{for } 0 < t_0 \le t \le T_0.$$

Notice that (3.2) reduces the computation of $D_x^\alpha\tilde{K}_h * u_h$ to finding the coefficients of $\tilde{K}_h$ and forming the convolutions between $\psi_{((r-2)\varepsilon)}$ and the basis functions in S_h.

REFERENCES

[1] J.H. BRAMBLE, J.A. NITSCHE, A.H. SCHATZ:"Maximum norm interior estimates for Ritz-Galerkin methods".Math. Comp. 29(1975), 677-688.

[2] J.H. BRAMBLE, A.H. SCHATZ: "Higher order local accuracy by averaging in the finite element method".To appear.

[3] J.H. BRAMBLE, A.H. SCHATZ, V. THOMEE, L.B. WAHLBIN: "Some convergence estimates for semidiscrete Galerkin type approximation for parabolic problems".To appear.

[4] J.H. BRAMBLE, V. THOMEE: "Discrete time Galerkin methods for a parabolic boundary value problem". Ann. Mat. Pura Appl. 101(1974), 115-152.

[5] J. DOUGLAS, T. DUPONT: "Galerkin methods for parabolic equations". SIAM J. Numer. Anal. 7(1970),575-626.

[6] J. DOUGLAS Jr., T. DUPONT, M.F. WHEELER: "A quasi-projection approximation method applied to Galerkin procedures for parabolic and hyperbolic equations". MRC Technical Summary Report No. 1465, Ocatober 1974.

[7] T. DUPONT:"Some L^2 error estimates for parabolic Galerkin methods". The Mathematical Foundations of the Finite Element Method with Applications to Partial Differential Equations, Ed. A.K. Aziz, Academic Press, New York, 1972, 491-504.

[8] H.P. HELFRICH:" Fehlerabschätzugen für das Galerkinverfahren zur Lösung von Evolutionsgleichungen".Manuscripta Math. 13(1974), 219-235.

[9] J.A. NITSCHE, A.H. SCHATZ: "Interior estimates for Ritz-Galerkin methods".Math. Comp. 28(1974), 937-958.

[10] H.S. PRICE, R.S. VARGA:" Error bounds for semi-discrete Galerkin approximations of parabolic problems with application to petroleum reservoir mechanics".Numerical Solution of Field Problem in Continuum Physics, SIAM-AMS Proc., vol. II, AMS, Providence R.I., 1970, 74-94.

[11] V. THOMEE: "Spline approximation and difference schemes for the heat equation".The Mathematical Foundations of the Finite Element Method with Applications to Partial Differential Equations, Ed. A.K. Aziz, Academic Press, New York, 1972, 711-746.

[12] V. THOMEE:" Some convergence results for Galerkin methods for parabolic boundary value problems .Mathematical Aspects of Finite Elements in Partial Differential Equations, Ed. C. de Boor, Academic Press, New York, 1974, 55-88.

SOME SUPERCONVERGENCE RESULTS IN THE FINITE ELEMENT METHOD

Milos Zlámal
Technical University in Brno, Czechoslovakia

ABSTRACT

Superconvergence of the gradient of finite element solutions of 2nd order problems is proved at Gaussian points when finite element spaces constructed by means of polynomials of the Serendipity family are used. Another superconvergence is proved for "moments" of the finite element solution of higher order equations at Gaussian points when finite element spaces constructed by means of Hermite bivariate polynomials are applied.

1. SECOND ORDER PROBLEMS

In recent years it has been observed (we refer to Hulme [7], de Boor-Swarz [2] and Douglas-Dupont [6]) that finite element approximate solutions of differential equations can converge at exceptional points at rates that exceed the possible global rates. Exceptional points for gradient of finite element solutions of second order problems are predicted by Strang and Fix in [8], p.168. In this section we find such exceptional points.

We consider the Dirichlet problem

$$- \frac{\partial}{\partial x}\left[\alpha(x,y)\frac{\partial u}{\partial x}\right] - \frac{\partial}{\partial y}\left[\beta(x,y)\frac{\partial u}{\partial y}\right] + \sigma(x,y)u = f(x,y) \text{ in } \Omega , \qquad (1)$$

$$u = 0 \text{ on } \Gamma \qquad (2)$$

Here Ω is a domain which is a union of rectangles with sides parallel to the coordinate axes and Γ is its boundary. To (1) and (2) there is associated the bilinear functional

$$a(u,v) = \int_\Omega \left[\alpha(x,y)\frac{\partial u}{\partial x}\frac{\partial v}{\partial x} + \beta(x,y)\frac{\partial u}{\partial y}\frac{\partial v}{\partial y} + \sigma(x,y)uv\right]dxdy . \qquad (3)$$

We assume that

i) α, β are Lipschitz continuous in $\bar{\Omega}$:

ii) σ is bounded in Ω $f \in L_2(\Omega)$ $\qquad (H_1)$

iii) $a(u,v)$ is H_0^1 - elliptic, i.e. $a(v,v) \geq c_1 \|v\|_1^2$, $c_1 = \text{const} > 0$

The following notation is used in the paper: H^m $(m = 0,1,\ldots)$ denotes the Sobolev space of functions which together with their generalized derivatives up to the order m inclusive are in $L_2(\Omega)$. The inner product is given by $(u,v)_m = \int_\Omega \sum_{|i| \leq m} D^i u D^i v \, dxdy$, the norm is denoted by $\|\cdot\|_m$; $(\cdot,\cdot)_0$ is, of course, the inner product in $L_2(\Omega)$. By $|\cdot|_m$ we denote the seminorm $|u|_m = \left[\int_\Omega \sum_{|i|=m} (D^i u)^2 dxdy \right]^{\frac{1}{2}}$. If instead of Ω a domain $D \neq \Omega$ is considered we attach D as a subsctipt (i.e. we write $\|\cdot\|_{m,D}$, $|\cdot|_{m,D}$). By H_o^m (m = 1,2,...) we denote the subspace of H^m which arizes by completing in the H^m - norm the set of infinitely differentiable functions with compact support in Ω.

The weak solution of the problem (1),(2) is a function $u \in H_o^1$ which satisfies

$$a(u,v) = (f,v)_o \quad \forall v \in H_o^1 . \tag{4}$$

Let Ω be covered by rectangular elements with sides parallel to the coordinate axes the union of which is Ω. We denote by $h(h')$ the greatest (smallest) length of sides of all rectangular elements and assume that

$$\frac{h'}{h} \geq c_2 , \quad c_2 = \text{const} > 0 \tag{5}$$

The condition (5) is sufficient that the family of the rectangular elements be regular in the sense of the definition by Ciarlet-Raviart [4]. The finite element space denoted by V_h consists of functions which on every rectangular element are incomplete polynomials of the third degree determined uniquely by eight values: values at cotners and at mid-points of the sides. Let us consider an element e with sides of length $h_e \geq k_e$. If the side of length h_e is parallel to the x-axis and if (x_o, y_o) are coordinates of the center of e then the mapping

$$x = x(\xi,\eta) \equiv x_o + \frac{h_e}{2}\xi, \quad y = y(\xi,\eta) \equiv y_o + \frac{h_e}{2}\eta \tag{6}$$

mapps $\bar{e}$ one-to-one on the unit square $K_1 : -1 \leq \xi \leq 1,\ -1 \leq \eta \leq 1$; the inverse mapping is

$$\xi = \xi(x,y) \equiv \frac{2}{h_e}(x-x_o), \quad \eta = \eta(x,y) \equiv \frac{2}{h_e}(y-y_o). \tag{7}$$

A function $v \in V_h$ is on $\bar{e}$ of the form

$$v(x,y) = r\left[\xi(x,y),\ \eta(x,y)\right], \quad r(\xi,\eta) = \sum_{i=1} v_i N_i(\xi,\eta) \tag{8}$$

v_i are values of v at the corresponding eight nodes and the functions $N_i(\xi,\eta)$ are easy to compute (see, e.g. Zienkiewicz [9], p.109). For our purpose the important thing is that the polynomial $r(\xi,\eta)$ contains these eight terms: $1, \xi, \eta, \xi^2, \xi\eta, \eta^2, \xi^2\eta, \xi\eta^2$. Therefore the set of functions $\hat{v}(\xi,\eta) = v[x(\xi,\eta),\ y(\xi,\eta)]$ contains all quadratic polynomials and two terms, ξ^3 and η^3, are missing from cubic polynomials.

The finite element solution $u_h \in V_h$ is defined by

$$a(u_h,v) = (f,v)_o \quad \forall v \in V_h . \tag{9}$$

For the discretization error $E = u - u_h$ it follows from results by Ciarlet-Raviart [4] that

$$\|E\|_1 \leq Ch^2 \|u\|_3 \quad ; \tag{10}$$

in general, (10) is the best possible rate.

Let us denote by G the set of all maps of Gaussian points, i.e.

the set of all maps of the points $\xi = \pm \frac{\sqrt{3}}{3}$, $\eta = \pm \frac{\sqrt{3}}{3}$ through mappings (6) (in general, by Gaussian points we mean the points (ξ_i, ξ_j), $i,j = 1,\dots,p$ where ξ_i, $i = 1,\dots,p$ are zeros of the Legendre polynopial $P_p(\xi)$ of degree p; ξ_i, $i = 1,\dots,p$, are nodes of the Gaussian quadrature formula of degree 2p-1). Superconvergence for gradient of the approximate solution u_h at these points follows from.

Theorem 1. Let the conditions (H_1) and (5) be satisfed and let $u \in H^4$. Then the arithmetic mean A of values $|\mathrm{grad}E(P)|, P \in G$, is bounded by

$$A \leq Ch^3(|u|_3 + |u|_4) \tag{11}$$

or, equivalently,

$$h^2 \sum_{P \in G} |\mathrm{grad}E(P)| \leq Ch^3(|u|_3 + |u|_4). \tag{12}$$

Proof. a) Equivalence of (11) and (12) follows from (5). So we prove (12).

Let u_I be the interpolate of u (i.e. the function from V_h which assumes the same values at the nodes as u itself). We will prove first that

$$\|u_I - u_h\|_1 \leq (Ch^3(|u|_3 + |u|_4). \tag{13}$$

From (4) and (9) we have

$$a(u_I - u_h, v) = a(u_I - u, v) \qquad \forall v \in V_h . \tag{14}$$

If we prove that

$$|a(u - u_I, v)| \leq Ch^3(|u|_3 + |u|_4)\|v\|_1 \tag{15}$$

then setting $v = u_I - u_h \in V_h$ in (14) and using (H_1)-iii) we get (13). Hence, it is sufficient to prove (15).

Ti this end we write

$$a(u-u_I,v) = \sum_e \int_e \left[\alpha \frac{\partial(u-u_I)}{\partial x}\frac{\partial v}{\partial x} + \beta \frac{\partial(u-u_I)}{\partial y}\frac{\partial v}{\partial y} + \sigma(u-u_I)v\right] dxdy$$

and estimate each term of the above sum. Using (6) and the notation $\tilde{w}(\xi,\eta) = w[x(\xi,\eta), y(\xi,\eta)]$ for any w we have

$$\int_e \left[\alpha \frac{\partial(u-u_I)}{\partial x}\frac{\partial v}{\partial x} + \beta \frac{\partial(u-u_I)}{\partial y}\frac{\partial v}{\partial y} + \sigma(u-u_I)v\right] dxdy = \frac{k_e}{h_e}\int_{k_1} \tilde{\alpha} \frac{\partial(\tilde{u}-\tilde{u}_I)}{\partial \xi}\frac{\partial \tilde{v}}{\partial \xi} d\xi d\eta$$
$$+ \frac{h_e}{k_e}\int_{k_1} \tilde{\beta} \frac{\partial(\tilde{u}-\tilde{u}_I)}{\partial \eta}\frac{\partial \tilde{v}}{\partial \eta} d\xi d\eta + \frac{1}{4} h_e k_e \int_{k_1} \tilde{\sigma}(\tilde{u}-\tilde{u}_I)\tilde{v} d\xi d\eta \tag{16}$$

Now if $\alpha^o = \alpha(x_o, y_o)$ then from (H_1) and (5) it follows

$$\int_{k_1} \tilde{\alpha} \frac{\partial(\tilde{u}-\tilde{u}_I)}{\partial \xi}\frac{\partial \tilde{v}}{\partial \xi} d\xi d\eta = \alpha^o \int_{k_1} \frac{\partial(\tilde{u}-\tilde{u}_I)}{\partial \xi}\frac{\partial \tilde{v}}{\partial \xi} d\xi d\eta + O(h)|\tilde{u}-\tilde{u}_I|_{1,k_1} \cdot |\tilde{v}|_{1,k_1} . \tag{17}$$

Let us consider the functional $F(\tilde{u}) = \int_{k_1} \frac{\partial(\tilde{u}-\tilde{u}_I)}{\partial\xi}\frac{\partial\tilde{v}}{\partial\xi}\,d\xi d\eta$. From the Sobolev lemma we have $|\tilde{u}_I|_{1,k_1} \leq C\|\tilde{u}\|_{2,k_1}$, hence $|\tilde{u}-\tilde{u}_I|_{1,k_1} \leq C\|\tilde{u}\|_{2,k_1}$ and $|F(\tilde{u})| \leq C|\tilde{v}|_{1,k_1}\|\tilde{u}\|_{4,k_1}$. We prove that $F(\tilde{u}) = 0$ if $\tilde{u}$ is a cubic polynomial. It is sufficient to consider $\tilde{u} = \xi^3$ and $\tilde{u} = \eta^3$ because if $\tilde{u} = a_1 + a_2\xi + \ldots + a_6\eta^2 + a_7\xi^2\eta + a_8\xi\eta^2$ then $\tilde{u}_I = \tilde{u}$ and $F(\tilde{u}) = 0$. For $\tilde{u} = \xi^3$ we have $\tilde{u}_I = \xi_1$. As $\int_{-1}^{1}\frac{\partial\tilde{v}}{\partial\xi}\,d\eta$ is a linear polynomial of ξ we get $F(\xi^3) =$
$= \int_{-1}^{1}(3\xi^2-1)(A+B\xi)\,d\xi = 0$ (notice that $3\xi^2-1$ is an multiple of $P_2(\xi)$).
For $\tilde{u} = \eta^3$ we have $\frac{\partial(\tilde{u}-\tilde{u}_I)}{\partial\xi} = 0$, hence $F(\eta^3) = 0$. From the Bramble-Hilbert lemma (see [3]) it follows (taking into account (5))

$$|F(\tilde{u})| \leq C|\tilde{v}|_{1,k_1}|\tilde{u}|_{4,k_1} \leq Ch^3|v|_{1,e}\,|u|_{4,e}\,.$$

From (17) we obtain (as $|\tilde{u}-\tilde{u}_I|_{1,k_1} \leq C|\tilde{u}|_{3,k_1} \leq Ch^2|u|_{3,e}$)

$$\left|\int_{k_1}\tilde{\alpha}\,\frac{\partial(\tilde{u}-\tilde{u}_I)}{\partial\xi}\frac{\partial\tilde{v}}{\partial\xi}\,d\xi d\eta\right| \leq Ch^3(|u|_{3,e} + |u|_{4,e})|v|_{1,e}\,. \tag{18}$$

Similalrly

$$\left|\int_{k_1}\tilde{\beta}\,\frac{\partial(\tilde{u}-\tilde{u}_I)}{\partial\eta}\frac{\partial\tilde{v}}{\partial\eta}\,d\xi d\eta\right| \leq Ch^3(|u|_{3,e} + |u|_{4,e})|v|_{1,e}\,,$$

$$\left|\int_{k_1}\tilde{\sigma}(\tilde{u}-\tilde{u}_I)\tilde{v}\,d\xi d\eta\right| \leq Ch|u|_{3,e}\|v\|_{0,e}\,.$$

Hence, from (16)

$$\left|\int_e\left[\alpha\,\frac{\partial(u-u_I)}{\partial x}\frac{\partial v}{\partial x} + \beta\,\frac{\partial(u-u_I)}{\partial y}\frac{\partial v}{\partial y} + \sigma(u-u_I)v\right]dxdy\right| \leq Ch^3(|u|_{3,e}+|u|_{4,e})\|v\|_{1,e}$$

and summing we easily get (15).

b) Denote $u-u_I = \varepsilon$. We have $\frac{\partial\varepsilon}{\partial x} = \frac{2}{h_e}\frac{\partial\tilde{\varepsilon}}{\partial\xi}$, $\frac{\partial\varepsilon}{\partial y} = \frac{2}{k_e}\frac{\partial\tilde{\varepsilon}}{\partial\eta}$. Consider the functional $G(\tilde{u}) = \frac{\partial\tilde{\varepsilon}(Q)}{\partial\xi}$ where $Q = \left(\pm\frac{\sqrt{3}}{3}, \pm\frac{\sqrt{3}}{3}\right)$. It holds $|G(\tilde{u}) \leq C\|\tilde{u}\|_{4,k_1}$ and $G(\tilde{u}) = 0$ if $\tilde{u}$ is a cubic polynomial (again, it suffices to consider $\tilde{u} = \xi^3$, $\tilde{u} = \eta^3$; if $\tilde{u} = \xi^3$ then $\frac{\partial\tilde{\varepsilon}}{\partial\xi} = 3\xi^2-1$, if $\tilde{u} = \eta^3$ then $\frac{\partial\tilde{\varepsilon}}{\partial\xi} = 0$). Therefore $|G(\tilde{u})| \leq C|\tilde{u}|_{4,k_1} \leq Ch^2|u|_{4,e}$, hence $\left|\frac{\partial\varepsilon(P)}{\partial x}\right| \leq Ch^2|u|_{4,e}$ if $P\in G$. Similarly $\left|\frac{\partial\varepsilon(P)}{\partial y}\right| \leq Ch^2|u|_{4,e}$, thus

$$|\mathrm{grad}\,\varepsilon(P)| \leq Ch^2|u|_{4,e} \quad \text{if } P\in G. \tag{19}$$

From (19) and (5) it easily follows

$$h^2 \sum_{P\in G} |\mathrm{grad}\,\varepsilon(P)| \leq Ch^3 |u|_4 . \tag{20}$$

c) For any quadratic polynomial $q(\xi,\eta)$ it holds $\max\limits_{\bar{K}_1} q^2 \leq C\int\limits_{K_1} q^2 d\xi d\eta$ where C is an absolute constant (both sides of this inequality are positive definite quadratic forms of the coefficients of q bounded from below and above uniformly for $(\xi,\eta)\in\bar{K}_1$) Hence, if $w(x,y)$ is piecewise of the form $w(x,y) = q[\xi(x,y), \eta(x,y)]$ we have $\max\limits_{\bar{e}} w^2 = \max\limits_{\bar{k}} \tilde{w}^2 \leq C\|\tilde{w}\|_{o,k_1} \leq Ch^{-2}\|w\|_{o,e}$, consequently $h \max\limits_{\bar{e}}|w| \leq C\|w\|_{o,e}$ and summing we get

$$h^2 \sum_e \max_{\bar{e}}|w| \leq C\|w\|_o . \tag{21}$$

Setting $w = \dfrac{\partial(u_I-u_h)}{\partial x}$ and $w = \dfrac{\partial(u_I-u_h)}{\partial y}$ in (21) we obtain

$$h^2 \sum_e \max_{\bar{e}}|\mathrm{grad}(u_I-u_h)| \leq C|u_I-u_h|_1 . \tag{22}$$

Now $u-u_h = \varepsilon + u_I-u_h$, hence (12) is an immediate consequence of (20), (22) and (13).

2. SOME EXTENSIONS OF THE RESULTS

1. The estimate (12) is also valide if Newton boundary condition $\frac{\partial u}{\partial \nu} + \omega u|_\Gamma = 0$ ($\frac{\partial u}{\partial \nu}$ is the conormal $\alpha\frac{\partial u}{\partial x}\ell_1 + \beta\frac{\partial y}{\partial y}\ell_2$; here ℓ_1, ℓ_2 are direction cosines of the outward normal to Γ) is considered and if the cerresponding functional $a(u,v) = \int\limits_R [\alpha\frac{\partial u}{\partial x}\frac{\partial v}{\partial x} + \beta\frac{\partial u}{\partial y}\frac{\partial v}{\partial y} + \sigma uv]dxdy + \int\limits_\Gamma \omega uv d\Gamma$ is H^1-elliptic. The only change in the proof is that we have to estimate one more term of $a(u-u_I, v)$, namely $\int\limits_\Gamma \omega(u-u_I)v d\Gamma$. We write this term as a sum of integrals over sides lying on Γ. Let s_e be such a side parallel to the x-axis. We have $|\int\limits_{s_e}\omega(u-u_I)vd\Gamma| \leq C\int\limits_{s_e}|u-u_I||v|dx =$

$= C\,\frac{1}{2}\,h_e \int\limits_{-1}^{1}|\tilde{u}-\tilde{u}_I||\tilde{v}|d\xi \leq Ch[\int\limits_{-1}^{1}(\frac{\partial^3\tilde{u}}{\partial\xi^3})^2 d\xi]^{\frac{1}{2}} [\int\limits_{-1}^{1}\tilde{v}^2 d\xi]^{\frac{1}{2}}$ (last inequality can be proved by means of Bramble-Hilbert lemma using the fact that the shape functions are quadratic polynomials in each variable).

Hence $|\int\limits_{s_e}\omega(u-u_I)vdx| \leq Ch^3 [\int\limits_{s_e} v^2 dx]^{\frac{1}{2}} [\int\limits_{s_e}(\frac{\partial^3 u}{\partial x^3})^2 dx]^{\frac{1}{2}}$.

A similar bound is true for sides parallel to the y-axis. Summing we get

$\left|\int_\Gamma \omega(u-u_I)v\,d\Gamma\right| \leq Ch^3\left[\int_\Gamma v^2 d\Gamma\right]^{\frac{1}{2}}\left[\int_\Gamma\left\{\left(\frac{\partial^3 u}{\partial x^3}\right)^2\left(\frac{\partial^3 u}{\partial y^3}\right)^2\right\}d\Gamma\right]^{\frac{1}{2}}$. As for any $w\in H^1$ it holds $\int_\Gamma w^2 d\Gamma \leq C\|w\|_1^2$ we get the final estimate $\left|\int_\Gamma \omega(u-u_I)v\,d\Gamma\right| \leq Ch^3(|u|_3 + |u|_4)\|v\|_1$.

2. Superconvergence at Gaussian points holds when we use quadratic one-and three-dimensional elements or linear one-two, and three-dimensional elements of the Serendipity family. The general statement is the following:

$$h^N \sum_{P\in G} |\mathrm{grad}E(P)| \leq Ch^{p+1}(|u|_{p+1} + |u|_{p+2}); \tag{23}$$

here N = 1,2,3 is the dimension and p = 1,2 is the degree of elements. The proof is the same as in case of quadratic two-dimensional elements with some obvious changes, so we leave it out.

For cubic two-and three-dimensional elements of the Serendipity family the argument a) of the proof of Theorem 1does not work. Also an analog of the estimate (19) is true if and only if P is a map of the point $\xi = 0$, $\eta = 0$, i.e. an analog of (20) is not true the set of all Gaussian points. A different situation occurs if suitable cubic one-dimensional elements are applied.Let us namely consider cubic polynomials $r(\xi)$ determined by their values at $\xi = \pm 1$ and $\xi = \pm\frac{\sqrt{5}}{5}$. An easy calculation shows that if $\tilde{u} = \xi^4$ then $\frac{d(\tilde{u}-\tilde{u}_I)}{d\xi} = \frac{8}{5}P_3(\xi)$.
As $\frac{d\tilde{v}}{d\xi}$ is a quadratic polynomial we have $\int_{-1}^{1}\frac{d(\tilde{u}-\tilde{u}_I)}{d\xi}\frac{d\tilde{v}}{d\xi}d\xi = 0.$

It follows

$$h\sum_{P\in G}\left|\frac{dE(P)}{dx}\right| \leq Ch^4(|u|_4 + |u|_5). \tag{24}$$

The result can be extended for polynomials of higher degree than the fifth with suitably chosen nodes.

3. There is a question if (12) is true in case of curved quadrilateral elements. By a curved quadrilateral (see Zienkiewicz [9] or Ciarlet-Raviart [5])we mean the following element: We choose eight points $a_i = (x_i, y_i)$, $i = 1,\dots,8$. Then under certain conditions the mapping $x = \sum_{i=1}^{8} x_i N_i(\xi,\eta)$, $y = \sum_{i=1}^{8} y_i N_i(\xi,\eta)$ maps the unit square $\bar{K}_1$ one-to-one on a curved quadrilateral with corners a_i, $i = 1,\dots,4$ and with "mid-side" points a_i, $i = 5,\dots,8$. On every such element the trial functions are of the form $v(x,y)=r[\xi(x,y),\eta(x,y)]$, $r(\xi,\eta) = \sum_{i=1}^{8} v_i N_i(\xi,\eta)$. We did not finish the proof, however we hope to prove that (12) is true if the curved quadrilaterals arise by a small distorsion of rectangles. By a small distorsion we mean that the points a_i are in a $O(h_e^2)$ distance from corresponding corners and mid-side points of a rectangle with sides of the lenghts h_e, k_e. In general, we have to assume that the rectangle has sides parallel to the coordinate axes. However, if the coefficients α and β of equation (1) are equal

then such an assumption is not necessary.

3. HIGHER ORDER EQUATIONS

We consider the variational problem to find $u \in V$ such that

$$a(u,v) = L(v) \qquad \forall v \in V. \tag{25}$$

Here V is a subspace of H^n, $n \geq 2$, $L(v)$ is a linear functional continuous on V and the bilinear functional $a(u,v)$ has the form

$$a(u,v) = \int_\Omega \sum_{|i|=n} a_{ij} D^i u D^i v dx + \int_\Omega \sum_{|i|,|j| \leq n-1} a_{ij} D^i u D^j v \tag{26}$$

$\left(D^i = \dfrac{\partial^{|i|}}{\partial x_1^{i_1} \partial x_2^{i_2}}, \; dx = dx_1 dx_2\right)$. The domain Ω is again a union of rectangles with sides parallel to the coordinate axes. We assume that

i) a_{ii}, $|i| = n$, are Lipschitz continuous in $\bar{\Omega}$,

ii) a_{ij}, $|i|,|j| \leq n-1$ are bounded in $\bar{\Omega}$, $\qquad (H_2)$

iii) $a(u,v)$ is V-elliptic, i.e. $a(v,v) \geq c_1 \|v\|_n \quad \forall v \in V$

The conditions i) and ii) guarantee the boundedness of $a(u,v)$.

As examples of the problem (25) let us introduce a clamped and simply supported plate. Due to special form of the domain Ω we can take in both cases $a(u,v) = \int_\Omega \left[\frac{\partial^2 u}{\partial x_1^2}\frac{\partial^2 v}{\partial x_1^2} + 2\frac{\partial^2 u}{\partial x_1 \partial x_2}\frac{\partial^2 v}{\partial x_1 \partial x_2} + \frac{\partial^2 u}{\partial x_2^2}\frac{\partial^2 v}{\partial x_2^2}\right] dx$, $L(v)$ is equal to $(f,v)_0$ and the space V is equal to H_0^2 for a clamped plate and to $\{v;\, v \in H^2,\, v|_\Gamma = 0\}$ for a simply supported plate.

Let Ω be again covered by rectangular elements with sides parallel to the coordinate axes the union of which is Ω and let the condition (5) be again satisfied. As shape functions we choose bivariate Hermite polynomials $H_n(x_1,x_2) = \sum_{r,s=0}^{2n-1} \alpha_{r,s} x_1^r x_2^s$ which are uniquely determined by values of $D^i H_n$, $i_1, i_2 \leq n-1$, at corners of the corresponding rectangle (see Birkhoff-Schultz-Varga [1]). The trial functions belong to C^{n-1}, hence also to H^n and the set of all trial functions which lie in V is denoted again by V_h. The finite element solution is a function $u_h \in V_h$ such that

$$a(u_h,v) = L(v) \qquad \forall v \in V_h. \tag{27}$$

From results by Birkhoff-Schultz-Varga [1] it follows the estimate

$$\|E\|_n \leq Ch^n \|u\|_{2n} \; ; \tag{28}$$

the exponent of h is best possible.

Let us denote by G the set of all maps of Gaussian points (ξ_i, ξ_j), $i,j = 1,\dots,n$ (we remind that ξ_i, $i = 1,\dots,n$, are zeros of the Legendre polynomial $P_n(\xi)$). We shall prove superconvergence of n-th order derivatives of the approximate solution u_h at points of G.

Theorem 2. Let the conditions (H_2) and (5) be satisfied and let $u \in H^{2n+1}$. Then it holds

$$h^2 \sum_{P \in G} \sum_{|i|=n} |D^i E(P)| \leq C h^{n+1} (|u|_{2n} + |u|_{2n+1}). \tag{29}$$

Remark. A special case of Theorem 2 is the following assertion: There is superconvergence of moments of a clamped or simply supported plate at maps of the points $\left(\pm\frac{\sqrt{3}}{3}, \pm\frac{\sqrt{3}}{3}\right)$ when the trial functions are piecewise bicubic Hermite polynomials.

Proof of Theorem 2. It is similar to the proof of Theorem 1.

a) We prove that

$$\|u_I - u_h\|_n \leq C h^{n+1} (|u|_{2n} + |u|_{2n+1}). \tag{30}$$

As before it is sufficient to show that

$$|a(u-u_I, v)| \leq C h^{n+1} (|u|_{2n} + |u|_{2n+1}) \|v\|_n \quad \forall v \in V_h . \tag{31}$$

To this end we first remark that the term of $a(u-u_I, v)$ coming from the second sum in (26) is bounded in absolute value by $C\|u - u_I\|_{n-1} \|v\|_{n-1} \leq C h^{n+1} |u|_{2n} \|v\|_n$ (the estimate $\|u-u_I\|_{n-1} \leq C h^{n+1} |u|_{2n}$ follows from results of Birkhoff-Schultz-Varga [1]; the easiest way to prove it is to use Bramble-Hilbert lemma). We write the term coming from the first sum in (26) as a sum of integrals over all elements and estimate these integrals I_e,

$$I_e = \int_e \left[\alpha \frac{\partial^n (u - u_I)}{\partial x_1^n} \frac{\partial^n v}{\partial x_1^n} + \ldots + \beta \frac{\partial^n (u - u_I)}{\partial x_2^n} \frac{\partial^n v}{\partial x_2^n}\right] dx$$

($\alpha = a_{ii}$, $i = (n,0)$, $\beta = a_{ii}$, $i = (0,n)$). We have (with the same notation as before)

$$I_e = O(h^{2-2n}) \int_{k_1} \left[\tilde{\alpha} \frac{\partial^n (\tilde{n}-\tilde{u}_I)}{\partial \xi^n} \frac{\partial^n \tilde{v}}{\partial \xi^n} + \ldots + \tilde{\beta} \frac{\partial^n (\tilde{u}-\tilde{u}_I)}{\partial \eta^n} \frac{\partial^n \tilde{v}}{\partial \eta^n}\right] d\xi d\eta =$$

$$O(h^{2-2n}) \int_{k_1} \left[\alpha^\circ \frac{\partial^n (\tilde{u}-\tilde{u}_I)}{\partial \xi^n} \frac{\partial^n \tilde{v}}{\partial \xi^n} + \ldots + \beta^\circ \frac{\partial^n (\tilde{u}-\tilde{u}_I)}{\partial \eta^n} \frac{\partial^n \tilde{v}}{\partial \eta^n}\right] d\xi d\eta +$$

$$+ O(h^{3-2n}) |\tilde{u} - \tilde{u}_I|_{n, k_1} |\tilde{v}|_{n, k_1} .$$

Further

$$I_e = O(h^{2-2n}) \int_{k_1} \left[\alpha^\circ \frac{\partial^n (\tilde{u}-\tilde{u}_I)}{\partial \xi^n} \frac{\partial^n \tilde{v}}{\partial \xi^n} + \ldots + \beta^\circ \frac{\partial^n (\tilde{u}-\tilde{u}_I)}{\partial \eta^n} \frac{\partial^n \tilde{v}}{\partial \eta^n}\right] d\xi d\eta$$

$$+ O(h^{3-2n}) |\tilde{u}|_{2n, k_1} |\tilde{v}|_{n, k_1} . \tag{32}$$

Let us consider the functional $F(\tilde{u}) = \alpha^\circ \int_{k_1} \frac{\partial^n (\tilde{u}-\tilde{u}_I)}{\partial \xi^n} \frac{\partial^n \tilde{v}}{\partial \xi^n} d\xi d\eta$ which is the first term of the right-hand side of (32). As before if we prove that $F(\tilde{u}) = 0$ when $\tilde{u}$ is a polynomial of the degree not greater than 2n we get

$$|F(\tilde{u})| \leq C|\tilde{u}|_{2n+1,K_1}|\tilde{v}|_{n,K_1}. \tag{33}$$

It is sufficient to show that $F(\tilde{u}) = 0$ for $\tilde{u} = \xi^{2n}$, $\tilde{u} = \eta^{2n}$. If $\tilde{u} = \xi^{2n}$ we assume $\tilde{u}_I$ to be a function of the variable ξ only.

We have $\left.\frac{d^s(\tilde{u}-\tilde{u}_I)}{d\xi^s}\right|_{\xi=\pm 1} = 0$, $s = 0,\dots,n-1$, hence $\tilde{u}-\tilde{u}_I = A_n(\xi^2-1)^n$ and as $\tilde{u}-\tilde{u}_I = \xi^{2n} + \dots$ we get $A_n = 1$. Therefore $\frac{\partial^n(\tilde{u}-\tilde{u}_I)}{\partial\xi^n} = \frac{d^n}{d\xi^n}(\xi^2-1)^n = 2^n n! P_n(\xi)$. As $\int_{-1}^{1}\frac{\partial^n\tilde{v}}{\partial\xi^n}\,d\eta$ is a polynomial of ξ of degree not greater than n-1 we get $F(\xi^{2n}) = 0$. If $\tilde{u} = \eta^{2n}$ then $\frac{\partial^n(\tilde{u}-\tilde{u}_I)}{\partial\xi^n} = 0$ and $F(\eta^{2n}) = 0$.

When $F(\tilde{u})$ is equal to one of the terms which are denoted by ... in (32) then $F(\tilde{u}) = 0$ for $\tilde{u} = \xi^{2n}$ and $\tilde{u} = \eta^{2n}$ because $\tilde{u}-\tilde{u}_I$ is a function of ξ or of η only. For the last term we can repeat the same argument we have applied for the first term. Therefore (33) it true for all terms of (32) and hence

$$I_e = O(h^{2-2n})|\tilde{u}|_{2n+1,k_1}|\tilde{v}|_{n,k_1} + O(h^{3-2n})|\tilde{u}|_{2n,k_1}|\tilde{v}|_{n,k_1} =$$

$$= O(h^{n+1})(|u|_{2n,e} + |u|_{2n+1,e})|v|_{n,e}.$$

Suming over all elements we see that (31) is true.

b) We take into consideration that $\frac{\partial^n(\tilde{u}-\tilde{u}_I)}{\partial\xi^n} = 2^n n!\,P_n(\xi)$ for $\tilde{u} = \xi^{2n}$ and we easily prove for $\varepsilon = u - u_I$

$$h^2 \sum_{P\in G}\sum_{|i|=n}|D^i\varepsilon(P)| \leq Ch^{n+1}|u|_{2n+1}. \tag{34}$$

c) We again use (21) where we set $w = D^i(u-u_I)$, $|i| = n$. We get

$$h^2 \sum_{e}\sum_{|i|=n}\max_{\bar{e}}|D^i(u_I-u_h)| \leq C|u_I-u_h|_n. \tag{35}$$

Now (29) follows from (34), (35) and (30).

REFERENCES

[1] G.Birkhoff, M.H.Schultz, R.S.Varga: *Piecewise Hermite interpolation in one and two variables with applications to partial differential equations.*
Numer . Math. 11 (1968), 232-256 .

[2] C.de Boor, B.Swartz: *Collocation at Gaussian points.*
SIAM J.Numer. Anal. 10 (1973), 582-606.

[3] J.H.Bramble, S.R.Hilbert: *Estimation of linear functionals on Sobolev spaces with applications to Fourier transforms and spline interpolation.*
SIAM J.Numer. Anal. 7 (1970), 112-124.

[4] P.G.Ciarlet, P.A.Raviart: *General Lagrange and Hermite interpolation in* R^n *with applications to finite element methods.*
Arch. Rat. Mech. Anal. 46 (1972), 177-199.

[5] P.G.Ciarlet, P.A.Raviart: *The combined effect of curved boundaries and numerical integration in isoparametric finite element methods ;* <u>*The Mathematical Foundations of the Finite Element Method with Applications to Partial Differential Equations.*</u>
Ed. by A.K.Aziz, Academic Press, New York and London, (1972).

[6] J.Jr.Douglas, T.Dupont: *Superconvergence for Galerkin methods for the two point boundary problem via local projections.*
Numer. Math. 21 (1973), 270-278.

[7] B.L.Hulme: *One-step piecewise polynomial Galerkin methods for initial value problems.*
Math. Comp. 26 (1972), 415-426.

[8] G.Strang, G.J.Fix: <u>*An Analysis of the Finite Element Method*</u>.
Prentice-Hill, Englewood Cliffs, New York, (1973).

[9] O.C.Zienkiewicz: <u>*The Finite Element Method in Engineering Science.*</u>
Mc-Graw hill, London, (1971).